Environmental Energy Sustainability at Universities

Environmental Energy Sustainability at Universities

Editors

Alberto Jesús Perea Moreno
Francisco G. Montoya

MDPI • Basel • Beijing • Wuhan • Barcelona • Belgrade • Manchester • Tokyo • Cluj • Tianjin

Editors
Alberto Jesús Perea Moreno
Department of Applied Physics,
Radiology and Physical
Medicine, University of Cordoba
Spain

Francisco G. Montoya
Department of Engineering,
Electrical Engineering Section,
University of Almeria
Spain

Editorial Office
MDPI
St. Alban-Anlage 66
4052 Basel, Switzerland

This is a reprint of articles from the Special Issue published online in the open access journal *Sustainability* (ISSN 2071-1050) (available at: https://www.mdpi.com/journal/sustainability/special_issues/Environmental_Energy_Sustainability_Universities).

For citation purposes, cite each article independently as indicated on the article page online and as indicated below:

LastName, A.A.; LastName, B.B.; LastName, C.C. Article Title. *Journal Name* **Year**, *Volume Number*, Page Range.

ISBN 978-3-03943-765-8 (Hbk)
ISBN 978-3-03943-766-5 (PDF)

Contents

About the Editors

Alberto Jesús Perea Moreno Associate Professor at the Department of Applied Physics, Radiology and Physical Medicine in the University of Cordoba (Spain), received his M.S. as an Agricultural Engineer and Ph.D. in Geomatics at the University of Cordoba (Spain). He has published over 40 papers in JCR journals (http://orcid.org/0000-0002-3196-7033), H-index 13. His main interests are Renewable Energy, Energy Saving, Biomass, Sustainability, and Remote Sensing. He was the Secretary of the Applied Physics Department (2017-2019) at the University of Cordoba. Awards: 2019 Winners Sustainability Best Paper Awards.

Francisco G. Montoya Professor at the Engineering Department and the Electrical Engineering Section at the University of Almeria (Spain), received his M.S. from the University of Malaga and his Ph.D. from the University of Granada (Spain). He has published about 75 papers in JCR journals and is the author or co-author of books published by MDPI, RA-MA, and others. His main interests are power quality, smart metering, smart grids, and evolutionary optimization applied to power systems and renewable energy. Recently, he has become passionately interested in Geometric Algebra as applied to Power Theory.

Preface to "Environmental Energy Sustainability at Universities"

The use of renewable energies and energy saving and efficiency are needs of global society and universities. Universities have a large responsibility and social impact, as they are an example and engine of social change. Universities, in the European context, must be at the forefront of ESA processes, seeking to be at the same level as, and preferably higher than, the rest of society, seeking a goal of 20% renewable energy for 2020 and, in the longer term, greater energy efficiency based on a diverse use of renewable energy and studying the feasibility of other energy processes (cogeneration, trigeneration, etc.). The application of renewable energies and energy efficiency allow universities to make significant savings in their costs and contribute to sustainable development and the fight against climate change. Actions in pursuit of these goals in addition to the objective of energy saving should promote research and form an example for the university community. This book aims to advance the contribution of energy saving and the use of renewable energies in order to achieve more sustainable universities.

Alberto Jesús Perea Moreno, Francisco G. Montoya
Editors

Editorial

Environmental Energy Sustainability at Universities

Francisco G. Montoya [1] and Alberto-Jesus Perea-Moreno [2],*

[1] Department of Engineering, University of Almeria, ceiA3, 04120 Almeria, Spain; pagilm@ual.es
[2] Department of Applied Physics, Radiology and Physical Medicine, University of Cordoba, Campus de Rabanales, 14071 Córdoba, Spain
* Correspondence: aperea@uco.es; Tel.: +34-957-212633

Received: 2 November 2020; Accepted: 4 November 2020; Published: 5 November 2020

Abstract: The use of renewable energies and energy saving and efficiency are needs of global society and universities. Universities have a large responsibility and social impact, as they are an example and engine of social change. Universities, in the European context, must be at the forefront of sustainability progress, seeking to be at the same level, and preferably higher than the rest of society, seeking the goal of 20% in renewable energy for 2020 and, in the longer term, greater energy efficiency based on a diverse use of renewable energy and studying the feasibility of other energy processes (cogeneration, trigeneration, etc.). The application of renewable energies and efficiency allow universities to make significant savings in their costs and contribute to sustainable development and the fight against climate change. Actions on these aspects in addition to the objective of saving should seek to promote research and form an example for the university community. This Special Issue aims to advance the contribution of energy saving and the use of renewable energies in order to achieve more sustainable universities.

Keywords: energy saving; renewable energy; universities; zero-energy buildings; energy efficiency; sustainability; bioclimatic architecture; sustainable transport; photovoltaic; energy saving in laboratories; energy saving in data processing centres

---

1. Introduction

Efficient energy consumption has now become one of the most important points on which society must raise awareness and work on it, because it is today, more than ever, when natural resources are scarcer and scientists are showing more evidence of climate change. Energy consumption is one of the main sources of environmental impact at the University, and also represents a significant economic expense. Likewise, from the environmental point of view, it is worth noting that through the energy savings, we will be contributing to a good use of energy and, in turn, we will be providing solutions that minimize the impact or energy footprint on society. In this sense, a lower use of resources and the promotion of renewable energies will result in an important contribution to reduce the evolution towards a negative climate change.

This Special Issue aims to advance the contribution of energy saving and the use of renewable energies in order to achieve more sustainable universities. This Special Issue seeks contributions spanning a broad range of topics related but not limited to:

- Solar energy
- The use of rooftops for energy generation
- Energy conversion from urban biomass or residues
- Energy management for sewage water
- Bioclimatic architecture and green buildings
- Wind energy cogeneration
- Public and private urban energy saving

- Policy for urban energy saving
- Electric meters
- Zero-energy buildings

2. Publication Statistics

The publication statistics of the call of papers for this Special Issue, regarding the articles published or rejected with respect to the total number of articles submitted, were:

- 14 articles submitted (100%)
- 4 articles rejected (28.6%)
- 10 articles published (71.4%)

The regional distribution of authors by countries for the published articles is presented in Table 1, in which it is possible to observe 24 authors from three countries. Note that it is usual for an item to be signed by more than one author and for authors to collaborate with others from different countries.

Table 1. Authors' countries.

Country	Authors
Spain	12
Mexico	9
Colombia	3
Total	24

3. First Affiliation and Country of the Special Issue Authors

Table 2 shows the affiliations of the authors who have participated in this Special Issue.

Table 2. Authors' affiliations of this Special Issue.

Author	First Affiliation	Country	Reference
Garrido-Yserte, R.	University of Alcalá	Spain	[1]
Gallo-Rivera, M.-T.	University of Alcalá	Spain	[1]
Perea-Moreno, M.-A.	Universidad Internacional de La Rioja (UNIR)	Spain	[2–4]
Hernandez-Escobedo, Q.	Universidad Nacional Autónoma de México	Mexico	[2–5]
Rueda-Martinez, F.	Universidad Nacional Autónoma de México	Mexico	[2]
Perea-Moreno, A.-J.	University of Cordoba	Spain	[2–4]
Ramirez-Jimenez, A.	University of Almeria	Spain	[3]
Dorador-Gonzalez, J.M.	Universidad Nacional Autónoma de México	Mexico	[3]
Grisales-Noreña, L.F.	Instituto Tecnológico Metropolitano	Colombia	[5]
Ramos-Paja, C.	Universidad Nacional de Colombia	Colombia	[5]
Gonzalez-Montoya, D.	Instituto Tecnológico Metropolitano	Colombia	[5]
Alcalá, G.	Universidad Veracruzana	Mexico	[5]
Manzano-Agugliaro, F.	University of Almeria	Spain	[4,6–8]
Chihib, M.	University of Almeria	Spain	[6,7]
Salmerón-Manzano, E.	Universidad Internacional de La Rioja (UNIR)	Spain	[6–8]
Novas, N.	University of Almeria	Spain	[7]
Leon, I.	University of the Basque Country UPV/EHU	Spain	[9]
Oregi, X.	University of the Basque Country UPV/EHU	Spain	[9]
Marieta, C.	University of the Basque Country UPV/EHU	Spain	[9]
Chavero-Navarrete, E.	Universidad Autónoma de Querétaro	Mexico	[10]
Trejo-Perea, M.	Universidad Autónoma de Querétaro	Mexico	[10]
Jáuregui-Correa, J.-C.	Universidad Autónoma de Querétaro	Mexico	[10]
Carrillo-Serrano, R.-V.	Universidad Autónoma de Querétaro	Mexico	[10]
Rios-Moreno, J.-G.	Universidad Autónoma de Querétaro	Mexico	[10]

4. Topics of Environmental Energy Sustainability at Universities

The research carried out by the different authors is classified according to the topics of the Special Issue in Table 3. It was noted that one "Environmental Energy Sustainability at Universities" topic dominated the rest: "Sustainability".

Table 3. Topics of Environmental Energy Sustainability at Universities.

Topics	Number of Manuscripts	Reference
Energy efficiency	2	[1,6]
Energy conversion from urban biomass or residues	2	[2,4]
Solar energy	2	[3,5]
Sustainability	3	[7–9]
Wind energy	1	[10]

Author Contributions: The authors all made equal contributions to this article. All authors have read and agreed to the published version of the manuscript.

Funding: This research received no external funding.

Conflicts of Interest: The authors declare no conflict of interest.

References

1. Garrido-Yserte, R.; Gallo-Rivera, M.-T. The potential role of stakeholders in the energy efficiency of higher education institutions. *Sustainability* **2020**, *12*, 8908. [CrossRef]
2. Perea-Moreno, M.-A.; Hernandez-Escobedo, Q.; Rueda-Martinez, F.; Perea-Moreno, A.-J. Zapote seed (*Pouteria mammosa* L.) valorization for thermal energy generation in tropical climates. *Sustainability* **2020**, *12*, 4284. [CrossRef]
3. Hernandez-Escobedo, Q.; Ramirez-Jimenez, A.; Dorador-Gonzalez, J.M.; Perea-Moreno, M.-A.; Perea-Moreno, A.-J. sustainable solar energy in mexican universities. case study: The National School of Higher Studies Juriquilla (UNAM). *Sustainability* **2020**, *12*, 3123. [CrossRef]
4. Perea-Moreno, M.-A.; Manzano-Agugliaro, F.; Hernandez-Escobedo, Q.; Perea-Moreno, A.-J. Sustainable thermal energy generation at universities by using loquat seeds as biofuel. *Sustainability* **2020**, *12*, 2093. [CrossRef]
5. Grisales-Noreña, L.F.; Ramos-Paja, C.A.; Gonzalez-Montoya, D.; Alcalá, G.; Hernandez-Escobedo, Q. Energy management in PV based microgrids designed for the Universidad Nacional de Colombia. *Sustainability* **2020**, *12*, 1219. [CrossRef]
6. Chihib, M.; Salmerón-Manzano, E.; Manzano-Agugliaro, F. Benchmarking energy use at University of Almeria (Spain). *Sustainability* **2020**, *12*, 1336. [CrossRef]
7. Chihib, M.; Salmerón-Manzano, E.; Novas, N.; Manzano-Agugliaro, F. Bibliometric maps of BIM and BIM in universities: A comparative analysis. *Sustainability* **2019**, *11*, 4398. [CrossRef]
8. Salmerón-Manzano, E.; Manzano-Agugliaro, F. The role of smart contracts in sustainability: Worldwide research trends. *Sustainability* **2019**, *11*, 3049. [CrossRef]
9. Leon, I.; Oregi, X.; Marieta, C. Contribution of university to environmental energy sustainability in the city. *Sustainability* **2020**, *12*, 774. [CrossRef]
10. Chavero-Navarrete, E.; Trejo-Perea, M.; Jáuregui-Correa, J.-C.; Carrillo-Serrano, R.-V.; Rios-Moreno, J.-G. Pitch angle optimization by intelligent adjusting the gains of a pi controller for small wind turbines in areas with drastic wind speed changes. *Sustainability* **2019**, *11*, 6670. [CrossRef]

Publisher's Note: MDPI stays neutral with regard to jurisdictional claims in published maps and institutional affiliations.

Article

The Role of Smart Contracts in Sustainability: Worldwide Research Trends

Esther Salmerón-Manzano [1] and Francisco Manzano-Agugliaro [2,*]

[1] Faculty of Law, Universidad Internacional de La Rioja (UNIR), Av. de la Paz, 137, 26006 Logroño, Spain; esther.salmeron@unir.net

[2] Department of Engineering, ceiA3, University of Almeria, 04120 Almeria, Spain

* Correspondence: fmanzano@ual.es

Received: 7 May 2019; Accepted: 26 May 2019; Published: 30 May 2019

Abstract: The advent and development of digital technologies has had a significant impact on the establishment of contracts. Smart contracts are designed as computer code containing instructions for executing user agreements, offering a technologically secure solution with numerous advantages and applications. However, smart contracts are not without their problems when we try to fit them into the traditional system of contract law, and their presumed benefits can become shortcomings. Bibliometric studies can help to assess the current state of science in a specific subject and support decision making and research direction. Here, this bibliometric study is used to analyze global trend research in relation to this novel contractual methodology, the smart contract, which seems to have experienced exponential growth since 2014. Specially, this analysis was focused on the main countries involved and the institutions that lead this research worldwide. On the other hand, the indexations of these works are analyzed according to major scientific areas and the keywords of all the works, to detect the subjects to which they are grouped. Community detection has been used to establish the relationship between countries researching in this area, and six clusters have been identified, around which all the work related to this topic is grouped. This work shows the temporal evolution of research related to smart contracts, highlighting that there are two trends—e-commerce and smart power grids. From the perspective of driving sustainability, smart contracts could provide a contribution in the near future.

Keywords: bibliometrics; community detection; energy; law; sustainability; smart contracts

1. Introduction

A contract is where individuals, groups, companies, institutions, and even governments enter into an agreement, where each of them is committed to fulfilling certain conditions. If the contract is traditional, it is written in language appropriate to the territory or legislation where the agreement is drafted, and if the parties involved agree, then they sign the document and legally agree to comply with it. All economic transactions between companies or individuals, for goods, services, or relations between the parties, are implemented by means of contracts; purchase and sale, lease, supply, loan, transport, and work are some of the most common examples. More modern examples include the contractual relationship between authors and publishers on copyright [1] and how insurance contract law differs widely between jurisdictions [2]. The performance of a contract is, ultimately, the will of the parties, and if one of them resolves not to comply with the law, it grants actions to the other signatory parties, and the appropriate judicial or arbitral process must be conducted. However, a question always arises at any time a contract is written, which is a tradeoff that must be addressed—whether or not to make a contract flexible but incomplete or rigid but comprehensive [3].

The digital age dominates world trade, so smart contracts can have a place in the foreseeable future. It is enough to mention that firms deploying computerized order systems are now responsible for more than 60% of the trading volume in U.S.-listed stocks [4]. The emergence of electronic and self-executing contracts is the inevitable consequence of the automation process of the Internet and the Internet of things. The legal regime integrates this contracting format without difficulty, but achieving a fully automated process implies, for example, resorting to network payment mechanisms, which are not always adapted to the current contractual type. The use of virtual currencies, such as bitcoin or electronic money, could cover this role, but the scarce or non-existent regulation of virtual currencies [5], and their dual character of unit of value and unit of account, hinder the functionality and legal security of the use of Blockchain technologies in standardized and automated contracting formats.

Blockchain technology is a distributed ledger that enables subscribers to enter and update records in the ledger, and cryptography assures that stored records will remain the same after they are added [6]. This ensures that no alterations can be made as changes would invalidate the whole register. The block network is represented by the nodes and virtual machines that are connected in peers, and each node involved has a ledger copy. The virtual machines run nodes. Once a new block has been agreed upon in the network, each node will refresh its record by appending the new block. All transactions are processed and sent from the involved nodes. All nodes in the network will agree on a consensus method for aggregating new records to the ledger [7]. As an example of a programming language, a JavaScript-like language called Solidity can provide a method for executing computer code on blockchain nodes. Computer programs that verify contracts digitally, enforce those contracts, and run on a blockchain network are called smart contracts [8].

So-called electronic contracts have been a known and enforced reality for several decades. A first question to be resolved is what is meant by a "self-executory" contract and smart contract. In our opinion, these terms allow us to approach the same reality from partially different perspectives. From a technical or informatic prism, a smart contract would be a sequence of code and data that carries out the operation in its foreseen case and does not constitute a contract in the legal sense, even though such a term appears in its name. From a legal standpoint, the term "smart contract" would refer to an existing agreement between parties for which the code sequence would be a portion or all of the same. In other words, the code itself does not constitute a contract but responds to an agreement that gives meaning to it, and that serves as its expression.

Some authors define smart contracts as self-executing digital transactions that use decentralized cryptographic mechanisms [9]. Although novel, this form of compromise is not new; it has been on the table for more than thirty years. Specifically, it was in 1994 that US computer scientist Nick Szabo proposed what was then a fanciful notion of computerized transaction protocols for intelligent contracts that executed the terms of a contract [10]. In this way, smart contracts proposed the combination of protocols with user interfaces to formalize and secure relationships across computer networks [11]. Recently, the development of the Blockchain and Bitcoin technologies has once again driven the approach to the potential of smart contracts [12]. In Figure 1, the process of creating a smart contract and the blockchain is represented in a schematic form.

Smart contracts are not like commonly understood contracts, particularly for legal scholars and practitioners. The difference, however, is that because these contracts are intelligent, they can be fulfilled automatically. Even if these contracts are fulfilled automatically, it is necessary for each of the members to do their part. The main differences between smart and traditional contracts are the ways they are written, their legal implications, and how the agreed conditions will be fulfilled. These distinctive characteristics are the ones that provide the advantages and disadvantages of both types of contracts, which are easily observable when understanding how they work.

However, there is a long history of self-executory contracts. Take the example of 'on demand' guarantees. While clearly contracts, on demand guarantees do not reflect any particularly general idea of what a contract is, but rather a highly specialized institutional context where, firstly, it is possible to codify a transaction so that self-executing rights have practical meaning and, secondly, there is a highly specific, narrow context of use, where the multiplicities generally implicit in a contract can be controlled.

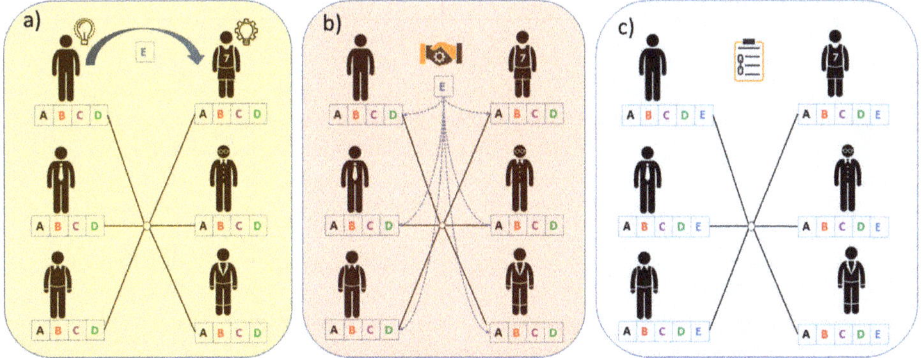

Figure 1. Smart contracts: (**a**) transaction idea, (**b**) smart contracts and blockchains, (**c**) transaction confirmed and added as a block to the blockchain.

A novel area of law is emerging around blockchain platforms and automated transactions [10]. The so-called Internet of Things (IoT) refers to a connection to the web for millions of devices. In IoT, fridges, washing machines, televisions, and vehicles can connect to the Internet and exchange data with the millions of other users or computers on the web. In this scenario, which is predicted for the near future, smart contracts could go beyond single-tract contracts and ensure the execution of successive-tract contracts. However, some authors, specialized in law, advise on the emerging risks in the use of smart contracts, which could certainly be a branch of research in this field [13].

Blockchain systems can be beneficial for non-centrally controlled storage, notarizing, and subsequent execution of intelligent contracts. However, fundamental problems can arise about the modification and termination of intelligent contracts. To simplify the modification of intelligent blockchain contracts, declarative language could be used, but compared to its imperative counterparts, it may not live up to expectations in terms of computational complexity and associated costs. For these reasons, we must emphasize that imperative and declarative approaches are not incompatible, but instead have the potential to complement each other, which can lead to interesting theoretical and practical opportunities [14].

However, "failed" smart contracts already exist. These contracts have even classified into prodigal, suicide, and greedy contracts [15]. Prodigal contracts are those which have fallen into the hands of hackers, thereby changing the direction the Ethers should go in this case. This fraud has caused crypto-currencies to reach a fraudulent address and become the property of the fraudster who had been placed between the contracting party and the actual recipient of the crypto-currency [16]. Suicide contracts are those that are closed when an exit requirement is activated by the person carrying out the attack. It may be that there is a wrongly implemented exit clause, as has already happened, and the consequence is quick to occur. Under the cover of a legal act, the wrong person ends up taking all the encrypted money that the smart contract entails [17]. It should also be noted that inadequate protection of the information in one of these contracts also ends up allowing funds to be moved to illegitimate places. Greedy contracts may be due to bad practice or miswriting, but the fact is that the contracting party will no longer have the legitimacy to receive its encrypted currency. It gets out of their control, and ends the contract. This is an example of economic loss due to vulnerability failure [18].

From a sustainability point of view, it is possible to find many works that show the potential of smart contracts [19]. Nikolakis et al. [20] studied how law, regulation, and private standards have evolved to enhance sustainability in value chains. As an example, they show how blockchains can improve sustainability by informing consumers about the origin of products, provide guarantees about the authenticity of information, and offer a mechanism for enforcing representations through the smart contract function of the blockchain. Park et al. [21] propose the implementation of an energy transaction platform based on P2P (peer–to–peer) blockchains to support energy efficient transactions between prosumers, which will encourage a more sustainable trading ecosystem between consumers and prosumers. Giungato et al. [22] propose the development of an Energy Internet, based on a new type of power grid structure based on the generation of renewable energy, distributed energy store devices, and the existent of the Internet [23]. Gatteschi et al. [9] propose a use for the insurance sector and give the example of B3i, the first blockchain-centered insurance consortium [24].

Other characteristics of smart contracts to expand their potential for sustainability are to accelerate and automate the exchange of information on the value of natural resources and environmental sustainability. Examples of sustainable supply chain traceability can be found as agrifood products [25], as forests (if the trees are cut without destroying natural forests) [26], or as payment for ecosystem services [26]. Another great smart contract approach is the application to improve logistics services and supply chains, such as in the pharmaceutical sector [27] or alimentary supply chain [28].

On the other hand, there are studies that warn about the problems of these technologies. For example, the advantages of blockchain technology can be overshadowed by the intentionally resource intensive nature of their transaction verification process, which now menaces the climate on which we depend to survive [17]. There is previous research that has studied the relationship of sustainability with "bitcoin", "digital currency", "cryptocurrency", and "virtual currency" [29], or the relationship of sustainability with the "Energy Internet" [30]. From a legal point of view, smart contracts, in contrast to traditional contracts, should address issues such as trial risks, enforcement risks, and jurisdictional risks. In fact, it would be useful to analyze the on-demand guarantee example to see in what kinds of institutional contexts they might be used. In this regard it is possible to find, as an example, contracts for insurance. Insurance contracts, or more specifically reimbursement in specific, narrowly defined loss scenarios, much more clearly provide a similar highly specialized institutional context comparable to the existing generally used self-executory contracts, such as on demand guarantees.

In short, this new technology has its advantages and disadvantages. Therefore, as it is a technology under development, work is continuing to optimize its operation to the maximum. On the other hand, there are investigations that alert to the problems of these technologies. For example, the advantages of blockchain technology can be overshadowed by the intentionally resource intensive nature of their transaction verification process, which now menaces the climate on which we depend to survive [31]. Until now, no systematic study of all published works related to smart contracts has been carried out. A bibliometric analysis is a useful tool, both for the study of the state of different scientific disciplines and for the scientific production of a given region, discipline, or topic. Its study aim is to physically represent the products of thought in documents. In other words, intellectual knowledge supported by material support—the publications.

Bibliometric analyses can determine which fields of research have been carried out, and which organisms and countries are the main ones applicable in researching this topic. Bibliometrics and the use of their indicators are necessary scientific tools because they allow the quantification of science in an objective way, as they show the current knowledge in a given scientific field and its compilation in bibliographic databases. The importance of bibliometric studies is carried out in all branches of science, such environment [32] and education [33]. In this context, the present work has the main objective of analyzing the global research trends on smart contracts, with special attention to analyzing the main areas in which efforts are being made by the scientific community.

2. Methods

One of the world's largest databases of scientific literature is Elsevier's Scopus, which contains approximately 18,000 titles from more than 5000 international publishers, including coverage of 16,500 peer-reviewed journals in the areas of Science, Technology, Medicine, and Social Sciences, including the arts and humanities. This is the methodological basis of this study, which has been used successfully in other bibliometric studies [34].

The coexistence of two large scientific databases, Scopus and Web of Science (WoS), raises the question of the stability of the statistics obtained by one or the other sources of information. Several studies have measured the overlap between databases and the impact of using different data sources for specific research fields on bibliometric indicators, demonstrating a larger number of journals indexed by Scopus compared to WoS [35]. Regarding the overlap, 84% of the WoS titles are also indexed in Scopus, while only 54% of the Scopus titles are indexed in WoS [36]. For example, some studies related to citations in the papers conclude that each database covered 90% of the citations in the other database when the citation period is limited to Scopus citation coverage for 1996 and beyond [37].

The methodology followed in this work is described in Figure 2. The search term is first consulted in the Scopus database (1). The term smart contract was used for the entire historical series up to 2018, where the exact search query was TITLE-ABS-KEY (smart AND contract). The resultant search is exported to Comma Separated Value (CSV) text format (2) for each of the fields studied (i.e., publications by year, type of publication, publications by category indexed by Scopus, publications by country, publications by institutions, and keywords and their frequency). Thirdly (3), the previously downloaded text files are imported into Excel, and all of them are represented without removing any data. Fourthly (4), the keywords are represented by means of the free online software Word Art (https://wordart.com/) to obtain a cloud words where the most important ones are highlighted. Fifth (5), the methodology was developed to analyze the scientific communities or clusters associated with this thematic. The exported information of the complete search was imported in csv format in the free and online bibliometric analysis software called VOSviewer (http://www.vosviewer.com/). Here, the relations between the countries, interpreted through the co-authors of every one of the works, were analyzed, and the research clusters of the works were analyzed, using the relations between all keywords of all the works.

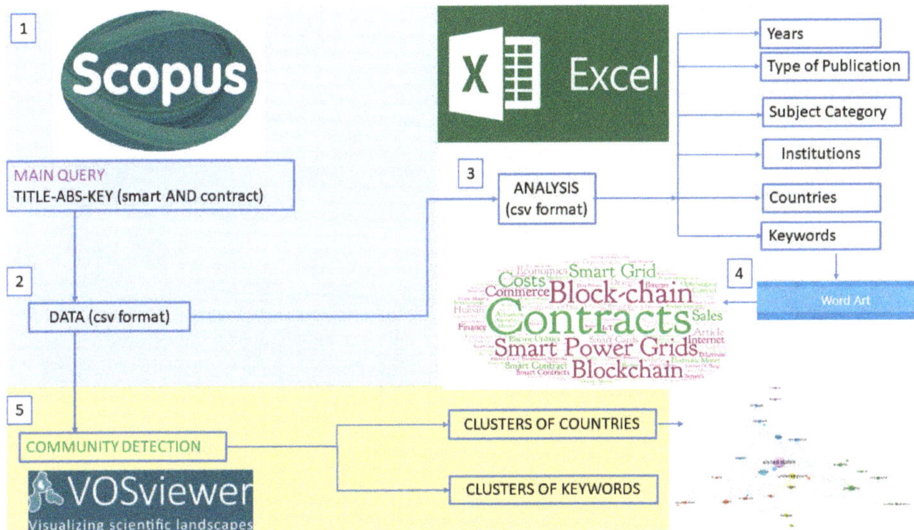

Figure 2. Methodology chart.

With respect to the chosen software, it should be noted that for the direct representation of results, bar charts, percentage distribution, or lines of evolution over time, a spreadsheet has been used via Microsoft Excel, which allows the direct import of the csv format exported by the Scopus database. For the clouds of words, the Word Art software has been chosen because it is free and online and allows the import of data from excel. Finally, the community detection software, we also opted for free software available online that allows the direct import of data in csv format exported from Scopus.

Finally, the community detection software used was the VOSviewer, which was also chosen for being free software available online that allows the direct import of data in the csv format exported from

Scopus and also allows the figures to be exported to a large range of graphical formats. The VOSviewer delivers three displays: Network visualization as clusters, overlay visualization as temporal evolution, and density visualization. In all cases, the parameters chosen for the analysis were: normalization method (association strength), layout (attraction 2, repulsion 0), clustering (resolution 1.00, minimum cluster size 1), and rotate (90 degrees).

3. Results and Discussion

3.1. Evolution of Scientific Production, Languages, and Types of Documents

The search yielded just over 1700 documents up to the year 2018, the evolution of which is reflected in Figure 3. This result shows that the first works published date back to the 1980s, but it was not until 2003 that they began to reach a certain relevance, with the volume of publications stabilizing at 50 works per year. However, the most remarkable increase can be seen from 2014 onwards, at which point the upward trend can be considered exponential.

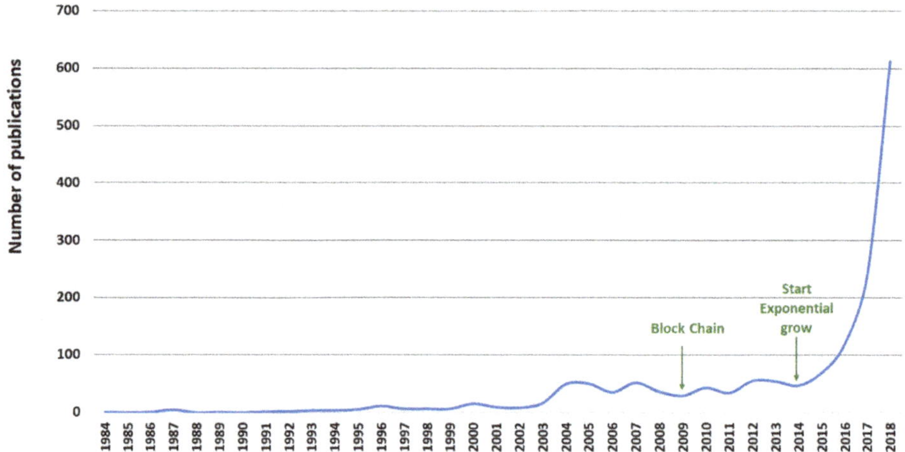

Figure 3. Evolution of scientific publications related to smart contracts.

If the results are analyzed according to the type of publication (Figure 4), it is observed that most results are communications presented at congresses (53%; 46% conference papers and 7% conference reviews), followed by articles in journals (38%; 33% articles and 5% reviews), as well as books and book chapters (2%). The rest is distributed among other formats such as editorials, notes, and letters. The high percentage of conferences in relative terms is because it is a very recent technology or area of interest. Thus, when the subject matter of research is consolidated, the percentage of books is higher and, above all, the number of articles in relation to number of congresses is higher. Note that scientific conferences are usually on very specific scientific topics and are aimed at sharing ideas among researchers. In short, these scientific events are key activities for the process of knowledge dissemination, for the presentation of new findings, and for the development of science in a community.

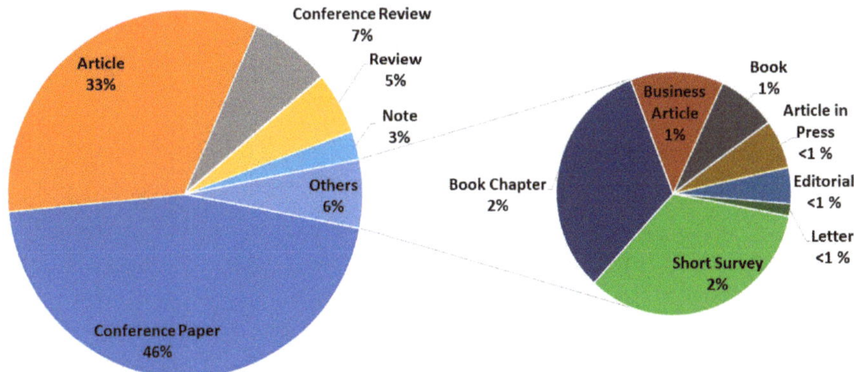

Figure 4. Distribution of publication type in relation to smart contracts.

These papers are written mostly in English, accounting for more than 97% of publications, as is usual when consulting international scientific databases. However, there are also works published in Chinese, French, Russian, German, Dutch, Portuguese, Spanish, and Japanese.

3.2. Distribution of Publications by Institution and Country

Regarding the countries which have carried out the most research on this subject, the USA stands out, with more than 22% of the total number of published papers, followed by China, with more than 6%, and finally the United Kingdom, Germany, and Italy, with slightly more than 5%.

The affiliations of the works do not essentially respond to the same order as the countries. Thus, it was shown that the ten institutions that have published the most papers are, in order, Danmarks Tekniske Universitet, Universita degli Studi di Trento, National University of Singapore, Tallinn University of Technology, Cornell University, Instituto Politecnico do Porto, Delft University of Technology, UC Berkeley, University of Electronic Science and Technology of China, and Tsinghua University. It can be seen that two are from the USA and two are from China. Table 1 lists the main institutions and their main keywords used.

Figure 5 shows the relationship between research in this area in different countries using the VOSviewer software. In the network visualization of Figure 5, countries are represented by a circle. The size of the label and the circle of an item is determined by the weight of the country. The higher the weight of a country, the larger the circle and the label of the country. These clusters or communities are shown in Table 2. There are six communities of scientific collaboration where, in principle, the apparent lack of affinity between countries is striking, except in the case of community 3, which corresponds to all European countries. The cluster name is selected by the country that has the greatest weight within the cluster. Figure 5 shows the great centrality of the USA in this area, and although they belong to other clusters, China and the UK also occupy important positions of centrality. The clusters are led by Japan, Germany, Italy, UK, USA, and China, which, as can be seen, are the most industrialized countries in the world, all of them belonging to the G8 countries except China. For example, in the case of United Kingdom, the relationship is concentrated mainly with Canada, Iran, and Spain.

Table 1. Main institutions and their keywords.

Institution	1	2	3
Danmarks Tekniske Universitet	Smart Power Grids	Energy Resources	Distributed Energy Resources
Universita degli Studi di Trento	Smart Cards	JAVA Card	Multiapplication Smart Cards
National University of Singapore	Electronic Money	Chromium Compounds	Cryptography
Tallinn University of Technology	E-governance	Decentralized Autonomous Organization	Government Data Processing
Cornell University	Cryptography	Electronic Money	Ethereum
Instituto Politecnico do Porto	Energy Resources	Smart Power Grids	Distributed Energy Resources
Delft University of Technology	Aggressive Environment	Anode Material	Cathodic Protection
UC Berkeley	Commerce	Smart Grid	Electricity Market
Chinese Academy of Sciences	Blockchain	Authentication	Network Security
University of Electronic Science and Technology of China	Blockchain	Big Data	Commerce
Tsinghua University	Blockchain	Bitcoin	Business Modelling

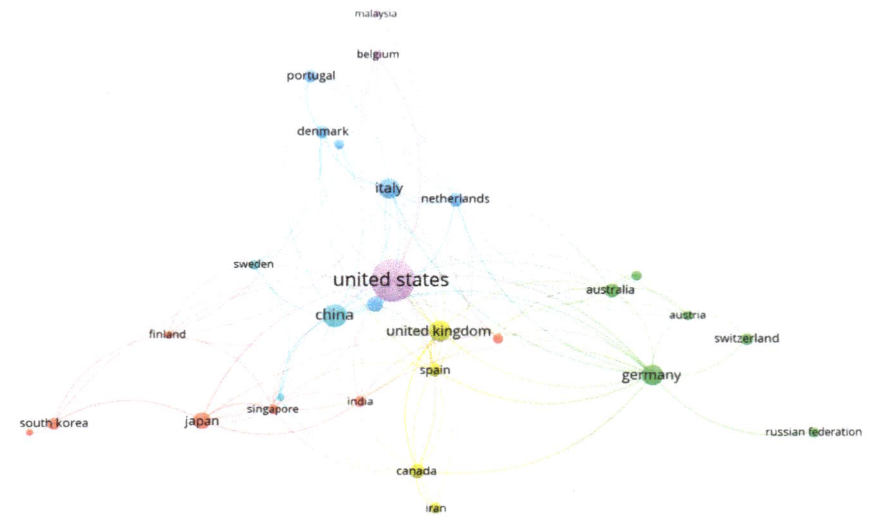

Figure 5. Relationship between countries publishing on smart contracts.

Table 2. The detected clusters of countries related to research on smart contracts.

Cluster	Color (Figure 5)	Main Countries	Main Country (by Number of Publications)
1	Red	Japan, Singapore, South Korea, Finland, India	Japan
2	Green	Australia, Austria, Germany, Russia, Switzerland	Germany
3	Blue	Denmark, Italy, Netherlands, Norway, Portugal	Italy
4	Yellow	Canada, Iran, Spain, UK	UK
5	Purple	Belgium, Malaysia, USA	USA
6	Cyan	China, Hong Kong, Sweden	China

3.3. Main Areas of Knowledge and Keyword Analysis

The analysis of the keywords with which the works are indexed is one of the most relevant aspects in bibliometric analysis [38,39]. If the results obtained are analyzed according to keywords, and a cloud word is made with on-line software (see Figure 6) a strong relationship can be observed between blockchain technology, smart grids (SGs), and smart energy grids, as well as virtual currencies or electronic money (Ethereum, Electronic Money). It is worth highlighting the main areas of knowledge in which research is being carried out.

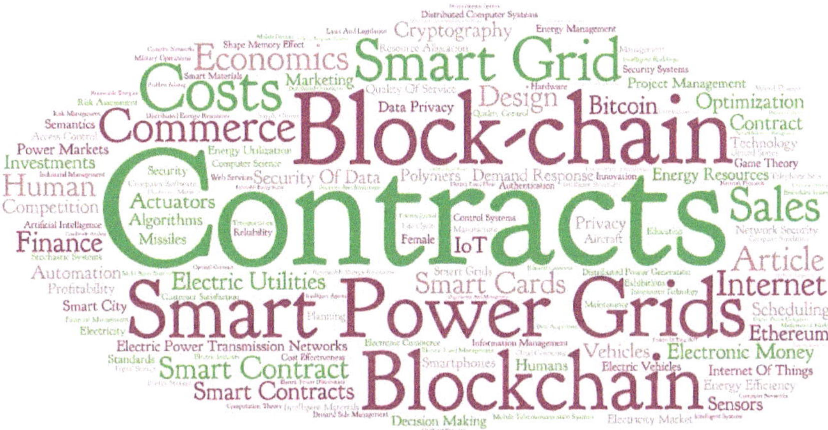

Figure 6. the cloud of keywords used in the work on smart contracts.

Figure 7 groups the keywords by large areas of knowledge, according to the Scopus indexation (subject area), and it can be seen that the first one, as was foreseeable, is computer science, followed by engineering via the theme of intelligent networks. The following areas of knowledge are those of energy and, later, of social sciences, among which the studies in law are framed. Indeed, if one looks at the keywords most commonly used in each country's publications (Table 3) one can see that blockchain and smart grid dominate in almost all the major countries with scientific publications on this subject.

There are also two keywords that, a priori, could go unnoticed, commerce and smart cards, which have a great relationship with the categories of social sciences and business, management, and accounting. This is particularly evident in the analysis of smart contracts in their focus on brokering, their transparent nature, the promise of greater commercial efficiency, lower legal and transaction costs, and, above all, the apparent advantage of anonymous transactions [40].

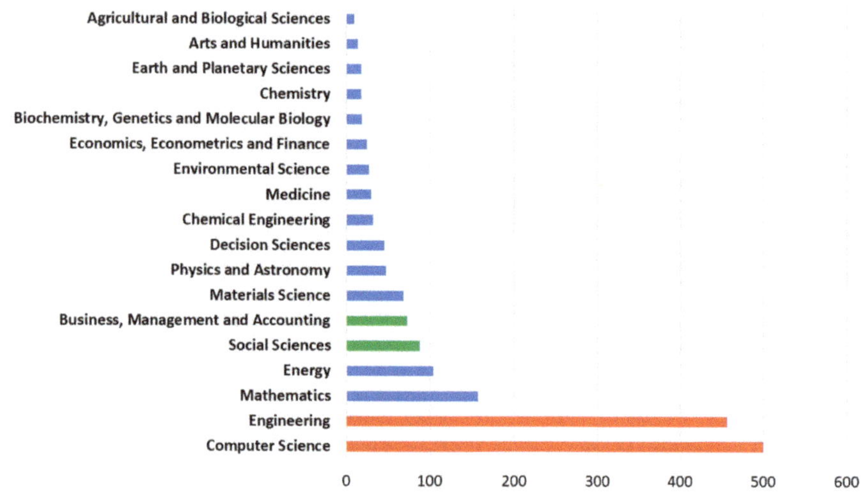

Figure 7. Thematic distribution of works related to smart contracts.

Table 3. Countries and their three main keywords related to smart contract.

Country	N	%	1°	2°	3°
United States	228	22.03	Blockchain	Smart Grid/ Smart Power Grids	Electronic Money
China	70	6.76	Blockchain	Smart Power Grids	Commerce
United Kingdom	61	5.89	Electronic Money	Blockchain	Commerce
Germany	55	5.31	Smart Power Grids	Electric Power Transmission Networks	Blockchain
Italy	55	5.31	Blockchain	Smart Cards	Commerce
Japan	36	3.48	Smart Power Grids	Commerce	Profitability/ Smart Grid
France	33	3.19	Automation	Costs/ Economics	Smart Cards
Canada	29	2.80	Blockchain	Electric Power Transmission Networks	Smart Power Grids
Netherlands	26	2.51	Commerce	Blockchain	Distributed Energy Resources
Australia	23	2.22	Blockchain	Smart Cards	Telephone Sets
Portugal	23	2.22	Smart Power Grids	Energy Resources	Smart Grid
Spain	22	2.13	Smart Power Grids	Distributed Energy Resource	Energy Resources
Denmark	20	1.93	Smart Power Grids	Electric Power Transmission Networks	Energy Resources
Switzerland	18	1.74	Blockchain	Demand Response	Smart Power Grids
South Korea	17	1.64	Smart Power Grids	Computer Science	Internet of Things
India	16	1.55	Computation Theory	Decision Making	Distributed Computer Systems
Iran	16	1.55	Smart Power Grids	Commerce	Costs

3.4. Community Detection: Analysis of the Interconnection Between Keywords

Considering a community as a system composed of multiple interdependent elements, with a very wide range of relationships and intensities that are highly variable and dependent on each other, we could accept, conceptually, that communities are made up of a highly cohesive central core and peripheral spheres with unions increasingly weaker compared to the center. The central core would be structured by the most significant elements of the community, in terms of granting a definable individuality, representing the links between its constituents, and the strongest and most significant elements within the entire community complex. Communities or clusters are usually groups that are more likely to relate to each other than to members of other groups. When this analysis was completed by collaboration between authors from different countries using what is known as community detection, Figure 8 was obtained. An available online application, called VosViewer, which was developed specifically for this type of analysis of scientific production, was used for this purpose [41].

Clusters, or communities in networks, are one of the most notorious aspects of leading bibliometric studies [42]. These communities are groups that are more likely to be interconnected with each other than with members of other groups [43]. By analyzing the keywords of all the works published on smart contracts with the application calibrated for community detection, Figure 8 was obtained, in which six clearly differentiated communities have been detected (Table 4).

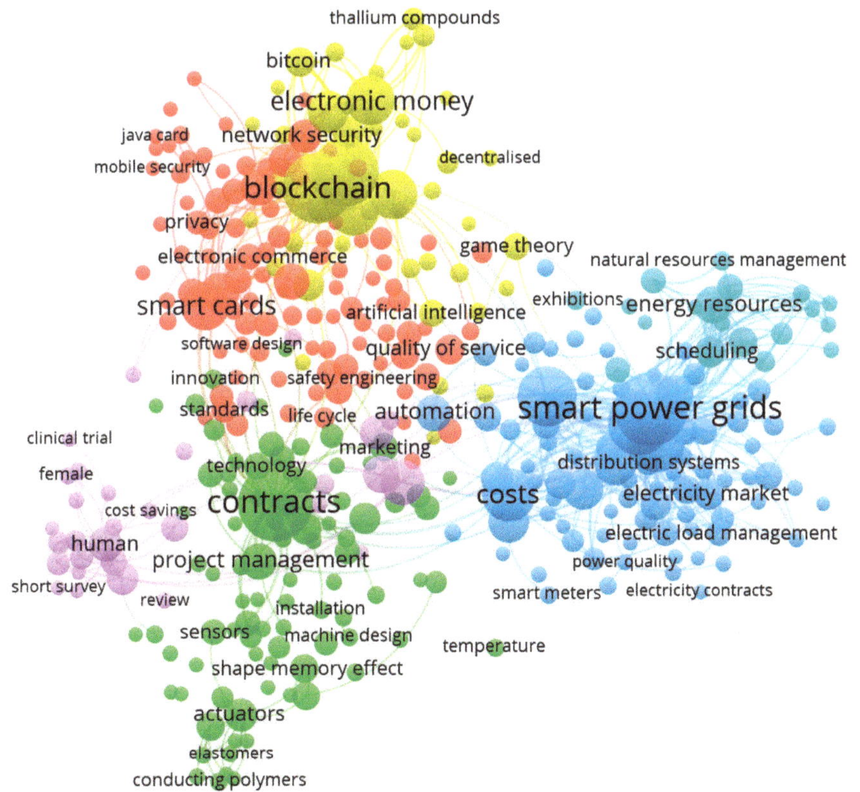

Figure 8. The relationship between keywords in the smart contract works.

Table 4. Detected clusters of keywords related to research on smart contracts.

Cluster	Color (Figure 8)	Keywords	Cluster Name	%
1	Red	Smart cards, security data, smart phone, internet, authentication, network security, internet of things, laws and legislation	Smart cards	26.9
2	Green	Contracts, computer software, project managements, marketing, societies and institutions	Contracts	23.6
3	Blue	Smart power grids, electric power transmission network, commerce, electric load management, costs	Smart power grids	23.3
4	Yellow	Blockchain, bitcoin, electronic money, smart contracts, cryptography	Blockchain	12.5
5	Purple	Economics, human, organization and management, decision making	Economics	8.5
6	Cian	Energy resources, distributed energy resources, virtual power players, natural resources managements	Energy	5.2

The largest volume is cluster 1 (red), which groups 26.9% of keywords. The main keyword is smart cards, which has the highest density relationship with authentication and internet within its cluster, as well as cryptography (cluster 4) and contract (cluster 2). The first works related to smart cards from 1998 are mainly for use in electronic commerce [44].

Cluster 2 (green), which groups 23.6% of keywords, is focused on contracts. Its clusters highlight its relationship with laws and legislation and trade (marketing, purchasing, or project management). The relationship with other clusters is mainly through economics (cluster 5), cost or sales (3), smart cards (cluster 1), and blockchain (cluster 4).

Cluster 3 (blue) is focused on smart power grids. In general, one could say that this cluster is fairly independent, since its main relationships are within its own cluster. Thus, its main relationships are with electric power transmission networks, electric utilities, electricity market, commerce, electric load, energy management, and wind power. With other clusters, its main links are through blockchain and game theory (cluster 4), economics (cluster 3), and energy resources (cluster 6).

Cluster 4 (yellow) is mainly grouped around blockchain, which has the highest density relationship with electronic money and bitcoin within its cluster. Blockchain is focused on programming and cryptocurrency, such as Bitcoin. This is because cryptocurrencies and smart contracts are based on the same technology (blockchain) [8]. Regarding the relationship with other clusters, we find a connection mainly with commerce (cluster 3), economics (cluster 5), and security data (cluster 1).

Cluster 5 is grouped around economics, human, organization and management, or decision making. The importance of this cluster is that although it is not particularly important in terms of weight (8.5% of keywords), it is centered on the network, which shows that it is a link-point for all this research. This result possibly suggests that, in the near future, smart contract applications will play an important role in the different domains of modern organizations [45]. The other clusters highlight their relationship with blockchain (cluster 4), contracts (cluster 2), and smart power grids (cluster 3). Within its cluster, the main relationship is with terms based on the social economy, where key words such as human, male, female, adult, investment, transparency, and decision-making stand out.

Based on the previous study of clustering of keywords by clusters, its temporal evolution can be analyzed (Figure 9). It can be seen how clustering has temporarily evolved since smart cards and contracts to mainly two lines—first, to electronic commerce or quality of service, and second, to costs; from the first line (electronic commerce) to blockchain or network security and electronic money, and from the second line to cost energy issues, mainly regarding smart power grids. This gives an idea about the transition of the worldwide research in this topic.

There is no doubt that, despite the objections that can and must be made, bibliometric studies facilitate the understanding of research activity in a given scientific field. In this research, it has been observed from the analysis of keywords and the scientific communities that support them, that there is not yet a community that dedicates itself to legislation and laws, despite being an essential aspect of the subject in which it is concerned. However, it should be noted that bibliometric analyses are generally valid in those areas in which scientific publications are an essential result of research. For this reason, the validity of bibliometric analyses is of maximum relevance to the study of basic areas, where scientific publications predominate, to a lesser extent in technological or applied areas, and to a much lesser extent in the areas of social and legislation. Therefore, comparisons between thematic areas and within these, scientific communities, should be made with caution, because the publication habits and productivity of authors differ according to knowledge areas. This is the case of this study, where differences were found between the areas of social sciences or business, management and accounting, and those of computer science or engineering.

In the absence of a greater degree of maturity and extension of use, smart contracts raise several issues from a legal standpoint. Our law does not contemplate them, nor do judicial precedents yet exist to help in this regard. However, it must be made clear that general contract law does provide criteria for verifying whether a smart contract can be legally valid and enforceable. The legal systems of our environment recognize the autonomy of the parties to freely reach legally enforceable agreements and contracts in the terms they consider, provided that the basic requirements of contract law are met, both in content (being a legal object and not a contravention of mandatory legal rules, ensuring the existence of valid consent of the parties, and obeying a legal cause) and in the manner of formalizing them.

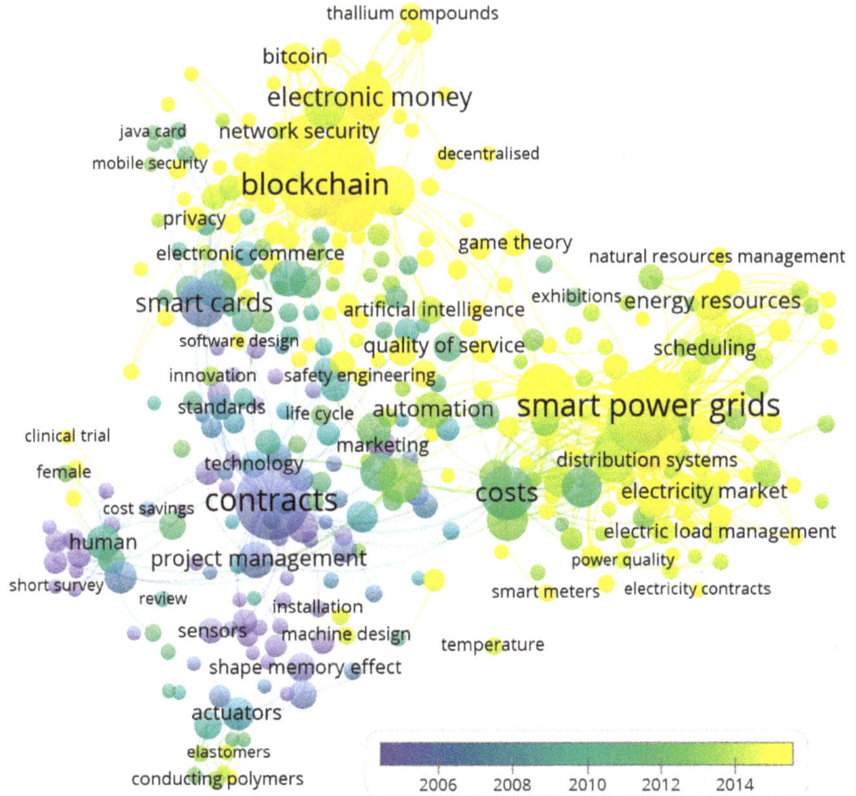

Figure 9. The relationship between keywords in smart contract works and its evolution.

4. Conclusions

This paper has analyzed the main trends in global research on intelligent contracts, highlighting the main countries that have made a scientific effort in this area, in order of importance—the USA, China, the United Kingdom, Germany, and Italy—and scientific collaboration between these countries does not necessarily respond to obvious or expected trade relations. For example, in the case of Spain, the relationship is concentrated mainly with Canada, Iran and the United Kingdom. The publication of research, mainly in the form of communications in congresses, shows that intelligent contracts are emerging technologies and still in the initial study phase, unlike other more established technologies that do the same work in the form of articles or even books. In order of importance, the main subject areas found were computer science, engineering, mathematics, energy, social sciences, business, management, and accounting. In the absence of a greater degree of maturity and widespread use, smart contracts raise several legal issues that need to be addressed. Consequently, this paper demonstrates that the Social and Legal Sciences occupy the fifth position in the number of published papers, clearly indicating that the technological aspect of this issue needs to be given a legal character, as its implications and scope are of the utmost interest for the international scientific community. Therefore, further legal progress must be made on the reality of this new form of contracts, especially to support the two main lines towards which they have evolved—electronic money and energy costs.

Finally, the analysis of the themes of all these works, by means of the analysis of the keywords, have shown that there are 6 clearly differentiated clusters around which all these works can be grouped. Their temporal evolution shows a tendency towards two main lines of research—the electronic money (cryptocurrencies, or bitcoins) and the smart power grids and electricity market (electricity consumption, electric industry). Given the growing scientific interest shown by the growing number of publications, smart contracts can overcome the problems pointed out by many detractors, and they could be a driver of sustainability. For environmental sustainability applications, smart contracts should pay particular attention to small communities, since it has been observed that small communities are still poorly covered by this particular technology, in spite of often being stewards of natural resources.

Author Contributions: E.S.-M. and F.M.-A. have contributed in the same way to the writing of this manuscript, both authors have read and approved the final manuscript.

Funding: This research received no external funding.

Acknowledgments: The authors of this manuscript wish to thank the CIAIMBITAL research center for its support of this research.

Conflicts of Interest: The authors declare no conflict of interest.

References

1. Bagheri, P.; Hassan, K.H. Access to information and rights of withdrawal in internet contracts in Iran: The legal challenges. *Comput. Law Secur. Rev.* **2015**, *31*, 90–98. [CrossRef]

2. Jackson, H.E. Variation in the intensity of financial regulation: Preliminary evidence and potential implications. *Yale J. Regul.* **2007**, *24*, 253.

3. Burden, K. "Cloud bursts": Emerging trends in contracting for Cloud services. *Comput. Law Secur. Rev.* **2014**, *30*, 196–198. [CrossRef]

4. Diamond, S.F.; Kuan, J.W. Are the stock markets "rigged"? An empirical analysis of regulatory change. *Int. Rev. Law Econ.* **2018**, *55*, 33–40.

5. Salmerón, E. Necesaria regulación legal del Bitcoin en España. *Rev. De Derecho Civ.* **2017**, *4*, 293–297.

6. Kuo, T.T.; Kim, H.E.; Ohno-Machado, L. Blockchain distributed ledger technologies for biomedical and health care applications. *J. Am. Med. Inform. Assoc.* **2017**, *24*, 1211–1220. [CrossRef]

7. Zheng, Z.; Xie, S.; Dai, H.N.; Chen, X.; Wang, H. Blockchain challenges and opportunities: A survey. *Int. J. Web Grid Serv.* **2018**, *14*, 352–375. [CrossRef]

8. Christidis, K.; Devetsikiotis, M. Blockchains and smart contracts for the internet of things. *IEEE Access* **2016**, *4*, 2292–2303. [CrossRef]

9. Gatteschi, V.; Lamberti, F.; Demartini, C.; Pranteda, C.; Santamaría, V. Blockchain and Smart Contracts for Insurance: Is the Technology Mature Enough? *Future Internet* **2018**, *10*, 20. [CrossRef]

10. Casado-Vara, R.; Prieto, J.; De la Prieta, F.; Corchado, J.M. How blockchain improves the supply chain: Case study alimentary supply chain. *Procedia Comput. Sci.* **2018**, *134*, 393–398. [CrossRef]

11. Sicari, S.; Rizzardi, A.; Grieco, L.A.; Coen-Porisini, A. Security, privacy and trust in Internet of Things: The road ahead. *Comput. Netw.* **2015**, *76*, 146–164. [CrossRef]

12. Lin, I.C.; Liao, T.C. A Survey of Blockchain Security Issues and Challenges. *IJ Netw. Secur.* **2017**, *19*, 653–659.

13. Briones, A.G.; Villaverde, D.V. Aspectos legales y riesgos emergentes en la utilización de Smart Contracts basados en Blockchain. In *FODERTICS 7.0: Estudios Sobre Derecho Digital*; Fundación Dialnet: La Rioja, Spain, 2019; pp. 245–252.

14. Governatori, G.; Idelberger, F.; Milosevic, Z.; Riveret, R.; Sartor, G.; Xu, X. On legal contracts, imperative and declarative smart contracts, and blockchain systems. *Artif. Intell. Law* **2018**, *26*, 377–409. [CrossRef]

15. Nikolic, I.; Kolluri, A.; Sergey, I.; Saxena, P.; Hobor, A. Finding the Greedy, Prodigal, and Suicidal Contracts at Scale. 2018. Available online: https://arxiv.org/abs/1802.06038 (accessed on 15 January 2019).

16. Giancaspro, M. Is a 'smart contract' really a smart idea? Insights from a legal perspective. *Comput. Law Secur. Rev.* **2017**, *33*, 825–835. [CrossRef]

17. Luu, L.; Chu, D.H.; Olickel, H.; Saxena, P.; Hobor, A. Making smart contracts smarter. In Proceedings of the 2016 ACM SIGSAC Conference on Computer and Communications Security, Vienna, Austria, 26–28 October 2016; pp. 254–269.

18. Atzei, N.; Bartoletti, M.; Cimoli, T. A survey of attacks on ethereum smart contracts (sok). In *Principles of Security and Trust*; Springer: Berlin/Heidelberg, Germany, 2017; pp. 164–186.

19. Mengelkamp, E.; Notheisen, B.; Beer, C.; Dauer, D.; Weinhardt, C. A blockchain-based smart grid: Towards sustainable local energy markets. *Comput. Sci. -Res. Dev.* **2018**, *33*, 207–214. [CrossRef]

20. Saberi, S.; Kouhizadeh, M.; Sarkis, J.; Shen, L. Blockchain technology and its relationships to sustainable supply chain management. *Int. J. Prod. Res.* **2019**, *57*, 2117–2135. [CrossRef]

21. Casino, F.; Dasaklis, T.K.; Patsakis, C. A systematic literature review of blockchain-based applications: Current status, classification and open issues. *Telemat. Inform.* **2018**, *36*, 55–81. [CrossRef]

22. de Vries, A. Renewable Energy Will Not Solve Bitcoin's Sustainability Problem. *Joule* **2019**, *3*, 893–898. [CrossRef]

23. Higgins, S. European Insurance Firms Launch New Blockchain Consortium. Available online: http: //www.coindesk.com/europe-insurance-blockchain-consortium/ (accessed on 11 November 2018).

24. Tian, F. An agri-food supply chain traceability system for China based on RFID & blockchain technology. In Proceedings of the 13th IEEE International Conference on Service Systems and Service Management (ICSSSM), Kunming, China, 24–26 June 2016; pp. 1–6.

25. Truby, J. Decarbonizing Bitcoin: Law and policy choices for reducing the energy consumption of Blockchain technologies and digital currencies. *Energy Res. Soc. Sci.* **2018**, *44*, 399–410. [CrossRef]

26. Kshetri, N. 1 Blockchain's roles in meeting key supply chain management objectives. *Int. J. Inf. Manag.* **2018**, *39*, 80–89. [CrossRef]

27. Casado-Vara, R.; González-Briones, A.; Prieto, J.; Corchado, J.M. Smart contract for monitoring and control of logistics activities: Pharmaceutical utilities case study. In Proceedings of the 13th International Conference on Soft Computing Models in Industrial and Environmental Applications, Donostia-San Sebastian, Spain, 6–8 June 2018; Springer: Cham, Switzerland, 2018; pp. 509–517.

28. Galvez, J.F.; Mejuto, J.C.; Simal-Gandara, J. Future challenges on the use of blockchain for food traceability analysis. *TrAC Trends Anal. Chem.* **2018**, *107*, 222–232. [CrossRef]

29. Khezr, S.; Moniruzzaman, M.; Yassine, A.; Benlamri, R. Blockchain Technology in Healthcare: A Comprehensive Review and Directions for Future Research. *Appl. Sci.* **2019**, *9*, 1736. [CrossRef]

30. Tapscott, D.; Tapscott, A. How blockchain will change organizations. *MIT Sloan Manag. Rev.* **2017**, *58*, 10.

31. Levy, K.E. Book-smart, not street-smart: Blockchain-based smart contracts and the social workings of law. *Engag. Sci. Technol. Soc.* **2017**, *3*, 1–15. [CrossRef]

32. Gimenez, E.; Salinas, M.; Manzano-Agugliaro, F. Worldwide research on plant defense against biotic stresses as improvement for sustainable agriculture. *Sustainability* **2018**, *10*, 391. [CrossRef]

33. Salmerón-Manzano, E.; Manzano-Agugliaro, F. The Higher Education Sustainability through Virtual Laboratories: The Spanish University as Case of Study. *Sustainability* **2018**, *10*, 4040. [CrossRef]

34. Salmerón-Manzano, E.; Manzano-Agugliaro, F. Worldwide scientific production indexed by Scopus on Labour Relations. *Publications* **2017**, *5*, 25. [CrossRef]

35. Gavel, Y.; Iselid, L. Web of Science and Scopus: A journal title overlap study. *Online Inf. Rev.* **2008**, *32*, 8–21. [CrossRef]

36. Salmeron-Manzano, E.; Manzano-Agugliaro, F. The electric bicycle: Worldwide research trends. *Energies* **2018**, *11*, 1894. [CrossRef]

37. Tibaná-Herrera, G.; Fernández-Bajón, M.T.; De Moya-Anegón, F. Categorization of E-learning as an emerging discipline in the world publication system: A bibliometric study in SCOPUS. *Int. J. Educ. Technol. High. Educ.* **2018**, *15*, 21. [CrossRef]

38. Garrido-Cardenas, J.A.; Mesa-Valle, C.; Manzano-Agugliaro, F. Human parasitology worldwide research. *Parasitology* **2018**, *145*, 699–712. [CrossRef]

39. Garrido-Cardenas, J.A.; Manzano-Agugliaro, F.; Acien-Fernandez, F.G.; Molina-Grima, E. Microalgae research worldwide. *Algal Res.* **2018**, *35*, 50–60. [CrossRef]

40. Turban, E.; McElroy, D. Using smart cards in electronic commerce. In Proceedings of the IEEE Hawaii International Conference on System Sciences, Kohala Coast, HI, USA, 9 January 1998; Volume 4, pp. 62–69.

41. Van Eck, N.J.; Waltman, L. Software survey: VOSviewer, a computer program for bibliometric mapping. *Scientometrics* **2010**, *84*, 523–538. [CrossRef] [PubMed]

42. Garrido-Cardenas, J.; Manzano-Agugliaro, F.; González-Cerón, L.; Gil-Montoya, F.; Alcayde-Garcia, A.; Novas, N.; Mesa-Valle, C. The Identification of Scientific Communities and Their Approach to Worldwide Malaria Research. *Int. J. Environ. Res. Public Health* **2018**, *15*, 2703. [CrossRef] [PubMed]

43. Chien, H.Y.; Jan, J.K.; Tseng, Y.M. An efficient and practical solution to remote authentication: Smart card. *Comput. Secur.* **2002**, *21*, 372–375. [CrossRef]
44. Szabo, N. Formalizing and Securing Relationships on Public Networks. 1997. Available online: http://ojphi.org/ojs/index.php/fm/article/view/548/469 (accessed on 15 January 2019).
45. Udokwu, C.; Kormiltsyn, A.; Thangalimodzi, K.; Norta, A. The State of the Art for Blockchain-Enabled Smart-Contract Applications in the Organization. In Proceedings of the 2018 2nd International Conference on Communication and Network Technology, Chandigarh, India, 29–30 March 2018.

Article

Bibliometric Maps of BIM and BIM in Universities: A Comparative Analysis

Mehdi Chihib [1], **Esther Salmerón-Manzano** [2], **Nuria Novas** [1] and
Francisco Manzano-Agugliaro [1,*]

1   Department of Engineering, CEIA3, University of Almeria, 04120 Almeria, Spain
2   Faculty of Law, Universidad Internacional de La Rioja (UNIR), Av. de la Paz, 137, 26006 Logroño, Spain
*   Correspondence: fmanzano@ual.es

Received: 27 May 2019; Accepted: 9 August 2019; Published: 14 August 2019

Abstract: Building Information Modeling (BIM) is increasingly important in the architecture and engineering fields, and especially in the field of sustainability through the study of energy. This study performs a bibliometric study analysis of BIM publications based on the Scopus database during the whole period from 2003 to 2018. The aim was to establish a comparison of bibliometric maps of the building information model and BIM in universities. The analyzed data included 4307 records produced by a total of 10,636 distinct authors from 314 institutions. Engineering and computer science were found to be the main scientific fields involved in BIM research. Architectural design are the central theme keywords, followed by information theory and construction industry. The final stage of the study focuses on the detection of clusters in which global research in this field is grouped. The main clusters found were those related to the BIM cycle, including construction management, documentation and analysis, architecture and design, construction/fabrication, and operation and maintenance (related to energy or sustainability). However, the clusters of the last phases such as demolition and renovation are not present, which indicates that this field suntil needs to be further developed and researched. With regard to the evolution of research, it has been observed how information technologies have been integrated over the entire spectrum of internet of things (IoT). A final key factor in the implementation of the BIM is its inclusion in the curriculum of technical careers related to areas of construction such as civil engineering or architecture.

Keywords: building information modeling (BIM); legal aspects; bibliometric; sustainability; clustering

---

1. Introduction

The growing requirements for establishing sophisticated buildings are making AEC (Architecture Engineering and Construction) projects more complex, while technological advances are helping the participants to collaborate more effectively during the construction process. In fact, the building information model (BIM) provides an intelligent model-based process that connects AEC professionals and helps them to design, build, and operate building infrastructure [1]. This tool allows professionals to design 3D models that incorporate data associated with physical and operational properties, which will help architects, engineers, and contractors to work on a coordinated digital model, giving everyone a better insight into how their work fits in the overall project [2]. BIM systems encourage greater cooperation between stakeholders through a unique integrated model during the design and construction stages. Adopting BIM in the construction industry will lead eventually to a better planning and preparation process by detecting conflicts between elements and improving coordination. In addition, it will help reduce time costing errors and help decision makers to increase their efficiency during the construction phase, and finally will help facilities management with future changes and renovation work.

BIM is the result of an international collaborative progress starting from Japan and moving to Europe and all the way to Northern America, so the history of this concept is not attributed to one name or one place alone. Although the BIM concept has existed since the 1970s, its development went through many steps until this term was first used officially to identify this notion [3]. The first commercial software known as computer-aided manufacturing (CAM) was developed by Dr. Patrick J. Hanratty in 1957. This numerical control machining technology has progressed to become computer-aided manufacturing [4]. Then, he immersed himself in computer-generated graphics and in 1961 developed DAC (Design Automated by Computer), which became later the first system that used CAM/CAD (Computer-Aided Manufacturing/Computer-Aided Design) interactive graphics [5]. The Augmenting Human Intellect paper where the BIM foundation was first documented was published by Douglas C. Englebart in 1962. With the incorporation of object-based systems, this BIM tool allowed architects to introduce several features and specifications for a building. This new advance made the fusion of parametric manipulation and a relational database possible and as a result the 2D illustration of the current design was formed [6]. Afterwards, 3D representations were developed with the Building Description System (BDS) illustrated by Charles Eastman et al. (1975). In their publication, they describe a generic prototype of BDS and consider the perspectives of parametric design and 3D representations with a "single integrated database for visual and quantitative analysis" [7].

After two years, the requirement to integrate building elements and monitor data accuracy was considered, in order to be used as a tool for estimating structural design costs. The Graphic Language for Interactive Design (GLIDE) tool was developed to implement this utility that allows for more reliable and accurate designs. However, both BDS and GLIDE have limited themselves to including only the design stage of the project, which would not allow the immersion of the different stages of the project life cycle [8]. By the year 1984, personal computers began building modeling programs, which included the first BIM (2D CAD) software used worldwide. However, this software wasn't operational until 1986, when Robert Aish used it in large and complex projects such as the renovation of the Heathrow airport terminal [9]. In the 1990s, several companies began to develop BIM tools, such as the Lawrence Berkeley National Laboratory [10]. Autodesk also began using the BIM concept in 2002 when it purchased the Texan company Revit Technology Corporation [11]. The Graphisoft company created the teamwork concept so that team members would be able to easily share BIM data with each other [12].

The complication of buildings and structures, increased construction and the imperative need to reduce design time, the increase in international design cooperation, and other factors led to the accelerated development of computer design tools.

By the early 2000s, objects and shapes have fully incorporated different type of data in the same file, meaning the designer, contractors, engineers and the owner could all work collaboratively on one centralized collaborative model. Objects and shapes had completely incorporated different types of data into the same file, allowing the designer, contractors, engineers, and owner to work collaboratively on a centralized collaboration model. BIM platforms such as the one shown in reference [13] have been created to incorporate parametric flexibility and sculpture geometry that supports NURBS (non-uniform rational B-spline) surfaces, and provides software that larger teams of architects and engineers can use to collaborate on an integrated model based on using a coherent system rather than a set of separate drawings. The new software works with all the information concerning the construction project, while the 3D model can include architectural, structural, electrical, sanitary, plumbing, Heating, Ventilating and Air Conditioning (HVAC) installation, and fire alarm system designs. All of these layers are merged into a BIM file that can be accessed by project holders at any time and from any location [13,14]. Advanced parametric techniques are then introduced into the BIM software [15], which can process complex and contemporary architectural shapes, enabling designers to create curved and complex architectural shapes. With the advancement of computational design technologies, more unique building designs will be realized [16–21].

BIM 4D modeling is employed in combination with the Geographic information system (GIS) to create a safe execution sequence of the process in order to enhance the visual tracking of construction supply chain management [22–24]. This can only prove that the BIM platform has a great potential to integrate various innovative operations related to construction. However, to improve the monitoring of the construction process [25,26], companies are using remote sensing technology to develop an approach called defect management for automatic quality inspection.

In the last decade, BIM technologies have improved greatly with the help of information modeling, meaning it is now possible to solve problems that were unimaginable years ago [27]. These technologies have introduced designer supervision, construction cost planning, risk management, etc. State-of-the-art architecture with unique structures and hazardous facilities, whose projects are subject to mandatory government expertise, can be addressed without great difficulty with the help of these powerful tools [28].

Northern European countries such as Finland adopted BIM regulations, e.g., Common BIM Requirement 2012 (COBIM). In 2016, the U.K became in the first country to legally mandat the use of BIM [29] for public funded projects. Germany is mandating BIM for all transportation projects so teams can collaborate and work in the same model [17,19], which will be useful in so many ways such as dealing with predictive risks and maintenance, improving g timelines and cost savings, as well as asset tracking and facilities management. Government agencies are using BIM software to plan and operate diverse forms of physical infrastructure, such as public sanitation, communication utilities, electricity grids, roads, bridges, and ports [3]. Several European and Asian countries, as well as Australia, and the USA have demanded the use of BIM in projects or have published formal standards of good practice [30]. A study at the Northumbria University campus used BIM to improve the collection of data and its accessibility for facilities management [31]. The digital representation of public infrastructure will not only help authorities to manage its current artworks but also will help them to plan better for future projects to avoid interference and unpredicted modifications.

A BIM execution plan for project implementation would help to explain the details of the necessary checklist and standards [32], such as ISO/TC 59/SC 13. These standards form the foundation for accurate and efficient communication and commerce that are required by the off-site construction industry [33]. Regrettably, this is not usually part of the contract. As such, that Northumbria University campus study had revealed insightful implications into significant legal aspects or contract provisions that need to be included in BIM contracts [34]. As an example, in the literature, it is possible to find engineering, procurement, and construction (EPC) contracting, which enables a contractor to be responsible for all works associated with the design, procurement, erection, and testing of a facility [35]. It is possible to find hydro-supported structures [36] being used as offshore wind turbines [37]. Some authors even propose the application of blockchain or smart contract [38] technologies as a possibility for this type of contract [39,40].

With the growing awareness of society and its contribution to maintaining sustainable systems, the construction industry has taken on this social concern and buildings are now designed with energy efficiency in mind. For this purpose, BEM has been combined with BIM. With this combination, the construction industry has the tools needed to solve problems related to the integrated energy analysis of buildings [41,42]. Another objective is the promotion for the construction of green buildings, the practice of building in a way that safeguards the natural environment [43].

Environmental sustainability concerns are frequently addressed as a complement to building design by pursuing ad hoc approaches to project implementation [44]. As a consequence, the most common problem in reaching a sustainable construction result is the lack of the right information at the right time to make crucial judgments. Furthermore, the design of these high performing buildings is a non-linear, complex, iterative, and multi-disciplinary process that requires efficient collaboration between interdisciplinary teams from the first stages to achieve sustainable outcomes [45]. Construction practitioners make extensive use of performance analysis tools to predict and quantify sustainability issues from the earliest stages of design and significantly improve quality and cost over the life cycle of a building [45]. There are very extensive and recent review studies on BIM and sustainability [46], noting that little work has been conducted about how it could be applied in refurbishment and demolition; but highlighting that BIM can improve social sustainability in two main areas: BIM provides a better facility design for a society's high standard of living. BIM transforms conventional practice, which is often highly fragmented, into a better collaborative effort that strengthens the working relationships among project participants. And this review [46] concludes that future policies of BIM for sustainability should consider improving the interoperability issue among BIM software and energy-simulation tools.

From an engineering point of view, it is important to reflect the complexity of new ways of working. This mixed set of knowledge, skills, and attitudes is essential to strengthen productivity, entrepreneurship, and the pursuit of excellent performance in an environment increasingly based on technologically advanced and sustainable outcomes [47]. A clear example of this involves developing effective measures using a "project-based learning" technique to improve student learning outcomes for the implementation of BIM in the area of sustainability [48]. Therefore, innovative concepts such as sustainability, green concepts, planning processes, or project execution are key aspects to evaluate the key benefits of BIM, and these concepts should be integrated in order to advance their curricula [49].

In this study, there are two objectives. to the first is analyze the background of the whole work published in relation to the BIM subject as bibliometric maps of this subject, and to see which clusters this scientific field is grouped around. And the second objective, given the importance it has for sustainability, is to determine which works of BIM are used in universities to contribute to their sustainability.

The previous published studies of bibliometric analysis on BIM focus on very specific aspects. For example, a study by Badrinath et al. (2016) [33] focuses on how the BIM is taught and then used for communication and visualization. That study was based on the bibliometric analysis of 445 BIM articles, but above all it was based on double keywords not covering the whole subject, these keywords were: "academic BIM education", "BIM curriculum", and "BIM course". Although the work is interesting, it is focused on a very specific area, which is academic BIM education. It was concluded that the case studies and experiences were the dominant type of publication.

In the literature there are other works focused on bibliometric studies of the BIM, but all of them make a subjective classification, and are based on databases other than those used of this study, for example in the WoS™ Core Collection [50] or Web of Science [51]. And although some of them open an important temporal window, from 1990 to 2016 only 567 publications appeared, and the ones which analyzed the relevant topic only numbered 445 [52].

Through Scopus, it is possible to find some bibliometric work but in a very short period of time (2006–2016) [53]. Although it uses community or cluster detection, it found only 4 clusters for 1031 available studies, it is notable that this work was looking for the specific topic of BIM-based Construction Networks (BbCNs). On the subject of the research, it was discovered that collaboration was a concept researched in isolation and without strong connections to other key areas of BIM research. Other works based on Scopus are state of the art revision works [54], and the database was therefore not used only to have a list of published works, but also to develop a bibliometric analysis. However, Scopus covers a wider range of journals in the area of construction project management than the WoS and contains more recent publications than other databases [55].

2. Research Methodology

The flowchart of the research methodology is shown in Figure 1. In this study, all the publications (4307) were collected from Elsevier Scopus database. This platform is useful for bibliometric studies because it allows you to download massive information for numerous bibliometric analysis. The search was conducted in March 2019 to extract research publications that include the following citations "BIM" or "BIM in University" and "Building Information Model" or "Building Information Model in University "in the title, abstract and/or keywords within the period (2003–2018). The following search queries were used: (TITLE-ABS-KEY ({BIM} OR {Building Information Model}), and another one specific for plant: (TITLE-ABS-KEY ({BIM} OR {Building Information Model} AND{Universit*}). Thus, in the first query all the published works on BIM have been found, and the second query acts as a filter on the first, where only those works that deal with the subject of the university remain. It should be noted that the subject categories are the result of the Scopus indexation.

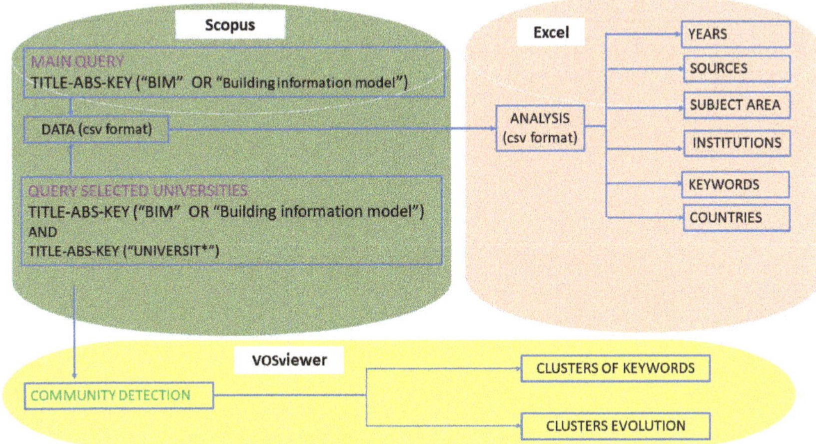

Figure 1. Flowchart of the information search and analysis.

2.1. Data Collection

The main query provides information such as the source of publications according to their authors, institutions, as well as their geographic locations. The publications acquired have been classified based on the following characteristics: the number of publications per year, distribution by subject area, document category, institution, and by country. The word frequency analysis is carried out to reflect the research field, and the core subject of BIM literature keywords are extracted and collected by filtering the main query. The records were consecutively processed using excel sheets, as well as the generation of the result corresponding graphics.

2.2. Community Detection

Community detection is a procedure that identifies geographic locations, trends, and other parameters of a large group of elements that interact with each other. This relationship between elements could vary in intensity that transcribes their dependency on each other. These multiple interdependent nodes evolve around one central core that is highly cohesive, while the density of interactivities decreases as we go far from the center. This structure is called a cluster, and the union of multiple clusters form from a complex network, which usually comes out in the form of a neural network. In this work we have proceeded by using Sw VOSviewer to detect the network community. This tool illustrates the most significant clusters based on the hierarchical connectivity algorithms [37]. This community detection software, VOSviewer, is free software available online that allows the direct import of data in the csv format exported from Scopus and also allows the figures to be exported to a large range of graphical formats. The VOSviewer delivers three displays: network visualization as clusters, overlay visualization as temporal evolution, and density visualization. In all of these cases, the parameters chosen for the analysis were: normalization method (association strength), layout (attraction 2, repulsion 0), clustering (resolution 1.00, minimum cluster size 1), and rotate (90 degrees).

3. Results and Discussion

3.1. Evolution of the Number of Publication Over the Years

We can observe in Figure 2 that the evolution of BIM in university publications is relatively weak compared to BIM global, which has a parabolic pattern over the period studied. The figure also shows that scientific production started to increase substantially since 2007. The total number of BIM and BIM in university publications are 4307 and 274, respectively (3.36%).

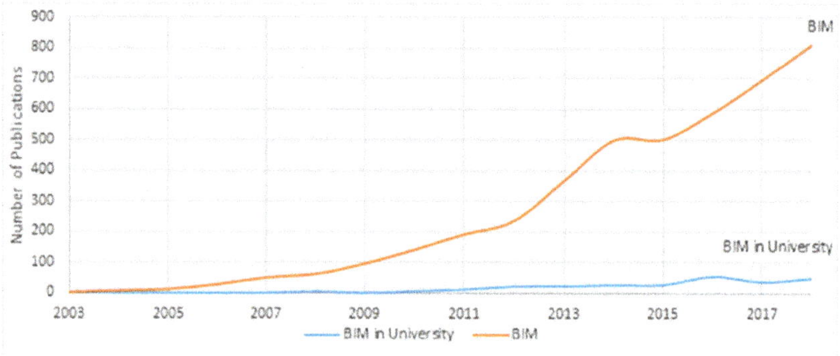

Figure 2. Evolution of the scientific production of BIM (2003–2018).

3.2. Distribution of Output in Subject Categories

Based on the Scopus classification, the distribution of publications on BIM research fields covered a total of 22 subject areas, see Figure 3. The largest number of documents corresponds to engineering (3234 records, 44%), computer science (1423 records, 20%), and business management and accounting (494 records, 7%) while the fourth largest number is for social sciences (389 records, 5%), the fifth is for mathematics (341 records, 5%) and the sixth is for environmental sciences (250 records, 3%). It is worth mentioning that a document can be related to more than one field of research at the same time. These six areas count for about 85% of all publications (Figure 3). The distribution of publications on BIM in the university research area reduced the number of subject areas, enclosing only 18 subject areas. The four first areas were the same as those showed by the global BIM research field. The first highest area according to number of publications was engineering (218 records, 43%), computer science was the second highest area (68 records, 13%), business management and accounting was the third highest (52 records, 10%), and social science was the fourth highest (48 records, 10%). Arts and humanities were in the fifth position (22 records, 4%), while energy accounted for the sixth position (17 records, 3%). BIM studies are mainly focusing on engineering and computer science, which involves architecture, mechanical structure design, and construction. It can be concluded that they are essentially the same categories that are given significance in both cases.

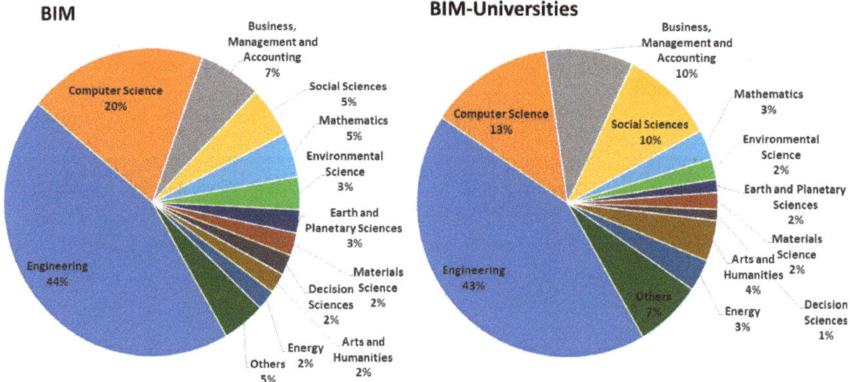

Figure 3. Distribution (%) of worldwide research of BIM and BIM in universities by subject area.

3.3. Types of Publications

The most common means used in scientific diffusion are journal articles. However, in the case of BIM, conference papers are the type of publication that counts for the biggest share with 49%, followed by scientific articles with 41% and articles in review with 3%, followed by books and book chapters with 3%. Figure 4 shows the percentage of the types of scientific production distributed on the building information model theme.

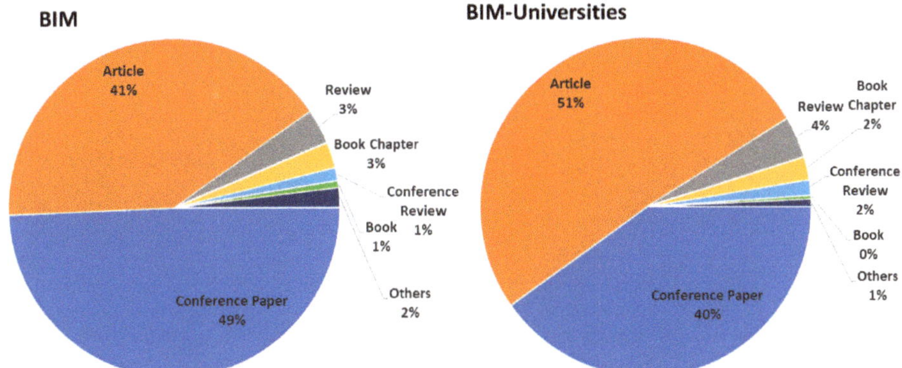

Figure 4. Distribution by publication type related respectively for global BIM and BIM in universities.

Regarding BIM in universities, scientific articles are the type of publication that counts for the largest share of BIM in university publications with 51%, followed by conference papers with 40%, articles in review with 3%, while books and book chapters had 2%. Figure 4 shows the percentages of the types of scientific production distributed on BIM in universities. It is remarkable that in the case of the universities the topic of BIM is predominant in the articles, while for BIM in general the topic is predominant in the conference papers. In general, the higher the percentage of conferences, the more novel the topic is. And when the percentage of books and book chapters is high, this indicates a topic that is scientifically established. BIM is therefore shown to be very novel, given the large percentage of conference paper in both cases.

3.4. Distribution by Countries and Institutions

If the distribution by country of the publications in BIM is represented (see Figure 5), it can be seen that the 10 highest countries are: the United States (20%), the United Kingdom (10%), China (9%), Australia (6%), South Korea (6%), Germany (5%), Canada (4%), Malaysia (3%), Italy (3%), and Taiwan (3%). It can be seen that almost 40% of publications are grouped in the first three countries.

BIM research has been produced in more than 160 institutions. Table 1 shows the top 20 the most productive institutions, with more than 4307 publications covering the BIM concept in the period studied. The first sixteen institutions (10% of total institutions) are from the USA, Australia, the UK, South Korea, China, Italy, Malaysia, Israel, and Germany. They are represented by the following affiliations: the Georgia Institute of technology (USA), Curtin University (UK), University of Florida (USA), University of Salford (UK), Kyung Hee University (South Korea), Pennsylvania State (USA), Hong Kong Polytechnic University (China), Politecnico di Milano (Italy), University Tongji University (China), Hanyang University (South Korea), Universiti Teknologi Malaysia (Malaysia), Cardiff University (UK), University College London (UK), Tsinghua University (China), Israel Institute of Technology (Israel), and the Technical University of Munich (Germany). Universities from the USA have the most representations with 988 publications, UK has the second highest production with 518 records, while China has the third highest rank with 460 publications. BIM in university publications have been produced in the same number of institutions as the global BIM.

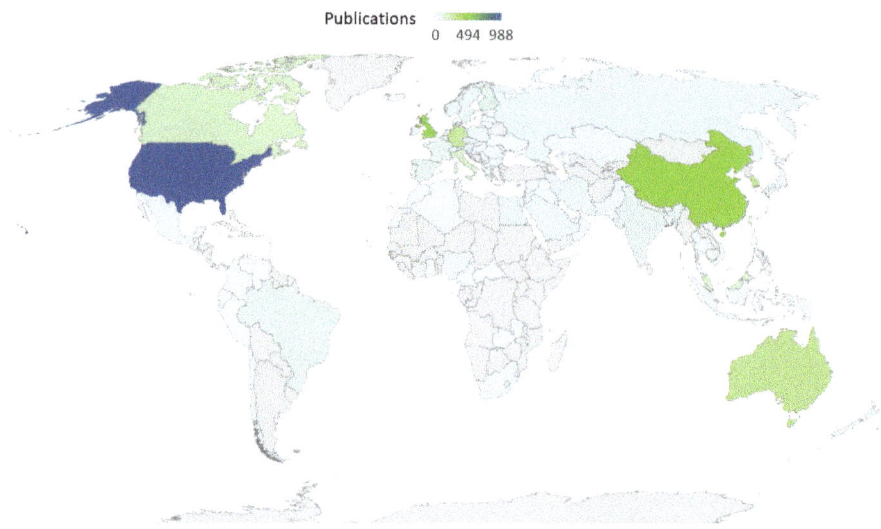

Figure 5. Distribution by countries of BIM publications.

Table 1. Classification of research institutions by record and countries.

BIM			BIM in University		
Affiliation	**Country**	**N**	**Affiliation**	**Country**	**N**
Georgia Institute of Technology	USA	101	Pennsylvania State University	USA	9
Curtin University	UK	98	Hong Kong Polytechnic University	China	9
University of Florida	USA	93	Tsinghua University	China	8
University of Salford	UK	68	Arizona State University	USA	6
Kyung Hee University	South Korea	56	Ceské vysoké ucení technické v Praze	Czech The Czech Republic	5
Pennsylvania State University	USA	55	National Taipei University of Technology	China	4
Hong Kong Polytechnic University	China	54	University of Southern California	USA	4
Politecnico di Milano	Italy	49	Vilniaus Gedimino technikos universitetas	Lithuania	4
Tongji University	China	49	Universidade de Lisboa	Portugal	4
Hanyang University	South Korea	48	Helsingin Yliopisto	Sweden	3
Universiti Teknologi Malaysia	Malaysia	47	University of Texas at San Antonio	USA	3
Cardiff University	UK	46	National Taiwan University	Taiwan	3
UCL (University College London)	UK	43	University of Salford	UK	3
Tsinghua University	China	43	University of Wyoming	USA	3
Technion—Israel Institute of Technology	Israel	42	Universitat d'Alacant	Spain	3
Technical University of Munich	Germany	41	International University of Florida	USA	3

However, the BIM in university research area had more than 274 publications in the studied period, although this number remains small in comparison to BIM global. The sixteen most productive institutions (10% of total institutions) have been developed in the following countries: the USA, the Czech Republic, the UK, China, Lithuania, Spain, Portugal, and Sweden, all of which are represented by the following universities: Pennsylvania State University (USA), Hong Kong Polytechnic University (China), Tsinghua University (China), Arizona State University (USA), Ceské vysoké Ucení technické v Praze (Czech Republic), National Taipei University of Technology (China), University of Southern California (USA), Vilniaus Gedimino technikos Universitetas (Lithuania), Universidade de Lisboa (Portugal), Helsingin Yliopisto (Sweden), University of Texas at San Antonio (U.S.A), National Taiwan University (Taiwan), University of Salford (UK), University of Wyoming (USA), Universitat d'Alacant (Spain), and the Florida International University (USA). Institutions from the USA are the most highly represented with 79 publications, the second country had 28 publications, while Chinese universities had the third rank with 18 publications.

3.5. Relationships between Countries

Figure 6 shows the labeled clusters with the relationships between the countries of the various publications. This is generally determined by the co-authorship of articles, i.e., authors from different countries who wrote the same article. Each element represents a country and the size of these elements is determined by the total number of publications of this country. The network counts 10 communities, their rank order respectively is the USA, the UK, China, Australia, Germany, South Korea, Italy, Spain, Finland, and Taiwan. It is observed that in the clusters of these countries, there is a major correlation with nearby or neighboring countries, which is not frequently found in scientific subjects. Examples of this include the USA with Canada, China with Hong Kong, Germany with Austria or Switzerland, the UK with Ireland, Finland with Norway, or the Czech Republic with Poland.

The countries that are in the middle of the cluster are the ones who are linked with the most nodes. Language plays a key role in the interconnections between countries. The largest community is the one that evolves around the USA. These publications are written mostly in English in more than 92% of cases, although the Chinese language has also appeared in 5.5% of publications since 2006, while other languages numbered less than 1% and included Japanese, Dutch, German, Polish, Russian, Spanish, and French.

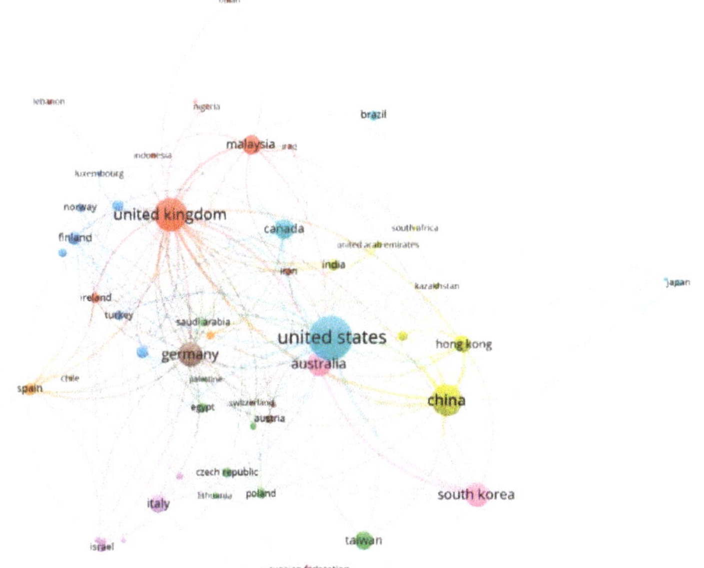

Figure 6. Countries network of BIM publications.

3.6. Keyword Analysis

The keywords analysis identifies the common interests of the researchers and their work. In this section we analyze the keywords acquired from the main query as well as their frequency of appearance in every article during the period studied. If the main keywords associated with the theme of the global BIM are analyzed, those of Table 2 are obtained, where the 15 main keywords have been selected. The words of the search itself, such as BIM or Building Information Modeling or Building Information Modelling, have not been taken into account (since it is written in both forms almost equally frequently). It is noted that the general main search keywords are also the first keywords of the particular search in universities. However, in the latter case there are, as expected, issues related to teaching: Students, Curricula, Teaching, Education, or Engineering Education.

Table 2. Main keywords related with both queries.

BIM	N	BIM in Universities	N
Architectural Design	3075	Architectural Design	168
Information Theory	1295	Information Theory	66
Construction Industry	935	Construction Industry	41
Buildings	671	Students	40
Construction	642	Buildings	39
Project Management	602	Curricula	38
Information Management	458	Project Management	37
Structural Design	458	Teaching	37
Life Cycle	377	Construction	35
Construction Projects	354	Education	35
Sustainable Development	291	College Buildings	34
Office Buildings	279	Information Management	30
Design	255	Engineering Education	27
Computer Aided Design	239	Surveys	20
Decision Making	224	Life Cycle	19

The more interesting data are the keywords College Buildings, which show that BIM is starting to also be applied for the construction of university buildings, while global BIM is mainly focused on office buildings. If a visual representation is made with clouds of the keywords, then Figure 7; Figure 8 are obtained. These figures show where a particular study has to be done for automation, sustainable development, or industry foundation classes (IFC).

Figure 7. Word cloud of keywords related to BIM (global query).

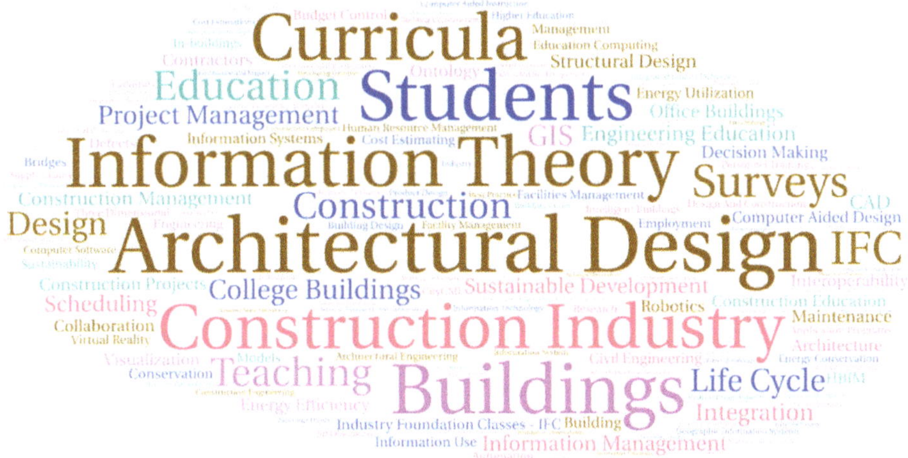

Figure 8. Word cloud of keywords related to BIM in universities.

Figure 9 shows the evolution of the following keywords: automation, sustainable development, and IFC. The three-keyword records were correlated during the greater part of the period studied. In addition, automation and sustainable development had approximately the same level of variance in the same value until 2013, when the gap started to grow wider between them. The automation record continued to increase remarkably until it reached 94 records in 2018. IFC had been scoring a lower score than sustainable development until the year 2018, when IFC obtained 53 records and sustainable development obtained 33 records.

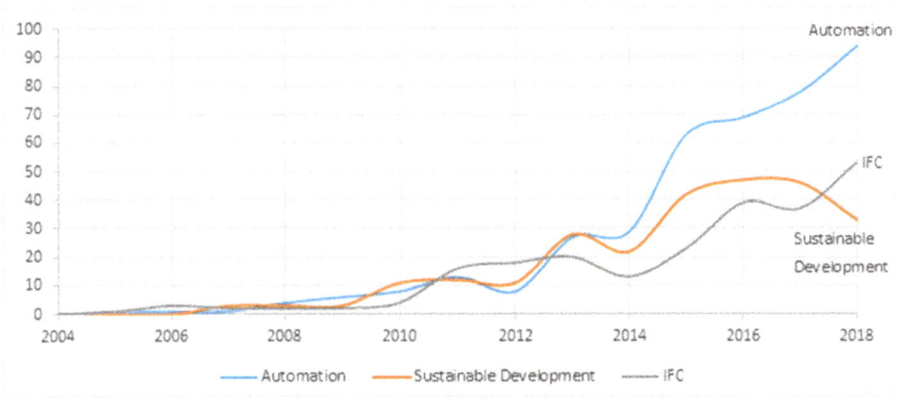

Figure 9. Countries network of BIM publishers and their community detection.

The major topics in Table 3 and Figure 10 constitute the structure of BIM. Management is the first topic that most relevant keywords evolve around, while the technology of BIM plays a crucial role in improving the interaction between different contributors and the ways in which they manage various aspects of the project development and BIM implementation during the building life cycle.

Table 3. Main keywords used by the communities detected in the topic BIM.

Cluster	Color	Main Keywords	Topic
1	Red	Construction management —Collaboration –Information technology—Bim adoption	Construction management
2	Green	Interoperability-Facility management—Industry foundation classes —Internet of Things	Documentation and Analysis
3	Blue	Architecture—Virtual Reality—Education—GIS	Architecture and Design
4	Orange	Lean construction—Implementation—Adoption —Benefits	Construction/Fabrication
5	Yellow	Energy efficiency—sustainable design—Leadership in Energy and Environmental Design—energy simulation—Building performance	Operation and maintenance

Figure 10 illustrates numerous keyword clusters in the form of a neural network with different colors, where the co-occurrence of keywords occurs at least 5 times. Each node is a keyword, and the link thickness between nodes represents the degree of connection. The BIM keywords analysis has identified six communities using a community detection algorithm and Table 2 shows their main clusters by their order of importance. The most significant clusters besides building information model are construction, interoperability, virtual reality, architecture, collaboration and construction management, visualization, and automation. These communities provide a general indication of the fields that are related to BIM, which are diverse and technology oriented.

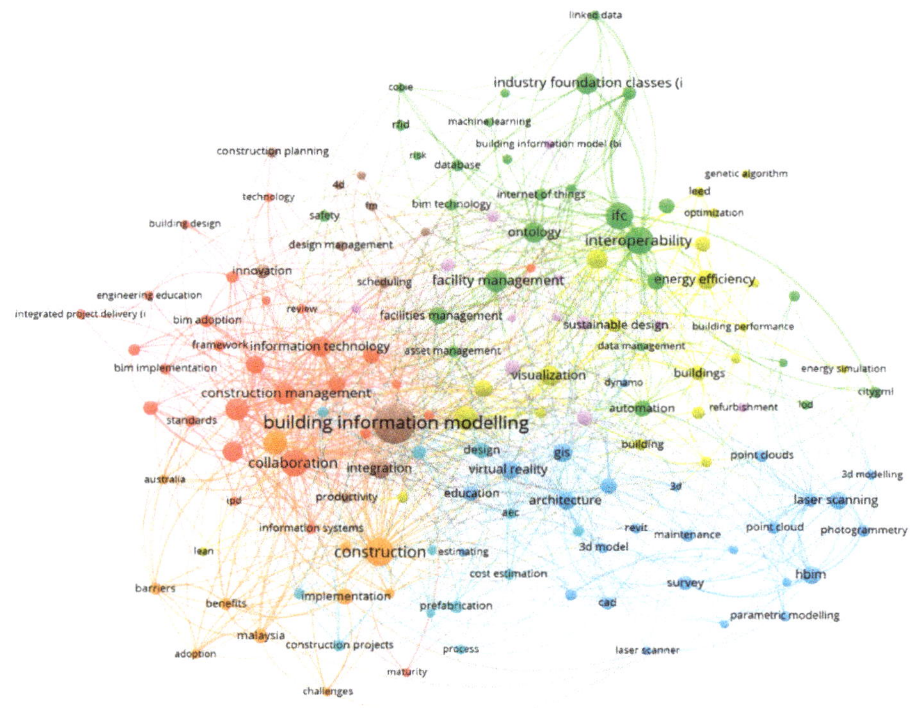

Figure 10. Keywords co-occurrence network related to BIM worldwide.

Analyzing the first cluster, in red in Figure 11, where only keywords with a minimum of 15 co-occurrences have been highlighted, it can be seen that it is related to construction management, collaboration, information technology, and BIM adoption. This shows that BIM is not only useful for geometric modeling of a building's performance but can also assist in the management of construction projects [56]. Some works highlight the synergy between facility management and BIM as a basis for multidisciplinary collaboration [57]. Within this community or cluster, the information technology that is developing object-oriented Computer Aided Design (CAD) tools compatible with BIM can be found, such as analysis tools, model verifiers, and facility management applications [58]. It should be remembered that the main difference between BIM technology and conventional 3D CAD is that the latter describes a building through independent 3D views, such as plans, while BIM uses all the information related to the building, including its physical and functional characteristics and information about the project life cycle, in a series of "smart objects" [59].

The second cluster, in green in Figure 12, where only keywords with a minimum of 15 co-occurrences have been highlighted, is shown to be associated with interoperability, facility management, industry foundation classes, and IoT. BIM is an expansive area of knowledge inside the architecture, engineering and construction industry [60]. An example of this is building automation, e.g., for on-site assembly services in prefabricated buildings with IoT, where the IoT-enabled platform can provide various decision support tools and services to different stakeholders [61]. As an example of industry foundation classes, it is possible to find works where various standards have been published, leading up to the 10-year development of industry foundation classes [62].

The cluster related to architecture, cluster 3 with the color blue in Figure 13, where only keywords with a minimum of 15 co-occurrences have been highlighted, has architecture; virtual reality, education, and GIS as the main related keywords. BIM offers the potential to achieve a lower project cost, increase productivity and quality, and reduce project turnaround time [58]. There are several great examples in the literature of the integration of architecture and GIS showing benefits such as reusability and extensibility [63]. Virtual reality is relevant because it allows us to make a 3D reconstruction of architecture appearance [64]. Both areas are very important in teaching, e.g., for the teaching of architecture through augmented reality [65,66].

The orange cluster four in Figure 14, is mainly related to the keyword's lean construction, implementation, and adoption benefits, cost/benefit analysis, awareness raising, and education and training, all of which are important activities to address the challenges of BIM usage. From the analysis of numerous works related to BIM, it was inferred that the benefit of BIM most frequently relates to cost reduction and control throughout the project life cycle, but significant time savings were also reported. The costs of BIM focused primarily on the use of BIM software [67]. Of course, the benefits are proportional to the form of implementation of the BIM [68].

The last cluster, in yellow in Figure 15, where only keywords with a minimum of 15 co-occurrences have been highlighted, focuses on the keywords: Energy efficiency, sustainable design, leadership in energy, environmental Design, energy simulation, and building performance. For a sustainable building, the use of its energy always involves customers and designers [69]. This includes important aspects of environmental design, e.g., to determine what the forecast CO_2 emissions are from the building and whether it will meet the performance criteria [70]. It should be noted that retrofitting of existing buildings offers significant opportunities for reducing global energy consumption and greenhouse gas emissions [71,72]. Therefore, a model based on BIM that can enhance the post-occupancy assessment processes and meet the industry standards for sustainable buildings would be useful.

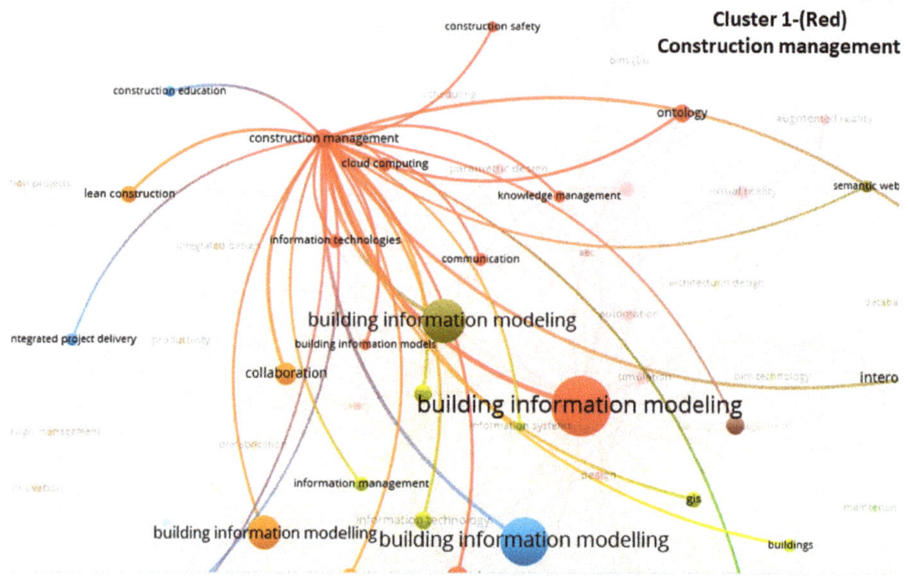

Figure 11. Keywords co-occurrence for cluster 1 (Construction management).

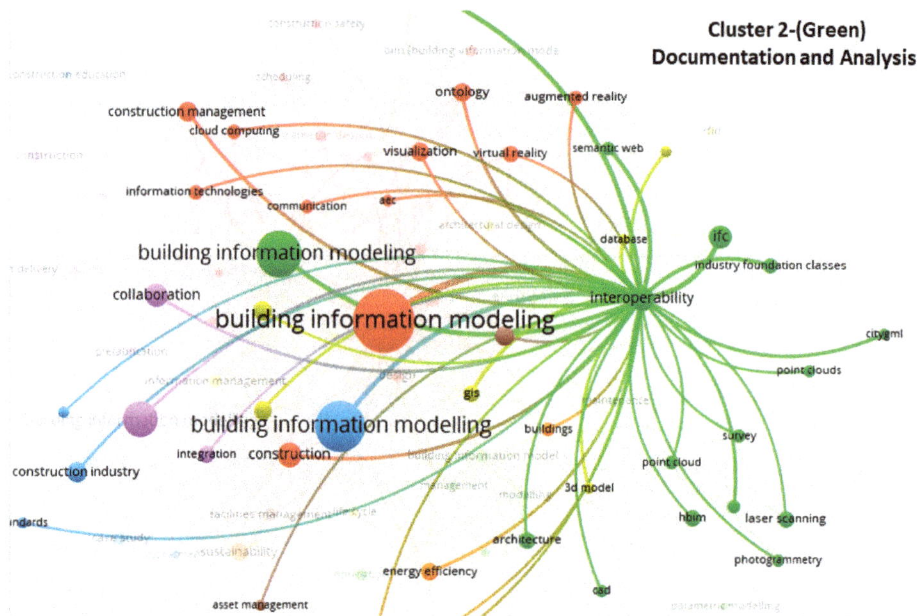

Figure 12. Keywords co-occurrence for cluster 2 (Documentation and Analysis).

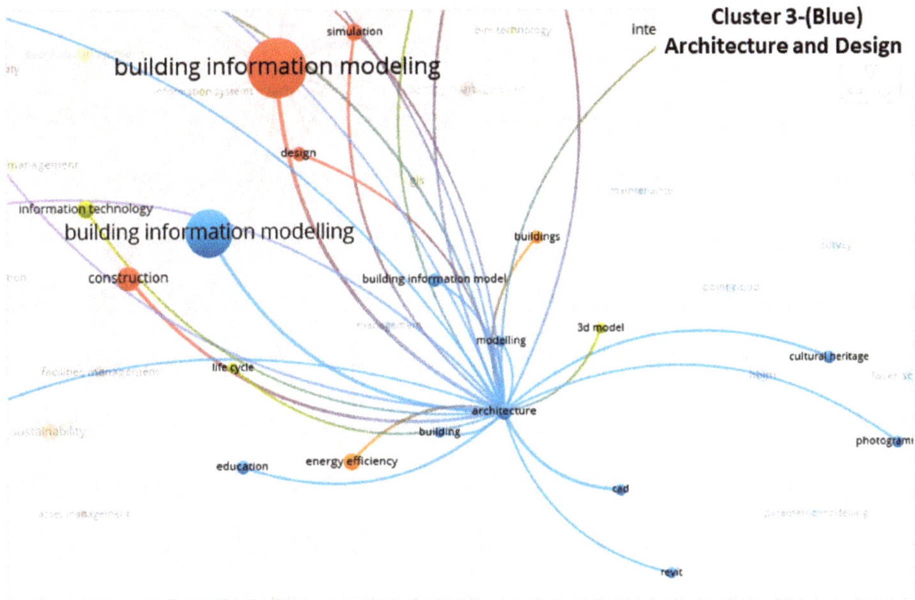

Figure 13. Keywords co-occurrence for cluster 3 (Architecture and Design).

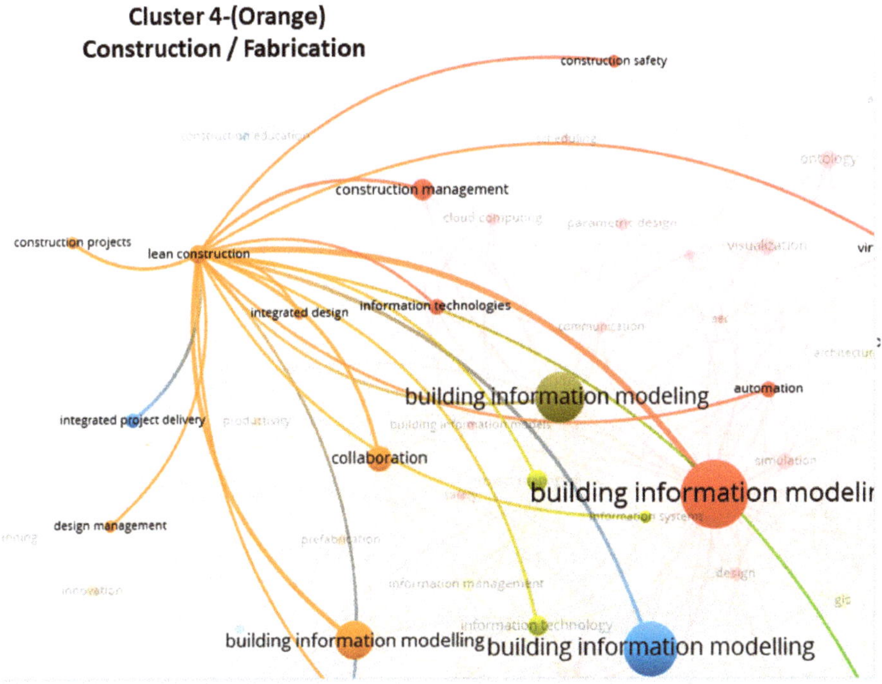

Figure 14. Keywords co-occurrence for cluster 4 (Construction / Fabrication).

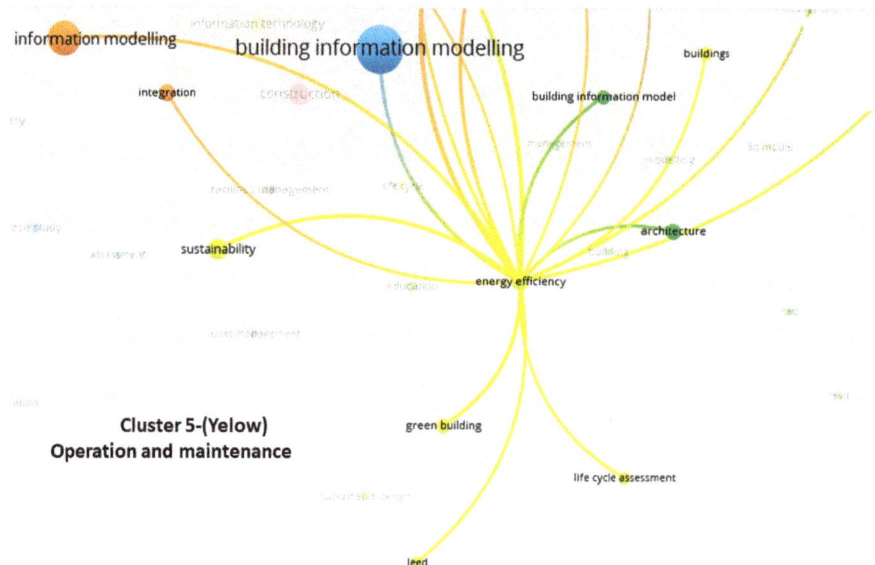

Figure 15. Keywords co-occurrence for cluster 5 (Operation and maintenance).

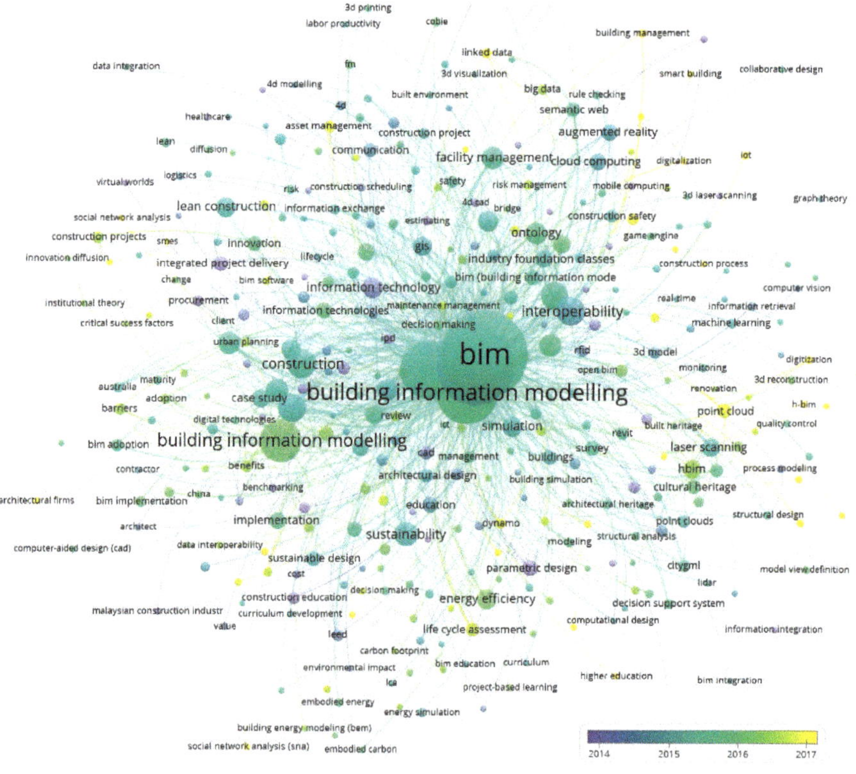

Figure 16. Evolution BIM research (2014–2017).

The network analysis illustrates the evolution of BIM within the period from 2014 to 2017 (Figure 16). It shows that all the major clusters spread from the central communities that represent BIM in different designations, while this analysis also reveals the structure of the current network and the composition of the future trends.

Of the whole studied period of BIM research, i.e., from 2003 to 2018, the period with the highest rate of evolution in BIM research found is from 2014 to 2017. In those years, the keywords were (indicated by purple in the year 2014): cad, parametric design, information technology, built heritage or integrated project delivery. In the years 2015–2016 (indicated in green) the main keywords were: lean construction, GIS, facility management, energy efficiency, or sustainability. The end of the evolution period (yellow, the year 2017) is shown by frequent keywords used: social network analysis, IoT, safety, SMES as the acronym for Small and Medium-sized Enterprises, and higher education. This evolution shows how information concepts and technologies have been incorporated into the research of the BIM as the IoT, obtaining importance in the curricula of students of higher education in careers such as engineering and architecture as a key factor for BIM implementation [73–75].

4. Discussion

Countries like China, the UK, Canada, South Korea, Germany, the USA, Australia and Italy are among the top publishers of both BIM global and BIM in universities. The geographic location plays a major role in the composition of most of the clusters. The UK's cluster is larger than more than six countries outside of the European continent combined. In addition, China is the leader in research about BIM within the Asian continent. The collaborating work of authors shapes the bibliometric map of BIM through numerous parameters, while citation network analysis of the cited references indicates a wide range of subjects in this field of research such as computer science, engineering, business, management and accounting. These different subjects show the diversity of this research area. The keywords analysis provided a list of diverse words related to themes like architectural design, construction management, interoperability, lean construction, virtual reality, visualization, robotics and sustainability development. The extensive amount of data that is generated to improve the facilities management requires multidisciplinary applications of BIM. Therefore, the use of advanced technology is emerging in order to be able to respond properly to market challenges. BIM applications are moving towards IoT, safety, digitalization, smart building, social network analysis, and point cloud. Thus, automation will play a significant role either in providing a highly accurate 3D model for the existing buildings, or in providing a system that measures, collects, and analyzes data of the key performance metrics based on the IoT concept. Furthermore, a digitally empowered framework will enable the decentralization of facilities management for single or multiple buildings [20,21], and could provide a finished product to end-users for cognitive building operation. In term of safety, professionals and researchers are working to develop an approach to integrate the risk factor in building an information model. The tool will be able to detect and quantify automatically any potential risk within the construction site and the life cycle of the project [17,19,26]. Several studies have applied social network analysis (SNA) to investigate major risks related to the act of building and to identify the network structure of all the contributor relationships [26]. Other research suggests using risk factors integration from an online application called the Safety in Design Risk Evaluator, which measures risks at the item-level in multistory buildings with a 4D building information model and a construction timetable [31]. Therefore, BIM trends as many other concepts are mainly heading towards the adoption of digital technologies, big data, IoT, smart models, and machine learning. The expertise areas extracted from the co-occurrence network include interoperability, IFC, lean construction, BIM implementation, energy efficiency and BIM education. Most of these fields are technology based, which has led to a fast-growing knowledge of BIM and its sub-areas that we can see in the evolution pattern. Therefore, BIM education should be constantly upgraded to deliver a valuable knowledge of this dynamic platform.

BIM should be understood as cycle where are all the phases related to the building industry. These phases are: programming, conceptual design, analysis, documentation, fabrication, construction, construction logistics or management, operation and maintenance, demolition, and renovation. The clusters obtained in the previous section reflect almost all of these phases of the BIM, but the cluster of demolition and renovation are missing. This gap in BIM research is already pointed out by some recent works [76]. This shows that these two fields of research within the BIM; although they are currently not fully developed.

5. Conclusions

This bibliometric approach can meaningfully contribute to the ongoing manual review of BIM. Conference papers are the main source of scientific publications, followed by scientific article and reviews. Experts and researchers mostly contribute to expanding BIM literature through these channels, and the rest are published through book chapters, conference reviews and article publications. The scientific contribution in this study refers to 4307 articles associated with BIM, where only 6.4% of these articles related are to BIM in universities and 46% is published in just three countries, which are the USA, UK, and China. The bibliographic records provide users with necessary data about the affiliation of different articles. Furthermore, four out of sixteen universities are present for both of the research topics BIM and BIM in universities. Georgia Institute of Technology and Pennsylvania State University are the leaders in this emerging area of research. It is also observed that the countries that made the usage of BIM mandatory in the regulation of construction are the ones which have the most interest in researching and developing this concept. The five clusters obtained in BIM research are those of the cycle in which all phases related to the construction industry are found: construction management, documentation and analysis, architecture and design, construction / fabrication, and operation and maintenance (related to energy or sustainability). However, the clusters of the last phases such as demolition and renovation are not present, which indicates a field that still needs to be developed and researched. With regard to the evolution of research, it has been observed how information technologies have been integrated with IoT. Finally, a key factor in the implementation of the BIM is its inclusion in the curriculum of technical careers related to construction such as civil engineering or architecture. Therefore, in order to remain up to date and meaningful, education in construction needs to take advantage of the opportunities and overcome the challenges presented by BIM. This bibliometric analysis provides a general overview of the subject in order to concentrate on the strategies that are still relevant and to open up promising new lines of research. This work opens new perspectives for the use of the BIM in universities, which has been found to be less extensively covered than BIM at a global level.

Author Contributions: M.C. dealt with literature review and article writing. E.S.-M. and N.N. analyzed the data and formatting. F.M.-A. Research idea, article writing. They share the structure and aims of the manuscript, paper drafting, editing and review. All authors have read and approved the final manuscript.

Funding: This research received no external funding.

Conflicts of Interest: The authors declare no conflict of interest.

References

1. Eastman, C.; Teicholz, P.; Sacks, R.; Liston, K. *BIM Handbook: A Guide to Building Information Modeling for Owners, Managers, Designers, Engineers and Contractors*; John Wiley & Sons: Hoboken, NJ, USA, 2011.
2. Hergunsel, M.F. Benefits of Building Information Modeling for Construction Managers and BIM Based Scheduling. Master's Thesis, Worcester Polytechnic Institute, Worcester, MA, USA, 2011.
3. Rynne, C. *Industrial Ireland 1750–1930: An Archaeology*; Collins Press: New York, NY, USA, 2006.
4. Sturdevant, J.R.; Bayne, S.C.; Heymann, H.O. Margin gap size of ceramic inlays using second—Generation CAD/CAM equipment. *J. Esthet. Restor. Dent.* **1999**, *11*, 206–214. [CrossRef]

5. Goodacre, C.J.; Garbacea, A.; Naylor, W.P.; Daher, T.; Marchack, C.B.; Lowry, J. CAD/CAM fabricated complete dentures: Concepts and clinical methods of obtaining required morphological data. *J. Prosthet. Dent.* **2012**, *107*, 34–46. [CrossRef]

6. Engelbart, C. *Augmenting Human Intellect: A Conceptual Framework*; Doug Engelbart Institute: Menlo Park, CA, USA, 1962.

7. Eastman, C. The use of computers instead of drawings in building design. *AIA J.* **1975**, *63*, 46–50.

8. Eastman, C.; Henrion, M. Glide: A Language for Design Information Systems. In *ACM SIGGRAPH Computer Graphics*; ACM: New York, NY, USA, 1977.

9. Aish, R. Three-dimensional input and visualization. In *Computer-Aided Architectural Design Futures*; Butterworth-Heinemann: Oxford, UK, 1986; pp. 68–84.

10. Van Nederveen, G.A.; Tolman, F.P. Modeling multiple views on buildings. *Autom. Constr.* **1992**, *1*, 215–224. [CrossRef]

11. Vysotskiy, A.; Makarov, S.; Zolotova, J.; Tuchkevich, E. Features of BIM implementation using autodesk software. *Procedia Eng.* **2015**, *117*, 1143–1152. [CrossRef]

12. Ulmanis, J. *CASE 6: Graphisoft: The Architecture of International Growth*; Entrepreneurial Icebreakers; Palgrave Macmillan: London, UK, 2015; pp. 193–209.

13. Mueller, V.; Smith, M. Generative Components and Smartgeometry: Situated Software Development. *Inside Smartgeometry Expand. Archit. Possibilities Comput. Des.* **2013**, *2013*, 142–153.

14. Watson, A. Digital buildings–Challenges and opportunities. *Adv. Eng. Inform.* **2011**, *25*, 573–581. [CrossRef]

15. Poole, M.; Shvartzberg, M. (Eds.) *The Politics of Parametricism: Digital Technologies in Architecture*; Bloomsbury Publishing: London, UK, 2015.

16. Wu, P.; Wang, J.; Wang, X. A critical review of the use of 3-D printing in the construction industry. *Autom. Constr.* **2016**, *68*, 21–31. [CrossRef]

17. Macher, H.; Landes, T.; Grussenmeyer, P. From point clouds to building information models: 3D semi-automatic reconstruction of indoors of existing buildings. *Appl. Sci.* **2017**, *7*, 1030. [CrossRef]

18. Ochmann, S.; Vock, R.; Klein, R. Automatic reconstruction of fully volumetric 3D building models from oriented point clouds. *ISPRS J. Photogramm. Remote Sens.* **2019**, *151*, 251–262. [CrossRef]

19. Wang, C.; Cho, Y.K.; Kim, C. Automatic BIM component extraction from point clouds of existing buildings for sustainability applications. *Autom. Constr.* **2015**, *56*, 1–13. [CrossRef]

20. Teizer, J.; Wolf, M.; Golovina, O.; Perschewski, M.; Propach, M.; Neges, M.; König, M. Internet of things (IoT) for integrating environmental and localization data in building information modeling (BIM). In Proceedings of the International Symposium on Automation and Robotics in Construction (ISARC), Vilnius Gediminas Technical University, Department of Construction Economics & Property, Taipei, Taiwan, 28 June–1 July 2017; Volume 34.

21. Pasini, D.; Ventura, S.M.; Rinaldi, S.; Bellagente, P.; Flammini, A.; Ciribini, A.L.C. Exploiting Internet of Things and building information modeling framework for management of cognitive buildings. In Proceedings of the 2016 IEEE International Smart Cities Conference (ISC2), Trento, Italy, 12–15 September 2016; Volumes 1–6.

22. Deng, Y.; Cheng, J.C.P.; Anumba, C. Mapping between BIM and 3D GIS in different levels of detail using schema mediation and instance comparison. *Autom. Constr.* **2016**, *67*, 1–21. [CrossRef]

23. Liu, X.; Wang, X.; Wright, G.; Cheng, J.C.P.; Li, X.; Liu, R. A state-of-the-art review on the integration of Building Information Modeling (BIM) and Geographic Information System (GIS). *ISPRS Int. J. Geo Inf.* **2017**, *6*, 53. [CrossRef]

24. Irizarry, J.; Karan, E.P.; Jalaei, F. Integrating BIM and GIS to improve the visual monitoring of construction supply chain management. *Autom. Constr.* **2013**, *31*, 241–254. [CrossRef]

25. Li, C.Z.; Hong, J.; Xue, F.; Shen, G.Q.; Xu, X.; Mok, M.K. Schedule risks in prefabrication housing production in Hong Kong: A social network analysis. *J. Clean. Prod.* **2016**, *134*, 482–494. [CrossRef]

26. Jin, Z.; Gambatese, J.; Liu, D.; Dharmapalan, V. Risk Assessment in 4D Building Information Modeling for Multistory Buildings. In Proceedings of the Joint CIB W099 and TG59 Conference, Coping with the Complexity of Safety, Health, and Wellbeing in Construction, Salvador, Brazil, 1–3 August 2018; pp. 84–92.

27. McGlinn, K.; Wagner, A.; Pauwels, P.; Bonsma, P.; Kelly, P.; O'Sullivan, D. Interlinking geospatial and building geometry with existing and developing standards on the web. *Autom. Constr.* **2019**, *103*, 235–250. [CrossRef]

28. Mishchenko, E.S.; Monastyrev, P.V.; Evdokimtsev, O.V. Improving the Quality of Training in Building Information Modeling. In *International Conference on Interactive Collaborative Learning*; Springer: Cham, Switzerland, 2018; pp. 453–459.

29. Alwan, Z.; Jones, P.; Holgate, P. Strategic sustainable development in the UK construction industry, through the framework for strategic sustainable development, using Building Information Modeling. *J. Clean. Prod.* **2017**, *140*, 349–358. [CrossRef]

30. Bradley, A.; Li, H.; Lark, R.; Dunn, S. BIM for infrastructure: An overall review and constructor perspective. *Autom. Constr.* **2016**, *71*, 139–152. [CrossRef]

31. Kassem, M.; Kelly, G.; Dawood, N.; Serginson, M.; Lockley, S. BIM in facilities management applications: A case study of a large university complex. *Built Environ. Proj. Asset Manag.* **2015**, *5*, 261–277. [CrossRef]

32. Fan, S.L.; Chong, H.Y.; Liao, P.C.; Lee, C.Y. Latent Provisions for Building Information Modeling (BIM) Contracts: A Social Network Analysis Approach. *KSCE J. Civ. Eng.* **2019**, *23*, 1427–1435. [CrossRef]

33. Badrinath, A.C.; Chang, Y.T.; Hsieh, S.H. A review of tertiary BIM education for advanced engineering communication with visualization. *Vis. Eng.* **2016**, *4*, 9. [CrossRef]

34. Fan, S.L.; Lee, C.Y.; Chong, H.Y.; Skibniewski, M.J. A critical review of legal issues and solutions associated with building information modeling. *Technol. Econ. Dev. Econ.* **2018**, *24*, 2098–2130. [CrossRef]

35. Lee, C.Y.; Chong, H.Y.; Wang, X. Enhancing BIM performance in EPC projects through integrative trust-based functional contracting model. *J. Constr. Eng. Manag.* **2018**, *144*, 06018002. [CrossRef]

36. Nawari, N.O. BIM Data Exchange Standard for Hydro-Supported Structures. *J. Archit. Eng.* **2019**, *25*, 04019015. [CrossRef]

37. Manzano-Agugliaro, F.; Calero, M.S.; Perea-Moreno, A.J.; De San Antonio Gomez, C. Ensayo de segregación en cimentaciones para la estructura soporte de aerogeneradores off-shore hechas por vertido en caída libre. *DYNA* **2018**, *93*, 221–227. [CrossRef]

38. Nawari, N.O.; Ravindran, S. Blockchain and the built environment: Potentials and limitations. *J. Build. Eng.* **2019**, *25*, 100832. [CrossRef]

39. Nawari, N.O.; Ravindran, S. Blockchain technology and BIM process: Review and potential applications. *J. Inf. Technol. Constr.* **2019**, *24*, 209–238.

40. Salmerón-Manzano, E.; Manzano-Agugliaro, F. The Role of Smart Contracts in Sustainability: Worldwide Research Trends. *Sustainability* **2019**, *11*, 3049. [CrossRef]

41. Maskil-Leitan, R.; Reychav, I. BIM's social role in building energy modeling. *Clean Technol. Environ. Policy* **2019**, *21*, 307–338. [CrossRef]

42. Gao, H.; Koch, C.; Wu, Y. Building information modeling based building energy modeling: A review. *Appl. Energy* **2019**, *238*, 320–343. [CrossRef]

43. Abanda, F.H.; Byers, L. An investigation of the impact of building orientation on energy consumption in a domestic building using emerging BIM (Building Information Modeling). *Energy* **2016**, *97*, 517–527. [CrossRef]

44. Zanni, M.A.; Soetanto, R.; Ruikar, K. Towards a BIM-enabled sustainable building design process: Roles, responsibilities, and requirements. *Archit. Eng. Des. Manag.* **2017**, *13*, 101–129. [CrossRef]

45. Bouchlaghem, D.; Shang, H.; Anumba, C.J.; Cen, M.; Miles, J.; Taylor, M. ICT-enabled collaborative working environment for concurrent conceptual design. *Archit. Eng. Des. Manag.* **2005**, *1*, 261–280. [CrossRef]

46. Chong, H.Y.; Lee, C.Y.; Wang, X. A mixed review of the adoption of Building Information Modeling (BIM) for sustainability. *J. Clean. Prod.* **2017**, *142*, 4114–4126. [CrossRef]

47. Zanni, M.A.; Soetanto, R.; Ruikar, K. Defining the sustainable building design process: Methods for BIM execution planning in the UK. *Int. J. Energy Sector Manag.* **2014**, *8*, 562–587. [CrossRef]

48. Crawley, D.B.; Hand, J.W.; Kummert, M.; Griffith, B.T. Contrasting the capabilities of building energy performance simulation programs. *Build. Environ.* **2008**, *43*, 661–673. [CrossRef]

49. Crawley, E.; Malmqvist, J.; Ostlund, S.; Brodeur, D.; Edstrom, K. Rethinking engineering education. *CDIO Approach* **2007**, *302*, 60–62.

50. Li, X.; Wu, P.; Shen, G.Q.; Wang, X.; Teng, Y. Mapping the knowledge domains of Building Information Modeling (BIM): A bibliometric approach. *Autom. Constr.* **2017**, *84*, 195–206. [CrossRef]

51. Zhao, X. A scientometric review of global BIM research: Analysis and visualization. *Autom. Constr.* **2017**, *80*, 37–47. [CrossRef]

52. Olawumi, T.O.; Chan, D.W.; Wong, J.K. Evolution in the intellectual structure of BIM research: A bibliometric analysis. *J. Civ. Eng. Manag.* **2017**, *23*, 1060–1081. [CrossRef]

53. Oraee, M.; Hosseini, M.R.; Papadonikolaki, E.; Palliyaguru, R.; Arashpour, M. Collaboration in BIM-based construction networks: A bibliometric-qualitative literature review. *Int. J. Proj. Manag.* **2017**, *35*, 1288–1301. [CrossRef]

54. Wong, J.K.W.; Zhou, J. Enhancing environmental sustainability over building life cycles through green BIM: A review. *Autom. Constr.* **2015**, *57*, 156–165. [CrossRef]

55. Aghaei Chadegani, A.; Salehi, H.; Yunus, M.M.; Farhadi, H.; Fooladi, M.; Farhadi, M.; Ale Ebrahim, N. A comparison between two main academic literature collections: Web of science and scopus databases. *Asian Soc. Sci.* **2013**, *9*, 18–26. [CrossRef]

56. Bryde, D.; Broquetas, M.; Volm, J.M. The project benefits of building information modeling (BIM). *Int. J. Proj. Manag.* **2013**, *31*, 971–980. [CrossRef]

57. Becerik-Gerber, B.; Jazizadeh, F.; Li, N.; Calis, G. Application areas and data requirements for BIM-enabled facilities management. *J. Constr. Eng. Manag.* **2011**, *138*, 431–442. [CrossRef]

58. Singh, V.; Gu, N.; Wang, X. A theoretical framework of a BIM-based multi-disciplinary collaboration platform. *Autom. Constr.* **2011**, *20*, 134–144. [CrossRef]

59. Azhar, S.; Khalfan, M.; Maqsood, T. Building information modeling (BIM): Now and beyond. *Constr. Econ. Build.* **2012**, *12*, 15–28. [CrossRef]

60. Succar, B. Building information modeling framework: A research and delivery foundation for industry stakeholders. *Autom. Constr.* **2009**, *18*, 357–375. [CrossRef]

61. Li, C.Z.; Xue, F.; Li, X.; Hong, J.; Shen, G.Q. An Internet of Things-enabled BIM platform for on-site assembly services in prefabricated construction. *Autom. Constr.* **2018**, *89*, 146–161. [CrossRef]

62. Howard, R.; Björk, B.C. Building information modeling–Experts' views on standardisation and industry deployment. *Adv. Eng. Inform.* **2008**, *22*, 271–280. [CrossRef]

63. Azhar, S. Building information modeling (BIM): Trends, benefits, risks, and challenges for the AEC industry. *Leadersh. Manag. Eng.* **2011**, *11*, 241–252. [CrossRef]

64. Kang, T.W.; Hong, C.H. A study on software architecture for effective BIM/GIS-based facility management data integration. *Autom. Constr.* **2015**, *54*, 25–38. [CrossRef]

65. Ning, X.; Wang, Y. 3D reconstruction of architecture appearance: A survey. *Comput. Inf. Syst.* **2013**, *9*, 3837–3848.

66. Kim, J.L. Use of BIM for effective visualization teaching approach in construction education. *J. Prof. Issues Eng. Educ. Pract.* **2011**, *138*, 214–223. [CrossRef]

67. Castro-Garcia, M.; Perez-Romero, A.M.; León-Bonillo, M.J.; Manzano-Agugliaro, F. Developing Topographic Surveying Software to Train Civil Engineers. *J. Prof. Issues Eng. Educ. Pract.* **2016**, *143*, 04016013. [CrossRef]

68. Volk, R.; Stengel, J.; Schultmann, F. Building Information Modeling (BIM) for existing buildings—Literature review and future needs. *Autom. Constr.* **2014**, *38*, 109–127. [CrossRef]

69. Migilinskas, D.; Popov, V.; Juocevicius, V.; Ustinovichius, L. The benefits, obstacles and problems of practical BIM implementation. *Procedia Eng.* **2013**, *57*, 767–774. [CrossRef]

70. Motawa, I.; Carter, K. Sustainable BIM-based evaluation of buildings. *Procedia Soc. Behav. Sci.* **2013**, *74*, 419–428. [CrossRef]

71. Habibi, S. The promise of BIM for improving building performance. *Energy Build.* **2017**, *153*, 525–548. [CrossRef]

72. AlFaris, F.; Juaidi, A.; Manzano-Agugliaro, F. Energy retrofit strategies for housing sector in the arid climate. *Energy Build.* **2016**, *131*, 158–171. [CrossRef]

73. Wong, K.D.; Fan, Q. Building information modeling (BIM) for sustainable building design. *Facilities* **2013**, *31*, 138–157. [CrossRef]

74. Ku, K.; Taiebat, M. BIM experiences and expectations: The constructors' perspective. *Int. J. Constr. Educ. Res.* **2011**, *7*, 175–197. [CrossRef]

75. Fridrich, J.; Kubečka, K. BIM–the process of modern civil engineering in higher education. *Procedia Soc. Behav. Sci.* **2014**, *141*, 763–767. [CrossRef]
76. Cheng, J.C.; Ma, L.Y. A BIM-based system for demolition and renovation waste estimation and planning. *Waste Manag.* **2013**, *33*, 1539–1551. [CrossRef]

Article

Pitch Angle Optimization by Intelligent Adjusting the Gains of a PI Controller for Small Wind Turbines in Areas with Drastic Wind Speed Changes

Ernesto Chavero-Navarrete [1,2], Mario Trejo-Perea [2,*], Juan-Carlos Jáuregui-Correa [2], Roberto-Valentín Carrillo-Serrano [2] and José-Gabriel Rios-Moreno [2]

[1] Centro de Tecnología Avanzada CIATEQ AC, Queretaro 76150, Mexico; ernesto.chavero@ciateq.mx
[2] Facultad de Ingeniería, Universidad Autónoma de Querétaro, Santiago de Querétaro 76010, Mexico; jc.jauregui@uaq.mx (J.-C.J.-C.); roberto.carrillo@uaq.mx (R.-V.C.-S.); riosg@uaq.mx (J.-G.R.-M.)
* Correspondence: mtp@uaq.mx; Tel.: +55-442-192-12-00 (ext. 6064)

Received: 29 October 2019; Accepted: 21 November 2019; Published: 26 November 2019

Abstract: The population growth demands a greater generation of energy, an alternative is the use of small wind turbines, however, obtaining maximum wind power becomes the main challenge when there are drastic changes in wind speed. The angle of the blades rotates around its longitudinal axis to control the effect of the wind on the rotation of the turbine, a proportional-integral controller (PI) for this angle achieves stability and precision in a stable state but is not functional with severe alterations in wind speed, a different response time is necessary in both cases. This article proposes a novel pitch angle controller based on auto-tuning of PI gains, for which it uses a teaching–learning based optimization (TLBO) algorithm. The wind speed and the value of the magnitude of the change are used by the algorithm to determine the appropriate PI gains at different wind speeds, so it can adapt to any sudden change in wind speed. The effectiveness of the proposed method is verified by experimental results for a 14 KW permanent magnet synchronous generator (PMSG) wind turbine located at the Universidad Autónoma de Querétaro (UAQ), Mexico.

Keywords: sustainability; renewable energy; wind turbine; pitch control; electric generation; universities

---

1. Introduction

Electricity is indispensable for the growth of a country, so it is essential to increase energy generation and meet the growing demand. Generating energy by traditional means such as petroleum products compromises the ecological balance of future generations. To take advantage of the natural resources that exist locally is to offer the population economic growth possibilities, wind energy is an alternative, using the force of the wind to convert it into electricity through a wind turbine [1]. Globally, wind energy has been booming, the Global Wind Energy Council (GWEC) in February 2018 reported that in 2017, more than 54 GW of wind energy were installed in more than 90 countries, nine of them with more than 10,000 MW installed and 29 that have now exceeded 1000 MW, increasing the accumulated capacity to 486.8 GW, 12.6% more than in 2016 [2]. In Mexico, the wind resource is currently being investigated [3], wind infrastructure has grown 300% in the last six years, the Asociación Mexicana de Energía Eólica (AMDEE) estimates that in 2024, 10,000 MW [4] will be reached and predicted that by 2031 14,000 MW [5] will be generated. The city of Querétaro has considerable wind potential [6], with an average annual wind speed of 7.3 m/s, NE/SW direction with a probability of the presence of 26% that defines it as a viable candidate for the incorporation of small wind turbines [7]. The Universidad Autónoma de Querétaro (UAQ) is working on the technological development of wind turbines with the design of its blades and the control of its systems, by now, three 14 KW wind turbines with two and three blades have been installed [8].

The amount of energy that can be obtained from a wind turbine is a function of the size of the rotor. The greater the length of the blades, the more energy is produced, so the capacity and size of wind turbines have increased exponentially in the last decade. The commonly used wind turbine has a diameter of 125 m sweeping area capable of producing up to 7.5 MW [9]. However, since 2016, it is more common to see wind turbines with a nominal power of 9.5 MW [10]. The V164-10 model capable of generating up to 10 MW is available for sale now and can be delivered for commercial installation until 2021 [11].

Technological advances in wind power generation systems focus on increasing the size of the rotor, which requires large plains, with a considerable wind resource. However, in complex terrain with hills, trees or buildings, where the wind resource is smaller, the installation of "small wind turbines" is necessary [12]. The international standard IEC 61400-1: 2014 defines "small wind turbines" those with an area of the rotor sweep less than or equal to 200 m^2 and a generation voltage less than 1000 V AC or 1500 V DC for both on-grid and off-grid applications [13].

The problem of installing wind turbines in this type of places is the randomness of the wind, it is necessary to use turbines that operate with different speeds to take full advantage of this natural resource, so it follows that for each wind speed there is an ideal rotation speed. This is called the optimum tip speed ratio (TSR), and it is different for each wind turbine according to its size and aerodynamic model [14]. The angle of inclination of the blade should be controlled in a wind turbine to maximize energy production, regulate the speed of rotation of the rotor, mitigate dynamic loads and ensure a continuous supply of energy to the network [15].

The pitch angle controller is based on rotating the blades simultaneously, with an independent or shared actuator. The angle used with the wind speed below the nominal value is zero, and then the angle increases when the wind speed is greater than the nominal speed [16]. The control method for the classic pitch angle is the PI [17,18]. The control strategy works correctly when the dynamics of the system is stable, however, the sensitivity of the generator's rotation speed to the pitch angle varies differently. If the wind speed is close to the nominal speed, the sensitivity of the generator shaft speed to the pitch angle is very small. Therefore, a higher response speed is required than at higher wind speeds, where a small change in the angle can have a large effect on the speed of the generator shaft. Nonlinear variation of the pitch angle versus wind speed implies the need for non-linear control, which means a constant change in the response speed of the controller according to the wind speed and the value of the change in the speed of the wind [19].

A literature review was carried out and it was found that there are different authors who have worked to solve the problem of nonlinearity for pitch control in a wind turbine [20]. In [21–23] a fuzzy logic control (FLC) was combined with proportional-integral-derivative control (PID), FLC is the means to change previously calculated PID gains according to the process variable error, if the error is negative or positive or if the measured value greatly exceeds. In [24], a control method based on an artificial neural networks (ANN) adapter was presented in which the parameters of the PID neural network were automatically regulated. The improved gradient descent method was used to optimize the weights of the networks and to avoid the weights of the neural network. In [25], a PID control model was designed, a gain programming procedure was planned using differential evolution optimization algorithm to apply to the appropriate controller as the operating point changes. In [17,26], the authors used a PI controller, however, the proportional gain K_p and the integral gain K_i were adjusted through the particle swarm optimization algorithm. In [27], a pitch controller based on the moth-flame optimization algorithm was proposed, the candidate solutions were moths and the PID parameters were the position of the moths in the search space. Therefore, moths can fly in a 3D space that represents the three parameters of the controller K_p, K_i, and K_d with the change of their position vectors.

In this document, we propose a PI controller, an optimization algorithm based on teaching–learning was developed that calculates the optimal gains for any change in wind speed. PI controller uses the error value between the measured speed of the generator shaft and the optimum value according to the wind speed, to determine the reference value of the inclination angle. The wind speed and the value of the magnitude of the change between each speed are evaluated in the proposed algorithm to calculate the value of the PI gains suitable for the adequate response time of the controller, so it can be adapted to any sudden changes in wind speed.

The implementation of the teaching–learning optimization method applied to the pitch control of a wind turbine is an important innovation for this type of application. Unlike the controllers mentioned above, this control model offers to researchers in this field a control alternative, with an algorithm that adjusts to wind conditions and converges on a solution in a shorter time because it uses the best solution of the iteration to change the existing solution in the population, in addition to using a minimum of computer resources since it only uses simple arithmetic operations such as sums and divisions, so the controller suppresses transitory excursions and achieve a good and fast regulation in the operation of the stable state.

Different simulations of the dynamic model were carried out to preset initial values of the control algorithm, and they were adjusted experimentally with a 14 KW wind turbine, two blades, 12 m in diameter, with a permanent magnet synchronous generator (PMSG) installed at the UAQ airport campus.

2. Wind Turbines

The turbines are composed of three main parts, the rotor of blades that converts the kinetic energy from the wind to mechanical energy, the gearbox that multiplies the speed of the rotor and transmits it to the rotational shaft of the generator, and the electric generator [28]. Additionally, some systems convert AC to DC using a rectifier and convert DC back to AC to match the frequency and phase of the network [29]. A diagram of a wind turbine is shown in Figure 1.

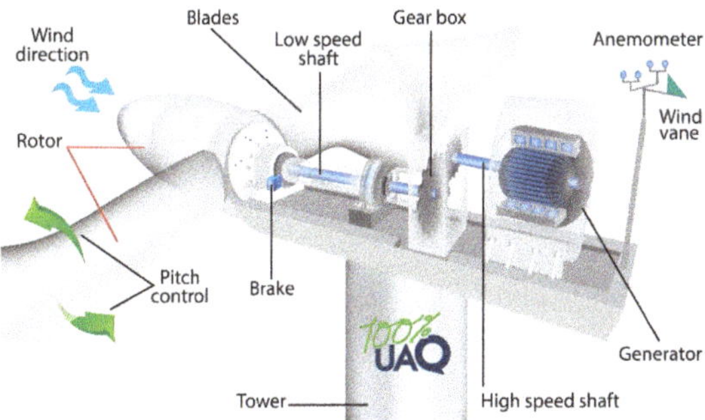

Figure 1. Parts of a wind turbine.

2.1. Mathematical Model

The mathematical model of the wind turbine is made by subsystems, the aerodynamic model of the rotor, the mechanical model and the electric model of the generator are determined.

To perform an aerodynamic analysis of how to extract the maximum power of the wind that passes through a turbine, the wind that crosses the sweeping area of the rotor is determined. The rotor power P_{rotor} extracted by the blades is described with the Equation (1) [30].

$$P_{rotor} = \frac{1}{2}\rho\,A\,V^3\,C_p \tag{1}$$

where the wind density is ρ, A is the swept area, V is the wind speed before the turbine. C_p is the power coefficient, which can be expressed in approximate method depending on tip speed ratio λ, that is the ratio between the tangential speed of the tip of a blade and the actual speed of the wind and on the pitch angle of the blade β, based on the characteristics of the turbine [31,32]. C_p is defined as:

$$C_p(\lambda,\beta) = 0.5176\left(\frac{116}{\lambda_i} - 0.4\beta - 5\right)e^{-\frac{21}{\lambda_i}} + 0.0068\lambda \tag{2}$$

where:

$$\lambda_i = \left[\left(\frac{1}{\lambda + 0.08\beta}\right) - \left(\frac{0.035}{\beta^3 + 1}\right)\right]^{-1} \tag{3}$$

$$\lambda = \frac{\Omega R}{V} = \frac{2\pi n R}{60V} \tag{4}$$

Ω is the rotation frequency in rad/sec, n is the speed of rotation in rpm and R is the radius of the rotor. Once the C_p has been calculated, it is possible to determine the torque of the rotor T_{rotor} with the following equation: [33,34].

$$T_{rotor} = \frac{1}{2}\rho\,\pi\,R^3 V^2\,C_t \tag{5}$$

where torque coefficient C_t is defined as:

$$C_t = \frac{C_p}{\lambda} \tag{6}$$

The mechanical system transmits the mechanical torque of the rotor, multiplies the speed of the rotor shaft n times towards the generator shaft. The mathematical model of the mechanical system can be simplified throughout the system into a two masses model, which is the most common model for wind turbine transmissions and can be used without losing accuracy [35,36]. In the mechanical model, the aerodynamic torque of the wind turbine rotor and the electromechanical torque of the generator act in opposition to each other and are the inputs to the model, while the rotation speeds are the output [37,38]. The mathematical model is represented by the Equations (7) y (8). The model of two masses proposed by [19] is shown in Figure 2.

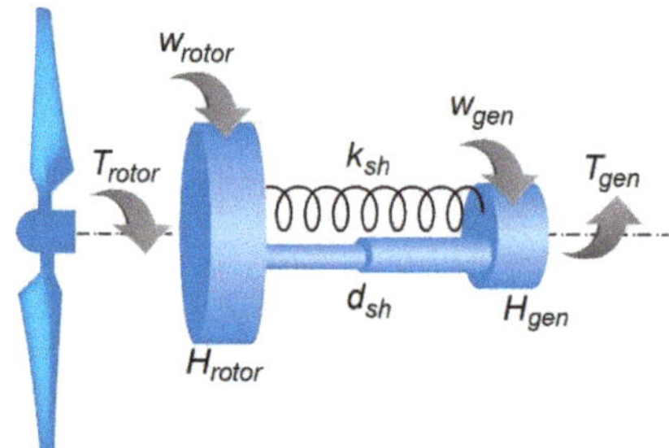

Figure 2. Two masses model of a wind turbine.

$$2H_{rotor}\frac{dw_{rotor}}{dt} = T_{rotor} - d_{sh}\left(w_{rotor} - w_{gen}\right) - k_{sh}\left(\theta_{rotor} - \theta_{gen}\right) \tag{7}$$

$$2H_{gen}\frac{dw_{gen}}{dt} = d_{sh}\left(w_{rotor} - w_{gen}\right) + k_{sh}\left(\theta_{rotor} - \theta_{gen}\right) - T_{gen} \tag{8}$$

where the inertial constant depends exclusively on the geometry and distribution of the mass of the element. The elasticity between adjacent masses is expressed by the spring constant k_{sh} and the mutual damping between adjacent masses is expressed by d_{sh} [36,37]. θ_{rotor} is the rotor angular position and θ_{gen} is the generator angular position. H_{rotor} and H_{gen} are the rotor inertia and generator inertia, respectively.

The inertial moments H_{rotor} and H_{gen} are calculated according to:

$$H_{rotor} = \frac{J_{rotor}\ w_{rotor}^2}{2\ P_n} \tag{9}$$

$$H_{gen} = \frac{J_{gen}\ w_{gen}^2}{2\ P_n} \tag{10}$$

J_{gen} is the inertia of the generator and is regularly provided by the manufacturer. In the case of the inertia of the rotor J_{rotor} it can be approximated according to:

$$J_{rotor} = \frac{1}{8} m_r R^2 \tag{11}$$

where m_r represents the mass of the rotor (includes the mass of the blades).

The generator is electromechanical equipment that converts mechanical power into electrical power, uses a stator and a rotor. The stator is a housing with coils mounted around it. The rotor is the rotating part and is responsible for producing a magnetic field, it can be a permanent magnet or an electromagnet. When rotating, its magnetic field is induced to the stator windings causing a voltage at the stator terminals [28]. According to the needs of the market, the synchronous permanent magnet generator (PMSG) is used commonly because of its adaptation to variable-speed turbines [38]. The PMSG has a high efficiency since its excitation is provided without any power supply. It requires the use of an AC/DC/AC power converter to adjust voltage and frequency to the supply network [39].

For the mathematical modeling of a PMSG, the three phases are transformed into an equivalent of two axes. This is because each one acts in a defined geometric space of the air gap. With the direct axis (d) in phase with the winding of the rotor field and the quadrature or displacement axis (q), 90 electrical degrees forward in a synchronous rotating d-q reference frame. Magnetic flux waves due to the winding of the stator are presented in two sinusoidal waves distributed rotating with a synchronous speed such that one is the maximum point on the axis d, and the other is the maximum point on the axis q. The stator output voltages d-q of this generator are given respectively by Equations (12) and (13) [40].

$$V_d = R_d I + L_d \frac{dI_d}{dt} - \omega_{gen} L_q I_q \tag{12}$$

$$V_q = R_q I_q + L_q \frac{dI_q}{dt} + \omega_{gen}\left(L_d I_d + \varphi_f\right) \tag{13}$$

where L are the inductances of the generator, R is the resistance and I is the currents in the axes d and q, respectively. φ_f is the permanent magnetic flux. ω_{gen} is the rotation speed of the PMSG.

$$\omega_{gen} = P_p \omega_{ref} \tag{14}$$

Pp is the number of pairs of poles. Electromechanical torque T_{gen} can be express as:

$$T_{gen} = \frac{3}{2} P_p \omega_{ref}\left(\left(L_q - L_d\right) i_d i_q + \varphi_f i_q\right) \tag{15}$$

ω_{ref} is the reference speed to control the speed of the generator shaft. The model reference speed is normally 120% but is reduced for power levels below 46% [41]. This behavior is represented in the model by using the following equation:

$$w_{ref} = -0.75 P^2 + 1.59 P + 0.63 \tag{16}$$

P is the electric power. In the controller, the speed reference is not directly a function of power, but the overall effect on the speed/power relationship is similar.

2.2. Control System

The complexity of modern wind turbines forces its control systems to ensure safe and efficient operation. The objective of control systems is to ensure a continuous supply of energy to the grid, maximize energy production, and mitigate dynamic and static mechanical loads. The pitch angle of the blade, the torque of the generator, and the frequency on the grid are the main parameters that must be controlled [42].

The definition of the control objectives depends on the operating regions of the wind turbine. These are closely related to wind speed, and one can identify four operating regions according to the wind speed. Region I represents the wind speed at which the rotor cannot move, so the rotation speed is zero. When the rotor begins to rotate, enters region II, this region is limited by the starting speed and wind speed where the generator rotates at its nominal speed. The objective of control in this region is to maximize energy production through MPPT strategies. Region III starts from the nominal speed to the stopping speed, which is the design speed limit and is required to stop rotation for safety. To maintain the constant nominal rotation speed, the pitch control is used. However, MPPT control is also used in this region to smooth out abrupt changes in wind speed, where the mechanical restrictions of the pitch system do not allow a rapid response. Finally, region IV is where the wind turbine must be stopped even with a mechanical brake [42,43].

The pitch control system in a wind turbine is used to regulate the power of the rotor, control its speed of rotation and stop the rotor out of the action of the wind. In region III, during high-burst winds, it is necessary to control the rotation speed of the rotor to protect the generator and electronic equipment from overloads. The inertia of the large rotors in acceleration and deceleration must be considered to decrease the dynamic mechanical stresses in the blades and the tower. The speed of rotation is variable with the angle of incidence. This angle is formed between the line of direction of the wind and line that marks the side of the blade. This angle of incidence is increased to take advantage of the wind speed and is reduced when the speed of rotation increases. The adjustment of this angle is made to keep the power captured by the wind constant. The angle of incidence changes with the pitch angle, which is the angle of rotation of the blade with respect to its axis. Consequently, the power coefficient is affected by the evolution of the pitch angle [44].

In Figure 3, the power coefficient curves with different values of the pitch angle are shown. When the angle value increases, the power coefficient decreases, the captured wind power is reduced, and there is a reduction in the rotor velocity.

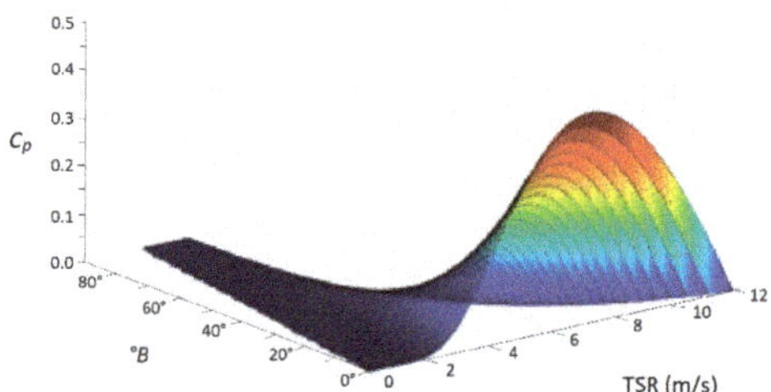

Figure 3. Power coefficient curves for different pitch angle values (β).

The control model used is a PI controller with feedback [17]. The purpose of a feedback control system is to reduce the error $e(k)$ to zero between the variable to control and its reference value as quickly as possible. The error is expressed as:

$$e(k) = \omega_{ref} - \omega_{rotor}(t) \tag{17}$$

The pitch control signal $u(t)$ to the plant is equal to the proportional gain K_p times the magnitude of the error plus the integral gain K_i times the sum of the errors of samples, k is the sample number from a total of k_{sim} samples.

$$u(t) = K_p e(k) + K_i \sum_{k=1}^{k_{sim}} e(t) \tag{18}$$

To select the parameters of the K_p and K_i controller that comply with the presumed behavior of the system, it is proposed to follow the tuning rules of Ziegler–Nichols [45] based on the experimental responses to a step input. However, there may be a large overshoot in your response that is unacceptable, so a series of fine adjustments is necessary until the desired result is obtained.

The pitch angle command activates a mechanic system back and forth for the pitch control system. The faster the settling, the lesser the mechanical stress on the turbine and structure.

3. Intelligent Search Algorithms Teaching–Learning

Teaching–learning-based optimization (TLBO) is based on the teaching and learning process in a classroom. In each generation, the best candidate solution in the population is considered the teacher, and the other candidate solutions are considered learners. The learners mostly accept instruction from the teacher, but also learn from each other. The score of an academic subject is analogous to the value of an independent variable or candidate solution feature [46,47]. The steps of the algorithm are described below.

Step 1: Define the optimization parameters.

Population size (Pn),
Number of generations (Gn),
Number of design variables (Dn),
Limits of design variables (L_U, L_L).
Objective function $f(x)$
Define the problem: Maximize $f(x)$, minimize $f(x)$.
X is a vector of design variables such that:

$$L_{U,i} \leq x_i \leq L_{L,i} \quad : \quad X_i \in x_i = 1, 2, ..., Dn$$

Step 2: Initialize the population.

Generate a random population according to the size of the population and the number of design variables. For TLBO, the size of the population indicates the number of students, and the design variables indicate the subjects that are offered. This population is expressed as:

$$Population = \begin{bmatrix} x_{1,1} & x_{1,2} & \cdots & x_{1,D} \\ x_{2,1} & x_{2,2} & \cdots & x_{2,D} \\ \vdots & \vdots & \vdots & \vdots \\ x_{Pn,1} & x_{Pn,2} & \cdots & x_{Pn,D} \end{bmatrix} \tag{19}$$

Step 3: Teacher Phase.

Calculate the average of the population, which will give the average for the subject as:

$$M_D = [m_1, m_2, m_3, \dots, m_D] \tag{20}$$

The best solution $T_{f(x)max,min}$ will be teacher $T_{teacher}$ for that iteration

$$T_{teacher} = T_{f(x)=max,\ min} \tag{21}$$

The teacher will change the mean of M_D to $X_{teacher}$, which will act as a new average for the iteration.

$$M_{new,D} = X_{teacher,D} \tag{22}$$

The difference between two means is expressed as:

$$Difference_D = r(M_{new,\ D} - T_F M_D) \tag{23}$$

r is a random number in a range [0,1]. The value of T_F is selected as 1 or 2. Teaching value that decides the value of the mean to be changed. No design value is random, given by the algorithm:

$$T_F = round[1 + rand(0,1)\{2-1\}] \tag{24}$$

The difference obtained is added to the current solution to update its values using:

$$X_{new,D} = X_{old,D} + Difference_D \tag{25}$$

Accept X_{new} if you give the function a better value.
Step 4: Learning Phase.
A student interacts randomly with other students. A student learns something new if the other student has more knowledge than he does. Randomly select two learners X_i and X_j, where $i \neq j$.
The knowledge of both students is compared.

$$IF \quad f(X_i) < f(X_j) \quad \rightarrow \quad X_{new,i} = X_{old,i} + r_i(X_i - X_j) \tag{26}$$

$$IF \quad f(X_i) > f(X_j) \quad \rightarrow \quad X_{new,i} = X_{old,i} + r_i(X_j - X_i) \tag{27}$$

When all the students have interacted, accept X_{new} if you give the function a better value.
Step 5: Termination criterion.
Stop if the maximum generation number is reached, otherwise, repeat from step 3.
Figure 4 shows a flow diagram of this algorithm.

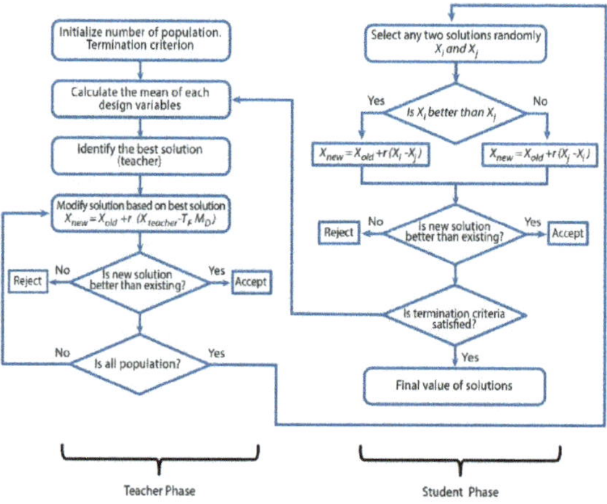

Figure 4. Flow chart for teaching–learning-based optimization.

This optimization method was selected since it converges on a solution in a shorter time compared to other known algorithms, this is because it uses the best iteration solution to change the optimal reference solution existing in the population, in addition to using a minimum of computer resources since it only uses simple arithmetic operations such as sums and divisions.

A performance analysis of the proposed algorithm was developed and compared with other intelligent search techniques such as genetic algorithm (GA) [48], simulated annealing (SA) [46], ant colony optimization (ACO) [46], differential evolution (DE) [46], and firefly algorithm (FA) [46], particle swarm optimization (PSO) [26], moth flame optimization (MFO) [27]. Test functions (28) suggested in [48] were used.

$$f(x_1, x_2) = (x_1 - 10)^3 + (x_2 - 20)^3 \tag{28}$$

$$for \ 13 \leq x_1 \leq 100 \ \& \ 0 \leq x_2 \leq 100 \tag{29}$$

This equation was selected due to its similarity to the objective function, since it has higher and non-linear terms of order, in addition to being initialized at a starting point (first solution). Another important feature is the mathematical minimization problem with two variables, as in the application of the wind turbine, it is necessary to find the value of the variables k_p, k_i that minimize the error in the controlled variable. Table 1 lists the specifications with which the test was performed.

Table 1. Test conditions.

Problem	Minimization
Variables	(x_1, x_2)
Restrictions	$13 \leq x_1 \leq 100 \ 0 \leq x_2 \leq 100$
Population size	5 for variable
Maximal interactions	1000
Acceptance criteria	100 interactions without change
Number of tests	5

From this analysis, the computation time for the search for each optimal solution was obtained. The computational resource was an HP Workstation i-7 processor and 32Gb RAM, 64bit, for all tests.

The optimum value is:

$$f(x_1, x_2) = -6961.8138 \tag{30}$$

$$for \ x_1 = 14.095 \ \& \ x_2 = 0.84296 \tag{31}$$

Table 2 shows this comparison, where it is evident that the teaching–learning algorithm obtained a better performance.

Table 2. Performance analysis of the teaching–learning algorithm compared with other intelligent search techniques.

Optimization algorithm	Run time (s)	Value	Error
PSO	2.44	−6875.94	|85.8738|
ACO	6.63	−6952.18	|9.63380|
FA	7.06	−5498.67	|1463.14|
GA	12.64	−6912.66	|49.1538|
SA	4.81	−6350.26	|611.553|
DE	1.71	−6942.59	|19.2280|
MFO	6.23	−5473.90	|1487.91|
TLBO	0.88	−6964.14	|2.32620|

According to the results presented in the previous table, the TLBO algorithm had better results in the computation time and obtained a minor error in relation to the exact known solution, this is due to the advantages of the algorithm. The advantages of the algorithm are that it does not require any parameters to adjust, which simplifies the implementation, uses the best solution of the iteration to change the existing solution in the population which increases the convergence rate. TLBO does not divide the population as other algorithms do, but it uses two different phases, the "teacher phase" and the "student phase", so that new random solutions can be evaluated, and no time is spent evaluating the same solutions among themselves. The disadvantage is that no measures are taken to handle the limitations of the problem. Solution selection is only done in a heuristic way by comparing two solutions.

4. Methodology

The proposed methodology for the optimization of the use of wind energy for the generation of electric energy through a PI controller with variable gains includes the following steps. First, know the available wind resource by making a preliminary study of the historical records of the climatic conditions at the installation site, as well as having physical knowledge of the wind turbine. This will allow us to characterize the wind behavior statistically to know the limits of wind conditions. Second, design a PI controller for the pitch angle, tuned for a response speed according to the sensitivity presented by the system, in a range of nominal wind speed values. This allows us to keep the controller in a stable state with typical winds. Third, once the response values of the controller are in their stable state, the limitations and operating ranges of the controller are proposed to establish the optimization parameters of the TLBO algorithm, which will improve the response of the controlled when atypical winds occur, such as bursts or turbulence.

4.1. Wind Resourse and Wind Turbine Specifications

The wind turbine used for experimentation is located at the UAQ, airport campus, road to Chichimequillas s/n, Ejido Bolaños, Querétaro, Qro. Z.C. 76140. The geographic location 20°37′24.1″ North and 100°22′06.0″ West and an altitude of 1969 m a.s.l. Figure 5 is an image of the airport campus of the autonomous university of Queretaro, where the wind turbine and the adjacent buildings are shown.

Figure 5. Universidad Autónoma de Querétaro, airport campus.

This research incorporates aerodynamic modeling based on a meteorological study with data collected in the weather station #76628 SMN-CONAGUA network, located in the place, with stored data from the last five years. The average annual recorded speed is 3.9 m/s, and the range of recorded wind speed values is between 0 and 15.69 m/s.

This wind turbine is a NACA 6812 airfoil with two blades of 6.4 m in length and 1.2 m at their widest point, constructed of fiberglass and polyester resin, weighing 260 kg each. The height of the tower is 18 meters. There is a multiplier box with a ratio of 1: 2.1, and a permanent magnet generator with a rated power of 14 KW at 14.6 rad/s speed.

4.2. Pitch Control

To form the plant to be controlled, the different mathematical models were integrated, in the dynamic model, the input data is wind speed that is the disturbance of the system and the pitch angle that is a variable that is originated in the controller, the output is the mechanical torque of the low-speed shaft. The information we obtained from the mechanical model is the rotation speed of the shaft at the output of the multiplier box or high-speed shaft. In the generator model, the generated electrical power was calculated and due to its electromagnetic properties, the torque of the generator, which in turn is a force opposing the rotor torque. The controlled variable is the speed of rotation of the rotor. The limitation in this control model is the speed of rotation of the blade, the maximum speed is 1.5 °/s because a 0.5 HP motor mechanically spins a reducer with a 60:1 ratio, which increases the torque to counteract the effects of air on the blade, but greatly decreases the speed. The control model that describes the operation of the systems is shown in Figure 6.

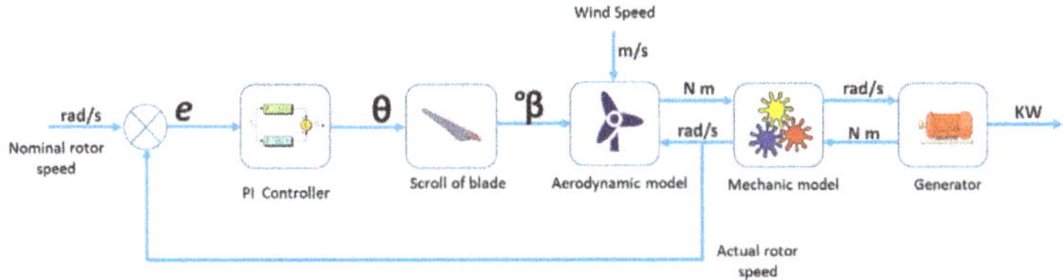

Figure 6. Proportional-integral (PI) pitch control model.

The controller gains were established empirically by performing various simulations in MatLab-Simulink R2018b V9.5.0.944444 program. The wind speed from 0 m/s to 15.69 m/s, maximum wind speed at the site, was used as an input variable. The setpoint for the controller is the rotation speed of the nominal generator shaft of 14.6 rad/s.

The system is limited to a wind speed between the starting speed and the cutting speed, within these limits the system must be stable. Tuning was performed by applying a step input with the value of the nominal wind speed, which means the maximum value of an accepted disturbance. With respect to the sensitivity of the system at different wind speeds, the system must not have an over impulse greater than 20%, so a range of PI gains from the controller that were within the limits of this condition was obtained. Therefore, the controller can minimize the error of the controlled variable under these disturbance conditions and ensuring that the system is stable. The gain values obtained using the Ziegler–Nichols method for PI controller and experimentally adjusted are $K_p = 2.2$ and $K_i = 0.1$, gain values $K_p = 10$ and $K_i = 0.01$ were also obtained for a system with 20% overshoot and $K_p = 1$ and $K_i = 1$ for an overdamped system.

Figure 7 shows the behavior of generator shaft speed with the gain values obtained for a stable system, a system with overshoot, and an overdamped system. Figure 8 contains the pitch angle movement for all cases.

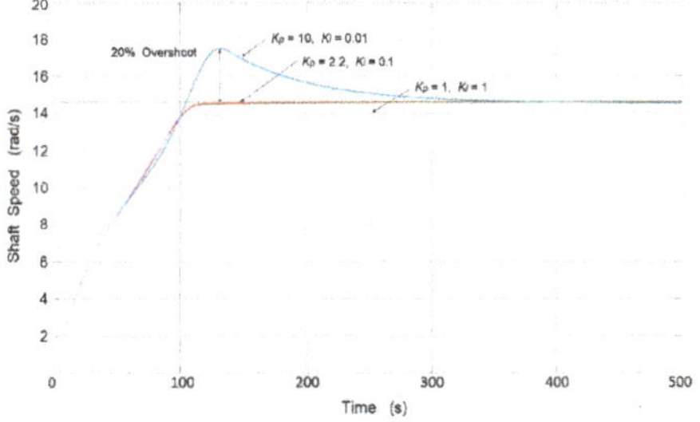

Figure 7. Gain values obtained for a stable system, a system with overshoot, and an overdamped system.

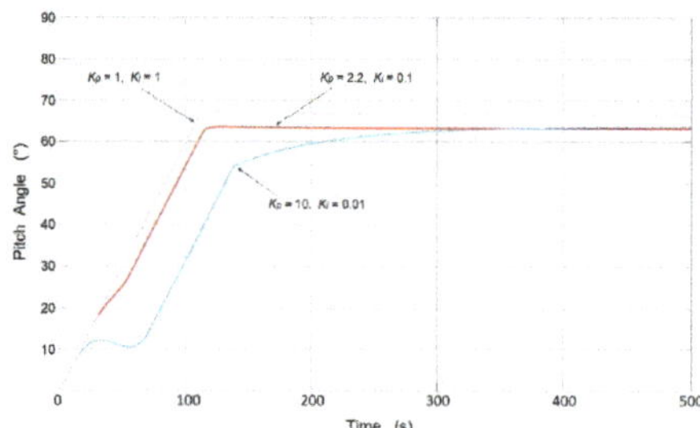

Figure 8. Pitch angle movement with different gains of PI controller.

4.3. TLBO Algorithm Application.

According to the analysis performed in the process of tuning a PI controller, in point 4.2, the optimization parameters were defined:

Population size (5), a small number of initial random solutions for each design variable is proposed, which reduces the convergence time, in addition, there is the possibility of increasing the search space since the switching between student-teacher phase is carried out faster and new random solutions can be evaluated and no time is spent evaluating the same solutions among themselves.

Number of generations (100), the maximum number of interactions was proposed after verifying that for this case study the convergence of a solution was obtained around 50 interactions.

Termination criteria: If there are more than 25 interactions without having a better profit proposal, the search process ends.

Design variables (K_p, K_i), controller gains.

Limits of design variables, $0 \leq K_p \leq 10, 0 \leq K_i \leq 1$ the limits were established based on knowing the optimal values of the design variables for the overshoot system and overdamped system, for nominal wind speed ranges.

Objective function, $f(x)$ = mathematical model, this model was described in Section 2 of this publication.

Define the problem: minimize $e(k)$, the objective is to find a solution that reduces the error between the nominal rotation speed of the generator shaft and the measured speed.

The calculation of the new gains was made every second, since the execution time of the algorithm was less than this time. MatLab-Simulink R2018b V9.5.0.944444 program was used to run the algorithm on an HP Workstation i-7 processor and 32 Gb RAM (64 bit). The experimentation was performed by programming a PIC16F87A using the Dev C ++ V5.0.0.4 software. The control model proposed proportional-integral with teaching–learning based optimization (PI-TLBO) is shown in Figure 9.

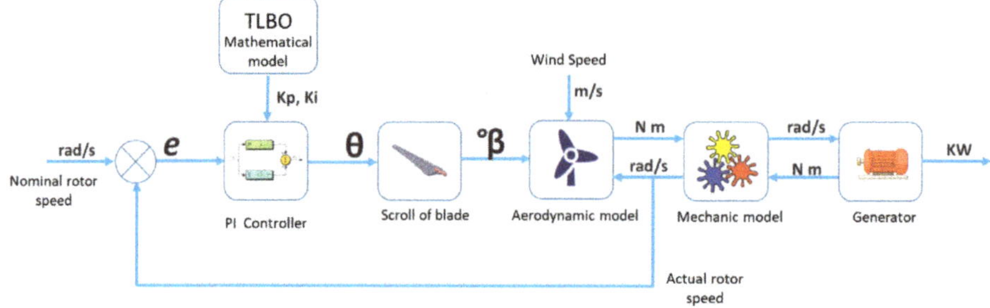

Figure 9. PI-TLBO model for pitch control.

5. Results

According to simulations performed to obtain the empirical values of the K_p and K_i gains, a maximum power factor of 4.47 and the wind speed of 6 m/s were determined under ideal conditions and shaft generator nominal speed of 14.6 rad/s. However, the magnitude of the change in wind speed makes the system unstable, since the thrust force provides different acceleration to the rotor rotation. Therefore, a PI controller with variable gains was proposed to obtain different response times in the controller and soften these changes. The effectiveness of a pitch control using the PI-TLBO algorithm is examined in real operating conditions. The recorded wind speed values are shown in Figure 10.

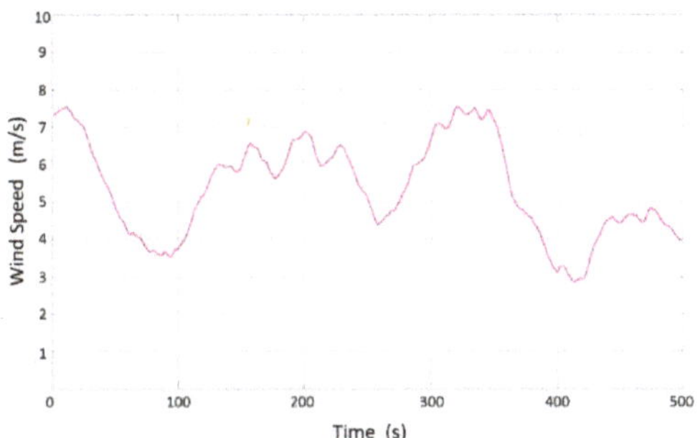

Figure 10. Wind speed recorded in the experiment with the algorithm.

The search time for each value of K_p and K_i is in a range of 0.4 and 0.75 seconds depending on whether the algorithm makes all interactions or not, so a variable adjustment is made every second. With this result, the convergence speed of the algorithm can be highlighted, where unlike other optimization algorithms, it uses the best iteration solution to update the value of the existing solution, in addition to reducing the processing time since they are only used simple arithmetic operations such as sums and divisions.

Figure 11 shows the development of the algorithm in (a) there is an error of the variable that decreases according to better values of K_p and K_i, in (b) and (c) we observe how the values K_p and K_i change respectively.

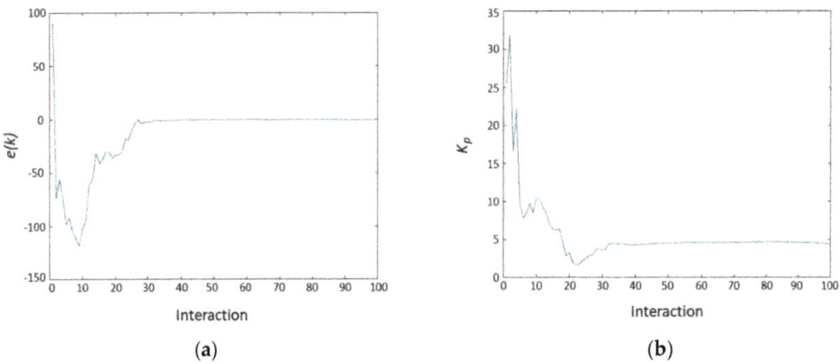

Figure 11. *Cont.*

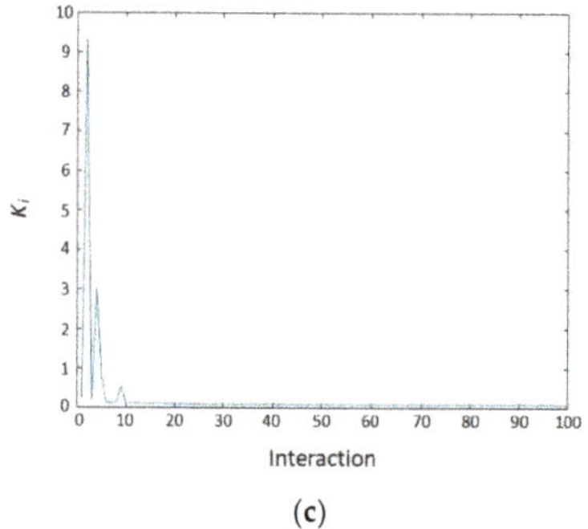

(c)

Figure 11. Algorithm development. (**a**) Decrease of the error value, (**b**) adjustment of the K_p value, and (**c**) adjustment of the K_i value.

The results of the experimentation show that the algorithm of optimization of gains K_p and K_i generate a better performance of a PI controller. In Figure 12, a comparison is made between the response of a PI controller and a PI-TLBO controller, the speed obtained from the generator shaft behaves more smoothly and close to the nominal speed, which represents a reduction in fatigue in the wind turbine structure and lower saturation in the PMSG.

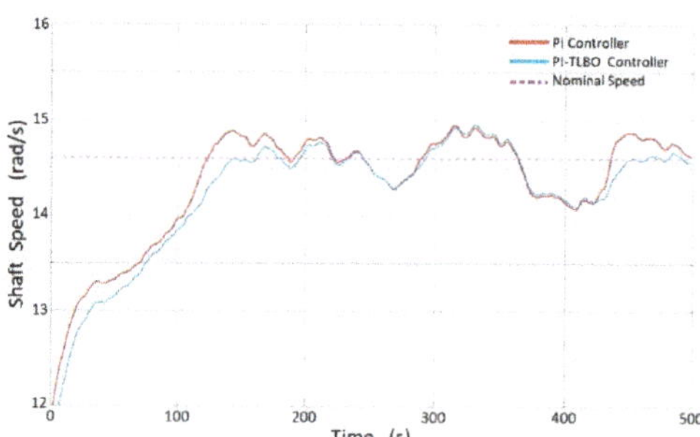

Figure 12. Comparison between the performance of the PI controller and the PI-TLBO controller.

It is also observed that the response of the PI_TBLO controller has a minor overshoot in the optimization range, the stabilization period is also reduced when a major disturbance occurs. Therefore, it can be suggested that after the TBLO optimization process for calculating the gains of the PI controller, the proposed control model is expected to control the pitch angle under various disturbances.

Figure 13 shows the movement of the K_p and K_i gains throughout the experimentation.

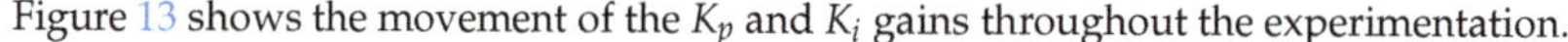
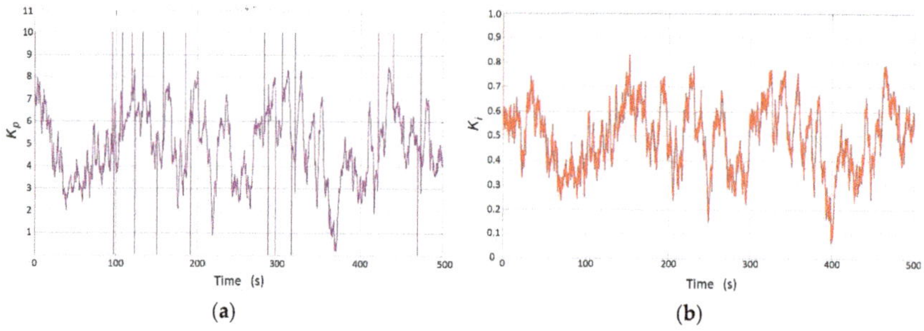

Figure 13. Evolution of gains (**a**) K_p and (**b**) K_i.

It is important to note the efficiency of the algorithm, a series of simulations were performed to validate the repeatability finding small variations in the response of the controller, this is due to the randomness of the data to generate new responses in the teacher phase. However, any of the responses of the PI-TLBO controller obtained better performance than that of the PI controller, this can be seen in Figure 14.

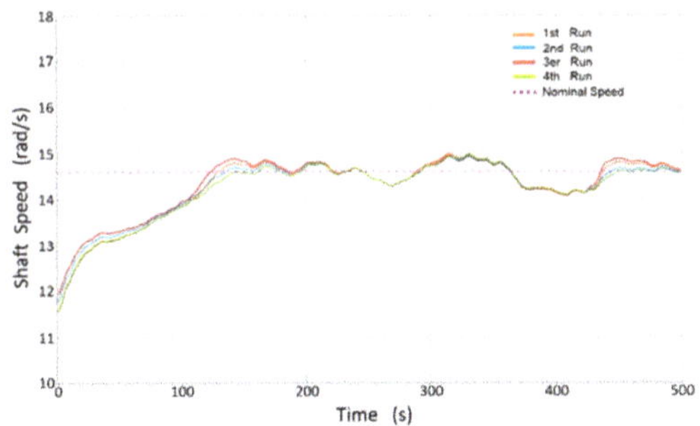

Figure 14. Repeatability of the PI-TLBO controller.

6. Discussion

The use of dynamic models to simulate the aerodynamic behavior of the system allows to know the behavior of a wind system, however, in this type of systems, it is not necessary to know the behavior of the wind speed or the magnitude of its changes that would be the variable input. The multiple variables involved in the output variable make the system nonlinear, which also becomes a problem to calculate the ideal parameters of the controller. That is why a controller that adjusts to various changes in the system variables is proposed, this controller is automatically adjusted by an optimization algorithm based on teaching–learning, which gave the system stability since the controller adjusted its control response according to wind conditions. A better response was obtained by constantly adjusting the controller with a method of close solutions to the optimal one, than by applying a single optimal deterministic method. This is since a suitable adjustment of the controller is required to the sensitivity of the system, since the closer it is to nominal wind speed, the system is more sensitive to wind changes, not so when the system is below the nominal speed.

A PI controller with dynamic gain adjustment and not a PID was used because the differentiation reacts as fast as the input changes with respect to time, the differentiation acts towards the future anticipating the overshoot trying to offer a response according to how quickly it increases or the input signal decreases, however, when applied in a system of dynamic gains, differentiation caused oscillations in the system, this is because it is not necessary to anticipate the future since the calculated gains act in the present. That is why it was decided to use only a PI controller, getting good results.

The presented control system affects the angle of inclination to limit wind energy, however, additional subsystems suggest the implementation of speed control to obtain an optimal electrical frequency and transfer the total electrical energy to the electricity grid.

It is convenient to use new forms of intelligent control, such as fuzzy systems, neural networks, and genetic algorithms. With this type of control, a predictive model for the expected climate would be achieved and with that would anticipate a better control response, which would reduce the frequent stops and starts of the system. In addition, the algorithms of these systems are programmed according to the knowledge and experiences that a human expert would have in the field. This is very useful since you cannot have an exact mathematical model of the meteorological parameters.

It is necessary and essential to prepare more professionals in universities with this new approach, which consists in considering the design of the wind turbine and its integration with the environment.

7. Conclusions

The use of wind energy in places where this natural resource is available should be an alternative for the generation of electrical energy, since it does not harm the environment and the wind does not end. There are places of difficult access due to their complicated orography surrounded by mountains where the wind is turbulent with various changes in wind speed.

A small deviation in the wind speed causes a large deviation in the output power of the wind turbine rotor due to the association of cubic links between these two parameters. This, in turn, translates into system vibration, mechanical fatigue, and an acceleration in the rotation speed that exceeds the nominal rotation speed of the generator.

Pitch control is the movement of the blades to receive wind power, conventional PI control is efficient in a stable state, but not in sudden changes in wind speed. This article proposes a new method of effectiveness for a PI controller based on the optimization of the gains used, which represents a better control response in different wind speed ranges. The proposed PI-TLBO algorithm proved to be efficient, since better performance was obtained compared to a conventional PI, it proved to be repetitive even though, in the teacher phase, the algorithm uses a random choice of parameters to generate new solutions. The algorithm has a rapid calculation speed due to the simplicity of its operations.

In general, with the PI-TBLO controller there is a lower overshoot and that the stabilization period is also reduced when disturbances occur, therefore it is highly efficient to control the pitch angle of a wind turbine under various atypical wind disturbances such as bursts and turbulence, reducing the transient effects of the controller and, consequently, the energy consumption in the actuator. Moreover, it was evident that the error between the nominal rotation speed of the generator shaft and the measured speed was considerably reduced, therefore, the generator is prevented from having losses due to magnetic saturation. According to the above, it can be ensured that the generation of energy in a wind turbine is increased with the use of an optimized PI-TLBO control algorithm to position the pitch angle.

The limitation of this work was the reaction speed of the mechanical system, the controller proved to be highly efficient, however, with extreme wind gusts, the actuator did not have enough speed to adjust the angle. An improvement to the design of the control model is the location of the wind speed measurement system, placing it at a specific distance, where the bursts of time are detected in advance and the controller can give a timely response, thereby the control transients that make the system unstable would be further reduced.

The use of algorithms that efficient the use of wind turbines is key to boost the use of wind energy with small wind turbines where the wind is not stable and the use of large commercial wind turbines is not possible or profitable.

Author Contributions: Conceptualization, E.C.-N. and J.-C.J.-C.; methodology, E.C.-N.; system control, E.C.-N. and R.-V.C.-S.; experiments, E.C.-N. and R.-V.C.-S.; software, R.-V.C.-S.; hardware, J.-C.J.-C.; validation, J.-G.R.-M.; formal analysis, M.T.-P.; investigation, E.C.-N.; resources, J.-C.J.-C..; data curation, J.-G.R.-M.; discussion, M.T.-P. and J.-C.J.-C.; writing—original draft preparation, E.C.-N.; writing—review and editing, M.T.-P. and J.-G.R.-M.; visualization, E.C.-N.; supervision, J.-C.J.-C.; project administration, J.-C.J.-C.; funding acquisition, J.-C.J.-C.

Funding: This research received no external funding.

Acknowledgments: This work was supported by Departamento de Investigación y Posgrado of the Facultad de Ingeniería and Universidad Autónoma de Querétaro.

References

1. Court, V. Energy Capture, Technological Change, and Economic Growth: An Evolutionary Perspective. *BioPhys. Econ. Resour. Qual.* **2018**, *3*, 12. [CrossRef]
2. Sawyer, S.; Dyrholm, M. *Global Wind Report Annual Market Update 2017*; GWEC Global Wind Energy Council: Brussels, Belgium, 2018.

3. Hernández-Escobedo, Q.; Perea-Moreno, A.J.; Manzano-Agugliaro, F. Wind energy research in México. *Renew. Energy* **2018**, *123*, 719–729. [CrossRef]

4. Coldwell, P.J. *Inaugural Speech in Wind Power Congress 2018*; Asociación Mexicana de Energía Eólica AMDEE: Ciudad de México, Mexico, 2018.

5. Rodriguez, L. *Inaugural Speech in Wind Power Congress 2018*; Asociación Mexicana de Energía Eólica AMDEE: Ciudad de México, Mexico, 2018.

6. Hernández-Escobedo, Q.; Manzano-Agugliaro, F.; Zapata-Sierra, A. Wind power in México. *Renew. Sustain. Energy Rev.* **2010**, *14*, 2830–2840. [CrossRef]

7. Rodríguez, F. Alternative energies a dissemination task of CINVESTAV-IPN. *NTHE Electron. J. Dissem. Dissem. Sci. Technol. Innov. State Querétaro* **2010**, *1*, 1–4.

8. Franco, J.A.; Jauregui, J.C.; Toledano, M. Optimizing Wind Turbine Efficiency by Deformable Structures in Smart Blades. *ASME. J. Energy Resour. Technol.* **2015**, *5*, 051206. [CrossRef]

9. Wang, L.; Liu, X.; Kolios, A. State of the Art in the Aeroelasticity of Wind Turbine Blades: Aeroelastic Modelling. *Renew. Sustain. Energy Rev.* **2016**, *64*, 195–210. [CrossRef]

10. González-González, A.; Jimenez-Cortadi, A.; Galar, D.; Ciani, L. Condition monitoring of wind turbine pitch controller: A maintenance approach. *Measurement* **2018**, *123*, 80–93. [CrossRef]

11. Nthe. Available online: http://www.concyteq.edu.mx/nthe1/pdfs/Nthe%201.pdf (accessed on 24 November 2018).

12. Hayter, S.J.; Kandt, A. Renewable Energy Applications for Existing Buildings. In Proceedings of the 48th AiCARR International Conference, Baveno-Lago Maggiore, Italy, 22–23 September 2011.

13. Culotta, S.; Franzitta, V.; Milone, D.; Moncada Lo Giudice, G. Small Wind Technology Diffusion in Suburban Areas of Sicily. *Sustainability* **2015**, *7*, 12693–12708. [CrossRef]

14. Bibave, R.; Kulkarni, V. A Novel Maximum Power Point Tracking Method for Wind Energy Conversion System: A Review. In Proceedings of the International Conference on Computation of Power, Energy, Information and Communication (ICCPEIC), Tamil Nadu, India, 28–29 March 2018.

15. Novaes-Menezes, E.J.; Araújo, A.M.; Bouchonneau da Silva, N.S. A review on wind turbine control and its associated methods. *J. Clean. Prod.* **2018**, *174*, 945–953. [CrossRef]

16. Menon, M.; Ponta, F. Dynamic aero elastic behavior of turbine rotors in rapid pitch-control actions. *Renew. Energy* **2017**, *107*, 327–339. [CrossRef]

17. Behera, S.; Subudhi, B.; Pati, B.B. Design of PI Controller in Pitch Control of Wind Turbine: A Comparison of PSO and PS Algorithm. *Int. J. Renew. Energy Res.* **2016**, *6*, 1.

18. Kumar, D.; Chatterjee, K. A review of conventional and advanced MPPT algorithms for wind energy systems. *Renew. Sustain. Energy Rev.* **2016**, *1*, 957–970. [CrossRef]

19. Amulya, M.; Prashanth, C.; Vijaya, M. Controlling flicker caused due to power fluctuations by using individual pitch control for a variable speed DFIG based wind turbine. *Int. Res. J. Eng. Technol.* **2017**, *4*, 286–293.

20. Chavero-Navarrete, E.; Trejo-Perea, M.; Jauregui-Correa, J.C.; Carrillo-Serrano, R.V.; Rios-Moreno, J.G. Expert Control Systems Implemented in a Pitch Control of Wind Turbine: A Review. *IEEE Access* **2019**, *7*, 13241–13259. [CrossRef]

21. Civelek, Z.; Lüy, M.; Çam, E.; Barışçı, N. Control of Pitch Angle of Wind Turbine by Fuzzy PID Controller. *Intell. Autom. Soft Comput.* **2015**, *22*, 463–471. [CrossRef]

22. Baburajan, S. Improving the efficiency of a wind turbine system using a fuzzy-pid controller. In Proceedings of the IEEE Advances in Science and Engineering Technology International Conferences (ASET), Abu Dhabi, UAE, 6 February–5 April 2018.

23. Xiao, Y.; Huo, W.; Nan, G. Study of Variable Pitch Control for Direct-drive Permanent Magnet Wind Turbines Based on Fuzzy Logic Algorithm. *J. Inf. Comput. Sci.* **2015**, *12*, 2849–2856. [CrossRef]

24. Kang, J.; Meng, W.; Abraham, W.; Liu, H. An adaptive PID neural network for complex nonlinear system control. *Neurocomputing* **2014**, *135*, 79–85. [CrossRef]

25. Taher, S.A.; Farshadnia, M.; Mozdianfard, R. Optimal gain scheduling controller design of a pitch-controlled VS-WECS using DE optimization algorithm. *Appl. Soft Comput.* **2013**, *13*, 2215–2223. [CrossRef]

26. Hodzic, M.; Tai, L.C. Grey Predictor reference model for assisting particle swarm optimization for wind turbine control. *Renew. Energy* **2016**, *86*, 251–256. [CrossRef]

27. Ebrahim, M.A.; Becherif, M.; Abdelaziz, A.Y. Dynamic performance enhancement for wind energy conversion system using Moth-Flame Optimization based blade pitch controller. *Sustain. Energy Technol. Assess.* **2018**, *27*, 206–212. [CrossRef]

28. Hansen, A.D. Wind Turbine Technologies. In *Wind Energy Engineering*, 1st ed.; Elsevier: Roskilde, Denmark, 2017; pp. 145–160.

29. Staggs, J.; Ferlemann, D.; Shenoi, S. Wind farm security: Attack surface, targets, scenarios and mitigation. *Int. J. Crit. Infrastruct. Prot.* **2017**, *17*, 3–14. [CrossRef]

30. Jamieson, P. *Innovation in Wind Turbine Design*, 3rd ed.; John Wiley & Sons Ltd: Garrad Hassan, UK, 2018.

31. Rezaei, M.M. A nonlinear maximum power point tracking technique for DFIG-based wind energy conversion systems. *Eng. Sci. Technol. Int. J.* **2018**, *21*, 901–908. [CrossRef]

32. Yang, B.; Yu, T.; Shu, H.; Zhang, Y.; Chen, J.; Sang, Y.; Jiang, L. Passivity-based sliding-mode control design for optimal power extraction of a PMSG based variable speed wind turbine. *Renew. Energy* **2018**, *119*, 577–589. [CrossRef]

33. Taoufik, M.; Abdelhamid, B.; Lassad, S. Stand-alone self-excited induction generator driven by a wind turbine. *Alex. Eng. J.* **2018**, *57*, 781–786. [CrossRef]

34. Giannakis, A.; Karlis, A.; Karnavas, Y.L. A combined control strategy of a DFIG based on a sensorless power control through modified phase-locked loop and fuzzy logic controllers. *Renew. Energy* **2018**, *121*, 489–501. [CrossRef]

35. Muyeen, S.M.; Hasan-Ali, M.; Takahashi, R.; Murata, T.; Tamura, J.; Tomaki, Y.; Sasano, E. Comparative study on transient stability analysis of wind turbine generator system using different drive train models. *IET Renew. Power Gener.* **2007**, *1*, 131. [CrossRef]

36. Abobkr, A.H.; El-Hawary, M.E. Evaluation of wind turbine characteristics built-in model in Matlab Simulink. In Proceedings of the IEEE Electrical Power and Energy Conference, Ottawa, ON, Canada, 12–14 October 2016.

37. Habibi, H.; Rahimi, H.; Howard, I. Power maximization of variable-speed variable-pitch wind turbines using passive adaptive neural fault tolerant control. *Front. Mech. Eng.* **2017**, *12*, 377–388. [CrossRef]

38. Honrubia-Escribano, A.; Gómez-Lázaro, E.; Fortmann, J.; Sørensen, P.; Martin-Martinez, S. Generic dynamic wind turbine models for power system stability analysis: A comprehensive review. *Renew. Sustain. Energy Rev.* **2018**, *81*, 1939–1952. [CrossRef]

39. Salih, H.W.; Wang, S.; Farhan, B.S.; Waqar, A. PSO tuned fuzzy based pitch blade controller of grid-tied variable speed wind turbine. In Proceedings of the IEEE 11th Conference on Industrial Electronics and Applications, Hefei, China, 5–7 June 2016.

40. Gong, J.; Xie, R. MPPT Control by Using a U-P Curve for PMSG Based Small Wind Turbines. *J. Energy Eng.* **2015**, *142*, 04015030. [CrossRef]

41. Eisa, S.A. Modeling dynamics and control of type-3 DFIG wind turbines: Stability, Q Droop function, control limits and extreme scenarios simulation. *Electr. Power Syst. Res.* **2019**, *166*, 29–42. [CrossRef]

42. Soued, S.; Ebrahim, M.A.; Ramadan, H.S.; Becherif, M. Optimal blade pitch control for enhancing the dynamic performance of wind power plants via metaheuristic optimizers. *IET Electr. Power Appl.* **2017**, *11*, 1432–1440. [CrossRef]

43. Asghar, A.B.; Liu, X. Estimation of wind turbine power coefficient by adaptive neuro-fuzzy methodology. *Neurocomputing* **2017**, *238*, 227–233. [CrossRef]

44. Villarrubia, L.M. *Ingeniería de la Energía Eólica*, 1st ed.; Marcombo: Barcelona, Spain, 2012.

45. Ogata, K. *Modern Control Engineering*, 5th ed.; Prentice Hall: Madrid, Spain, 2010.

46. Simon, D. *Evolutionary Optimization Algorithms, Biologically-Inspired and Population-Based Approaches to Computer Intelligence*, 1st ed.; John Wiley and Sons, Inc.: Hoboken, NJ, USA, 2013; pp. 441–445.

47. Rao, R.V. *Teaching Learning Based Optimization Algorithm and Its Engineering Applications*; Springer: New York City, NY, USA, 2016; Volume XVI, pp. 9–39.

48. Michalewicz, Z.; Schoenauer, M. Evolutionary algorithms for constrained parameter optimization problems. *Evolut. Comput.* **1996**, *4*, 1–32. [CrossRef]

 *sustainability*

Article

Contribution of University to Environmental Energy Sustainability in the City

Iñigo Leon [1], Xabat Oregi [1,*] and Cristina Marieta [2]

[1] Department of Architecture, University of the Basque Country UPV/EHU, Plaza Oñati 2, 20018 San Sebastián, Spain; inigo.leon@ehu.eus
[2] Department of Chemical and Environmental Engineering, Faculty of Engineering, University of the Basque Country UPV/EHU, Plaza Europa 1, 20018 San Sebastián, Spain; cristina.marieta@ehu.eus
* Correspondence: xabat.oregi@ehu.eus; Tel.: +34-943-01-5891

Received: 23 December 2019; Accepted: 19 January 2020; Published: 21 January 2020

Abstract: The environmental energy sustainability of universities has aroused great interest in recent years. In this study, environmental impact assessment tools are used to analyse the environmental impacts of the University of the Basque Country (UPV/EHU) since 2015 and to identify reform scenarios to make the university more sustainable. University campuses can be considered to be small cities that impact the environment of the cities where they are located. The environmental impacts of the UPV/EHU Gipuzkoa campus and the impacts on the city of Donostia-San Sebastián in which the university is located are analysed. The environmental impacts are calculated using simulation tools based on three-dimensional models of the university campus and the city. These results are compared with actual impact results from monitoring. The simulation results differ from the monitoring results but provide a rapid determination of the best future scenarios for a more sustainable university by taking the impacts on the city into account. This study enables the university to align its efforts with the Covenant of Mayors for Climate and Energy.

Keywords: university environmental impact; urban planning; sustainability assessment; covenant of Mayors

---

1. Introduction

The environmental energy sustainability of universities cannot be separated from the large-scale overarching problem that affects the entire world. There is growing evidence that the situation of the global environment has become critical in several aspects. Thus, problems, such as the depletion of natural resources, global warming, or the depletion of the ozone layer, have received considerable media coverage and have significant social repercussions.

It is estimated that over 50% of the population lived in urban settlements, in 2016, which will increase to over 60% by 2030; that is, two out of every three persons in the world will live in cities [1–5]. This problem is magnified if this densification is considered in conjunction with recent assessments that two-thirds of the world's primary energy consumption can be attributed to urban areas, which in turn means that 71% of the world's direct greenhouse gas (GHG) emissions are energy-related [6]. In the European Community, in particular, buildings and the construction sector in general are responsible for 40% of energy consumption and 25% of CO_2 emissions. [7–9].

This situation presents a clear demand, both from the public and private sectors, to forecasters and urban planners for greater environmental awareness in project implementation. This new awareness must encompass many interrelated problems [10], such as the consumption of resources, waste production, water consumption, GHG emissions and the protection of biodiversity and air quality. Most of these problems cannot be addressed at the level of a building or a facility. The urban scale is an extremely relevant scale at both the city and university campus levels [11]. The environmental and energy impacts at the building scale are magnified at the urban scale [12]. For this reason, urban planners are more frequently employing environmental and energy efficiency parameters in the design of new urban development spaces and in the creation of regeneration projects for city districts [13]. Therefore, it is absolutely necessary to use tools in the simulation, measurement and evaluation of parameters that can facilitate the sustainable development of cities [14–19]. It is crucial to include these parameters at the design stage to be able to choose the most sustainable urban proposals among different projects. Different neighbourhood sustainability assessment (NSA) tools are available depending on the objective of an urban development project. Each tool has its own particularities and produces various forms of environmental assessments for different data.

Several studies have analysed different systems [20]. Lee presents an in-depth review of five representative qualitative assessment tools at the building scale: BREEAM, LEED, CASBEE, BEAM Plus, and ESGB [21]. Reith and Orova [22], compare different NSA tools at three levels of detail for different indicators. Sharifi and Murayama [23], analyse seven systems of sustainability assessment, showing a clear difference in focus among the tools. An in-depth case study in Spain is used to demonstrate substantial differences among thirteen assessment tools for sustainability at the urban scale at the international level [24]. Lastly, Xabat et al. conducts a study on different recent development tools for the energy assessment of cities at the district level [25]. Among the available options, the dynamic energy atlas can be used to solve energy problems in a geographical context, with some drawbacks [26]. Another tool analysed in the study is CitySim, which can be used to simulate energy scenarios at an urban scale, although considerable expertise in energy simulations is required [27].

The NEST (neighbourhood evaluation for sustainable territories) tool is used in this study [28,29]. This recently developed and simply managed tool starts with a life cycle analysis at the building level to evaluate of the design of a district of a city and proposes improvement scenarios for a more sustainable evolution of the city [19]. Using NEST, three-dimensional (3D) models are used to evaluate a series of indicators to analyse the main environmental problems affecting sustainable design at an urban scale. This agile tool generates 3D graphical solutions that are very easy to interpret. The environmental energy sustainability of universities can be successfully studied at the university campus scale, and the impact of the university campus can be related to the city where it is located. This tool was initially developed to evaluate different sustainable scenarios for a city. However, present-day universities can be considered to be small cities because of their large size and population and the complexity of multiple campus activities, which directly or indirectly impact the environment [30,31]. These impacts are mainly related to the consumption of energy and materials related to operations and activities related to research, teaching, administration and services, and transportation to users' homes [32]. Thus, there is a growing demand for projects on sustainability in universities [33–37]. That is why the evaluation of the environmental impact of University of the Basque Country (UPV/EHU) has been carried out using the NEST tool [38].

A global evaluation can be performed using NEST at different scales of analysis, where the impacts of the university campus and the city are evaluated in an interrelated manner, and the links between impacts can be analysed. As a result, improvement scenarios can be proposed that do not focus on organisational or local policies but consider the best intentions of municipal policies in relation to university guidelines, thereby combining efforts to achieve common objectives for worldwide energy improvement.

2. Methods

The goals of this research project are as follows: (i) To compare the environmental impacts of the university in relation to the impacts of the city in which the university is located; (ii) to analyse the difference between scenarios simulated using the assessment tool and actual monitoring data to rapidly validate refurbishment scenarios; and (iii) to establish a scenario of joint reform between the university and the city in response to the city's environmental improvement plan to comply with international environmental commitments.

2.1. Study Case: The University Campus and the City

As discussed above, the aim of the project is to analyse the potential of the university to establish synergies with the city to meet environmental sustainability commitments at the international level, that is, the new Global Covenant of Mayors for Climate and Energy that was signed by the city of Donostia-San Sebastián in 2017. After separately evaluating the environmental impacts for the city and the university, a scenario for sustainable reform is proposed based on a joint plan, specific to the city, called the Sustainable Energy Action Plan (SEAP) of Donostia-San Sebastián. The environmental impacts for University of the Basque Country UPV/EHU are obtained by modelling and analysis of the Donostia-San Sebastián campus. The city of Donostia-San Sebastián is studied considering the main districts in relation to municipal plans and its pacts for sustainability at the international level. A 3D model of the university campus and the districts of the city is used to evaluate the environmental impact of the "building" and "transport" sectors, which have the most significant impacts. The "industry" and "waste" sectors are not considered. The industry that was located in what is now considered the centre of the Donostia-San Sebastián moved to the periphery and other places in the province in the 1980s and was replaced by the university campus, which did not maintain the industrial character. In terms of university waste management, different faculties have specific plans for the selective collection of waste (paper-cardboard, plastic-packaging, organic, batteries, toner, pens, computers, etc.), and the quantity of waste collected is monitored; however, there is no directive common to the entire university. In addition, in the comparison with the city there is a difference of criteria in the selection of conflicting waste to be monitored. The university and the city have different main activities: For example, the municipality is focused on reducing waste, such as baby diapers, whereas the university has begun to realise the significance of the impact of computers that have become obsolete increasingly quickly.

In summary, 3D modelling of the university campus and the main districts of the city is performed using the NEST tool to determine baseline scenarios of energy assessment. These results are compared with actual consumption to evaluate the accuracy of the simulation method relative to monitoring. Then, NEST is used to simulate scenarios of joint and interrelated energy improvement for both the university and the city and to analyse the feasibility of meeting the proposed goal of compliance with the Global Covenant of Mayors for Climate and Energy. To describe the characteristics and dimensions of the case study, we briefly present the two elements in the comparative analysis of the different phases of the proposed method.

2.1.1. City of Donostia-San Sebastián

The city of Donostia-San Sebastián can be mainly characterised by an approximate area of 60.89 km^2 and a population of 186,665 (year 2018) (Figure 1). According to the Köppen climate classification [39], the climate of Donostia-San Sebastián is an oceanic climate (Cfb), which is a climate with cool summers and cool (but not cold) winters and with a relatively narrow annual temperature range.

Figure 1. Aerial image of the city of Donostia-San Sebastián. The University campus is represented to the northwest of the city. Source: Google Earth.

Donostia-San Sebastián has been consistently committed to climate change over the past 20 years. The city has made the following significant commitments, among others: The Carta de Aalborg (Aalborg Letter) (1998); the First Local Plan to Combat Climate Change 2008–2013 (2008); the Safe and Sustainable Mobility Plan 2008–2024 (2008); a municipal ordinance on energy efficiency in buildings (2009); signature of the Covenant of Mayors (2011); Sustainable Energy Action Plan (2011); the environmental strategy Hiri Berdea 2030 (2014); Mayors for Adaptation to Climate Change (2014); a Plan of Action III of Local Agenda 21 2015–2022 (2015); the Adaptation to Climate Change Plan (2017); Plan de Accion Clima 2050 DSS (2017); and adherence to the new Global Covenant of Mayors for Climate and Energy (2017).

2.1.2. Campus of the University of the Basque Country in Donostia-San Sebastián

The University of the Basque Country is located in the three provinces of the autonomous community: Gipuzkoa (1997 km^2), Bizkaia (2217 km^2), and Álava (3030 km^2) [38]. The major university campuses are located in the three provincial capitals, Donostia-San Sebastián, Bilbao, and Vitoria-Gasteiz. The chosen campus for this study is the campus located in Donostia-San Sebastián (Figure 2), which will be evaluated and compared with the city that contains it.

The university campus in Donostia-San Sebastián can be considered to be an urban campus. The various faculties are located amidst pleasant green areas on a total area of 170,000 m^2 (Figure 1; Figure 3). The different university faculties and schools were originally scattered around different parts of the city, until Donostia-San Sebastián urban planning created a common campus for the development of new faculties and obtaining university degrees. Approximately 25% of the students of the entire university pursue higher education and the administration and services staff (PAS) and teaching and research lecturers (PDI) are located on this campus. This area is located northwest of the city and is crossed by an urban avenue with broad tracts of trees that allow access and exit of vehicles between the city and other towns in the province. There is a well-maintained public transportation network of trains and buses, as well as a network of cycling roads that connect practically the entire city.

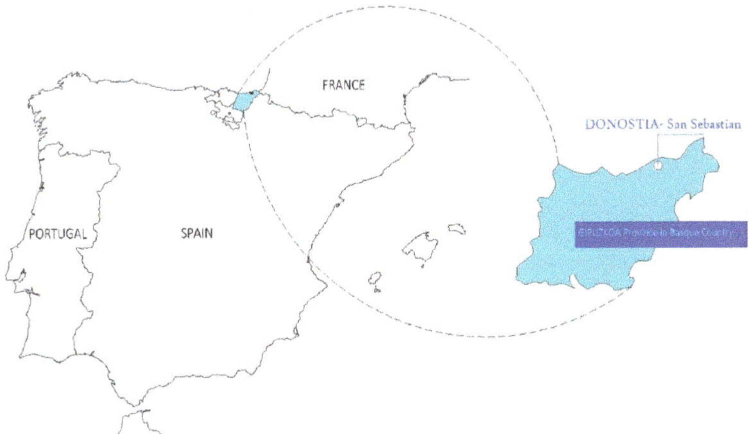

Figure 2. Location of the Basque Country and the city of Donostia-San Sebastián.

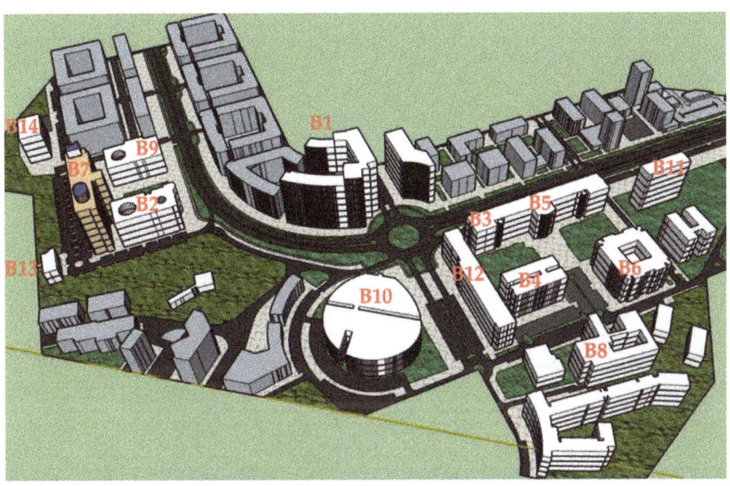

Figure 3. The different university buildings studied on the Donostia-San Sebastián campus (information about each building is provided in Appendix A).

The sample used in the study consists of higher education institutions (faculties) and other buildings that are necessary for the teaching, research, or management dynamics of the university (Figure 3). The faculties include the School of Engineering of Gipuzkoa; the Faculty of Economics and Business; the School of Education; Philosophy and Anthropology; the School of Computing; the School of Psychology; the School of Chemistry; the Technical School of Architecture; the Faculty of Law and the School of Education. Buildings with other uses are the Carlos Santamaria Centre (the central campus library), the Joxe Mari Korta Centre (RDI), the Ignacio Mª Barriola Centre (which contains the campus lecture hall consisting of 32 classrooms, rooms with other uses, an auditorium, and houses the majority of campus services); the Villa Julianategui (Campus Vice-Rectorship) and the most recent building, the Polyvalent Training and Innovation Centre (Centro Elbira Zipitria Zentroa). The aforementioned buildings are modelled using the NEST tool. To clearly investigate the area, private residential buildings adjoining the campus of Donostia-San Sebastián are also modelled.

2.2. Assessment Tool: NEST

NEST is a tool for the environmental, economic, and social analyses of an urban space and uses the Trimble SketchUp 3D modelling software; one of the most commonly used 3D graphic design programmes by designers and urban planners. NEST can be used to directly analyse a digital 3D model of a part of a city of interest. The tool is used to evaluate a series of indicators that have been developed on a scientific basis.

One of the great virtues of NEST is its graphical interface, which is very ergonomic because data can be simply entered into the geometric model to perform an easy, rapid, and effective analysis of real scenarios and proposed scenarios that are theoretically more sustainable. NEST takes into account four main elements of urban planning: (1) Buildings, (2) land use (roads, parking lots, green spaces, etc.), (3) infrastructure (street lighting), and (4) the mobility of the users of the urban space under study. NEST data can be entered or extracted in four different ways: (i) Manually (MA), (ii) manually through the NEST drop-down menu (MN), (iii) automatically by NEST (A), and (iv) imported using the software program Integrated Environmental Solutions (IES) [40]. Oregi et al. [29], and also, Leon et al. describe the assessment process, indicators, assessment scope, and hypothesis considered by NEST [38].

2.3. Scenarios and Strategies

To determine the improvement produced by a more sustainable university campus in conjunction with the environmental proposals in the strategic plan of the city, a baseline scenario or the prior-current situation must be analysed. Data for both the campus and the city can then be used to propose a refurbishment scenario to comply with the Covenant of Mayors guidelines at the international level (Figure 4).

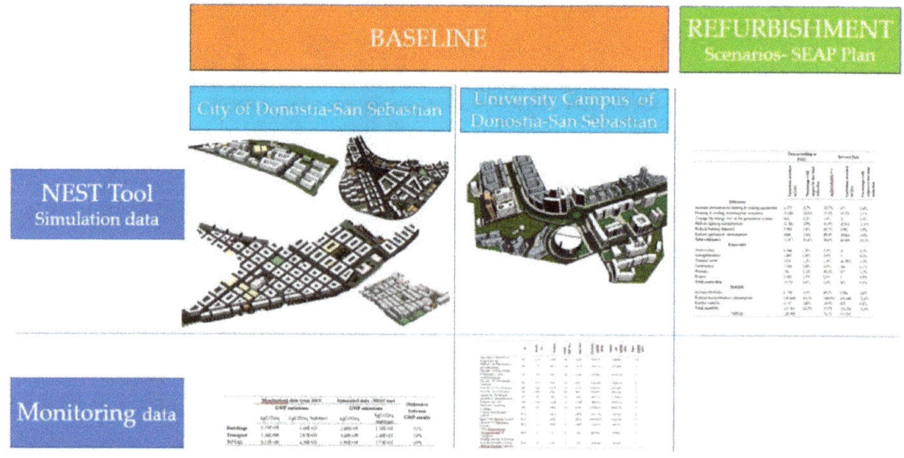

Figure 4. Scheme of the different scenarios analysed.

2.3.1. Baseline

To evaluate the importance of integrating simulation tools into this type of study, we use two different methodologies to study the city of Donostia-San Sebastián and the university campus: (1) Monitoring data and (2) simulation using the NEST software.

2.3.2. Refurbishment Scenarios

The refurbishment scenario is based on the SEAP plan established by the municipality. The SEAP is the official municipal document that summarizes the way the city of Donostia-San Sebastián has decided to go in the field of energy management including all the fields and activities in which the city can directly act or influence. The SEAP includes two parts. The first part is the energy diagnosis, which is the picture of all CO_2 emissions produced by energy consumption. The second part is the SEAP itself, which includes the environmental and energy targets and the list of actions to be implemented during next years, including a program for the implementation.

3. Results

The different results of the baseline scenario and the proposal of refurbishment strategies that affect both the city and the university will be broken down.

3.1. Baseline Scenario

The baseline is defined as the current scenario of the city of Donostia-San Sebastián and of the university campus related to the total global warming potential (GWP) emissions and emissions per habitant or user in 2015 year. The results of the city and the university are analysed separately for subsequent discussion.

3.1.1. City of Donostia-San Sebastián

Data on the energy consumption of the city and its emissions are obtained from different publications of the municipal, regional, and Basque Country authorities [41–47]. A comparative analysis of the information from the different data sources is used to compose a sample of the most relevant outputs or results for this study (Table 1). The main districts that make up the city of Donostia-San Sebastián are modelled in NEST to determine the CO_2 emissions of the sample (Table 1). The city of Donostia-San Sebastián is modelled in NEST (Figure 5), using information from different origins. The building geometry in the model is defined using DXF files provided by the city planning department and from cadastral information.

Table 1. Data analysis for the city of Donostia-San Sebastián.

	Monitored Data (Year 2015)		Simulated Data—NEST Tool		Difference between GWP Results
	GWP Emissions		GWP Emissions		
	Kg CO_2-eq	Kg CO_2-eq/habitant	Kg CO_2-eq	Kg CO_2-eq/habitant	
Buildings	2.79×10^8	1.49×10^3	2.48×10^8	1.33×10^3	−11%
Transport	5.36×10^8	2.87×10^3	4.48×10^8	2.40×10^3	−16%
TOTAL	8.15×10^8	4.36×10^3	6.96×10^8	3.73×10^3	−15%

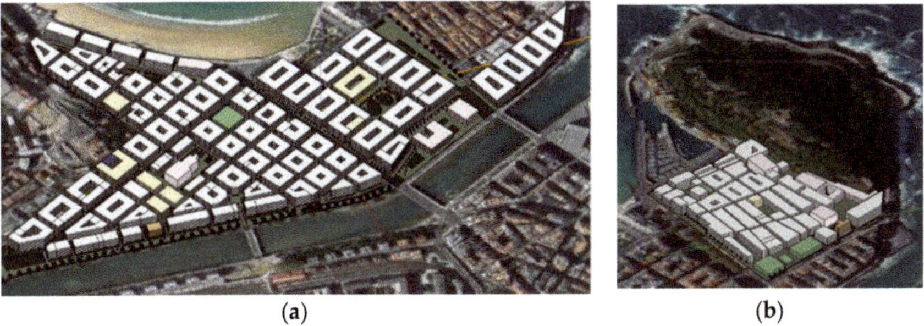

(a)   (b)

Figure 5. Two of the districts modelled in the neighbourhood evaluation for sustainable territories (NEST) for the city of Donostia-San Sebastián. (a) "Ensanche Cortazar" district; (b) "Parte Vieja" district.

Information on the energy performance of each building in the city is obtained from previous studies [29], and the Register of Energy Performance Certificates of the Basque Country [48]. The following statistics for the Donostia-San Sebastián inhabitants are determined from the mobility plan: 30% travel in private cars, 30% by bus, 3% by train, 25% by bicycle, and 17% by foot [46]. Based on Ecoinvent v3.0, NEST defines the environmental impact of the different mobility systems. The conversion factor from car, bus, tram, train, bicycle, and walking to GWP will be 0.29, 0.10, 0.09, 0.08, 0.00, and 0.00 (kg CO_2-eq/(user·km)), respectively.

A comparison of the results of the two calculation methodologies for the city of Donostia-San Sebastián shows that the monitoring global warming potential (GWP) data are 11% and 16% higher than the simulation results for the building and transportation sectors, respectively. Note that the main causes of this difference are the uncertainty in many of the hypotheses used in the two calculation processes and different quantification measures, such as the scope of the parameters. However, this difference is acceptable because the simulation can be used to rapidly evaluate the most sustainable option among different proposals. The identification of the most sustainable option remains valid after comparing the simulation and monitoring results. It is important to bear in mind the percentage deviation between the results of the two methodologies while acknowledging that simulation is indispensable.

In spite of the energy impact that it may suppose, the SEAP of Donostia-San Sebastián does not consider as input the energy consumption of the municipal sewage. Furthermore, the SEAP does not propose any strategy to reduce the environmental impact related to this process. Therefore, the municipal sewage will be out of the scope of this study.

3.1.2. Campus of Donostia-San Sebastián

Several parameters are monitored and quantified for the university campus in Donostia-San Sebastián for each of the campus buildings from 2015–2017 (see Appendix A). Note that the monitoring is limited to inventorying the different energy consumption points. Through the correct definition of "conversion factor" values, the energy consumption is transformed into environmental impact. For the natural gas source, the related impacts were deduced from the Ecoinvent database, applying the "Heat production, natural gas, at boiler modulating" process. The conversion factor from natural gas applied by this study to GWP will be 0.2 (kg CO_2-eq/kWh). For the oil source, the related impacts were deduced from the Ecoinvent database, applying the "heat production, light fuel oil, at boiler 100 kW, non-modulating" process and its conversion factor applied by this study to GWP will be 0.34 (kg CO_2-eq/kWh). Finally, the conversion factor from electricity (Spain 2016) applied during this case study to GWP will be 0.3 (kg CO_2-eq/kWh).

Information for the buildings that compose the campus in Donostia-San Sebastián are obtained from different UPV/EHU documents [49], to compile Table 2, which shows the environmental impacts associated with the mobility of users (workers, teachers, and students) of the Donostia-San Sebastián campus for 2015.

Table 2. Emissions from the university campus after the correction of the baseline scenario.

| | Monitored Data (Average 2015–2017 Years) | | Simulated Data—NEST Tool | | Difference between GWP Results |
| | GWP Emissions | | GWP Emissions | | |
	Kg CO_2-eq	Kg CO_2-eq/habitant	Kg CO_2-eq	Kg CO_2-eq/habitant	
Buildings	4.87×10^6	4.40×10^2	5.56×10^6	5.02×10^2	14%
Transport	4.94×10^6	4.46×10^2	4.12×10^6	3.72×10^2	−21%
TOTAL	1.01×10^6	9.11×10^2	9.68×10^6	8.74×10^2	−4%

The first revision of the campus model is developed in parallel in NEST (Figure 6), based on a model developed by Leon et al. [38]. However, the monitoring data show that the initial simulation model needs to be calibrated in two regards. First, the number of campus users is adjusted, because 12,248 users were used in Leon et al.'s study [38], whereas a corresponding mean value of 11,066 is determined for 2015–2017 from the monitoring process. Second, regarding transportation, the information in the documents show a new hypothesis for the mode of displacement of the campus users [49]: 36% travel in a private car, 30% by bus, 12% by train, 15% by bicycle, and 7% by foot. Table 2 shows the GWP emissions of the Donostia-San Sebastián campus that are obtained after defining and modelling all of the hypotheses for each building and the transport scenario in NEST.

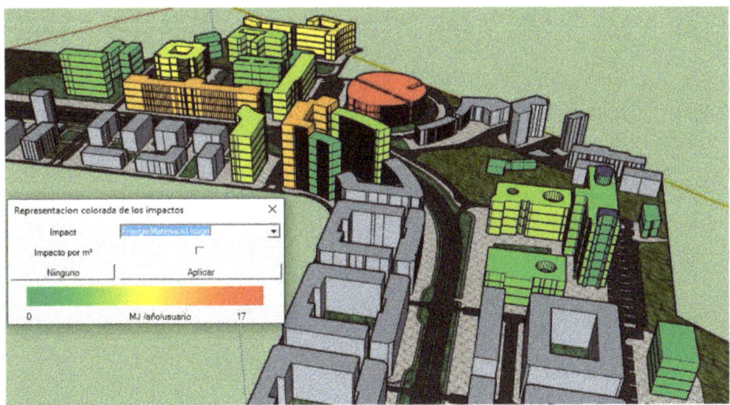

Figure 6. Graphical evaluation of the simulation of impacts for the campus with NEST.

A comparison of the results of the two calculation methodologies for the Donostia-San Sebastián campus shows that the monitored GWP data are 14% lower for the building sector and 21% higher for the transport sector than the simulation results. However, the total difference between the two methodologies is only 4%.

Considering the differences in the results for the buildings, the simulation process of NEST is based on a series of default assumptions to assess the energy consumption and environmental behaviour of the buildings. Considering these assumptions, it is understood that there will be certain variation between the building simulation result and the real performance of the building [50–52]. The reasons for the performance gap in a particular building can be several but in general, the performance gap happens due to the accuracy of the default values in the building simulation, variation of the weather data, or the influence of user, understood as user behaviour. Regarding transportation, it is very difficult to completely match the hypotheses simulated in the initial model with the monitored data, because the latter are based on questionnaires given out to campus users. This difference in the hypotheses results in an estimated impact for the transportation sector that is 21% higher for the university survey data than that calculated by NEST simulation.

3.2. Refurbishment Scenarios

3.2.1. Joint Plan Scenarios

In previous studies by Oregi et al. [29] and Leon et al. [38], theoretical rehabilitation scenarios associated only with the university were proposed and were not related to the action plans of the city of Donostia-San Sebastián. By contrast, in this study, the values and strategies defined by the SEAP of Donostia-San Sebastián [42], are used as a starting point to align the strategies of the university at the general level with those of the municipality at the local level. The SEAP of Donostia-San Sebastián comprises four strategic lines of action: (1) Energy efficiency, (2) renewable energies, (3) mobility, and (4) waste. The guidelines proposed by the city plan for adequate waste management (boosting second-hand markets (in particular), general awareness campaigns to promote reuse, promotion of reusable diapers, creating an infrastructure for territorial composting, taking advantage of surplus stores, etc.) are not aligned with university waste management strategies. Therefore, waste-related improvement actions are outside the scope of this study.

The first part of the study results is based on SEAP data (Table 3) and indicates the GWP emissions resulting from the application of 100% of the strategies for each strategic line. The relevance of each strategy group for a 100% reduction of GWP emissions is presented. However, all of the strategies have not been and will not be applicable to the city of Donostia-San Sebastián. Thus, in collaboration with different public stakeholders, the authors provide a critical review in the "revised data" section. This new section reflects three types of data: (1) The percentage of application or applicability of each strategy group; (2) the reduction in GWP emissions for this percentage of implementation; and lastly, (3) the relevance of each strategy group to 100% reduction in GWP emissions after reviewing the applicability of the strategies (more information about each SEAP strategy can be found in Appendix B). The authors use these revised values to determine the strategies for consideration in this study in terms of realisable actions implemented between 2011 and 2019 in the municipality of Donostia-San Sebastián.

Table 3. Summary table of improvement strategies according to the SEAP of Donostia-San Sebastián.

	Data according to SEAP		Revised Data		
	Emissions Avoided (tCO_2)	Percentage with Respect to the Total Reduction	Applicability (%)	Emissions Avoided (tCO_2)	Percentage with Respect to the Total Reduction
Efficiency					
Increase performance heating and cooling equipment	6272	2.7%	10.7%	671	0.4%
Heating and cooling consumption reduction	32,043	14.0%	33.5%	10,721	6.1%
Change the energy mix of the generation system	864	0.4%	2.9%	25	0.0%
Reduce lighting consumption	22,561	9.9%	92.9%	20,952	12.0%
Reduce heating demand	5963	2.6%	83.7%	4,992	2.9%
Reduce appliances consumption	4168	1.8%	80.0%	3,334	1.9%
Total efficiency	71,871	31.4%	56.6%	40,696	23.3%
Renewable					
Photovoltaic	6344	2.8%	0.3%	16	0.0%
Aero generators	1887	0.8%	0.0%	0	0.0%
Thermal solar	3234	1.4%	1.4%	44	0.0%
Geothermal	1856	0.8%	8.8%	164	0.1%
Biomass	754	0.3%	50.0%	377	0.2%
Biogas	5682	2.5%	0.0%	0	0.0%
Total renewable	19,757	8.6%	3.0%	601	0.3%
Mobility					
Increase biofuels	6776	3.0%	88.3%	5,984	3.4%
Reduce transportation consumption	126,448	55.2%	100.0%	126,448	72.4%
Electric vehicle	4117	1.8%	20.0%	823	0.5%
Total mobility	137,341	60.0%	97.0%	133,255	76.3%
TOTAL	228,969		76.2%	174,553	

According to the SEAP data, the strategic line of mobility is the sector in which up to 60% of total GWP reduction can be obtained. Thus, the main action group is focused on the "Reduce transportation consumption", whose application contributes 55.2% to the total reduction in GWP emissions. The strategic line of energy efficiency contributes 31.4% to the total reduction in GWP and includes action groups such as "Heating & cooling consumption reduction" and "reduce lighting consumption", whose application would contribute 14.0% and 9.9%, respectively, to the total reduction in GWP emissions. Within this line of efficiency, there are 28 other strategies with an overall influence below one percent (see the data in Appendix B). Lastly, the strategic line of renewable systems contributes 8.6% to the total reduction in GWP. This line includes action groups such as "Photovoltaic" and "Biogas", whose application would contribute 2.8% and 2.5%, respectively, to the total reduction in GWP emissions.

The estimated reduction in GWP emissions changes completely under an objective and critical review of the application or applicability of these strategies. With the support of different public stakeholders, the authors have conducted an exhaustive study on the application of each of the strategies in the municipality of Donostia-San Sebastián, tracking all of the actions based on different public sources and data from the energy department of the city of Donostia-San Sebastián.

An immediate conclusion that can be drawn is that it is essential to maintain the implementation of mobility strategies because of their 97% applicability. An opposite conclusion is drawn for the implementation of renewable technologies in Donostia-San Sebastián, for which only three percent of the SEAP proposed objectives has been implemented over the last eight years. Lastly, the applicability of most of the strategies associated with the strategic line of efficiency is projected to exceed 50%, and these strategies should therefore be considered in the study.

The strategies considered in this study based on a critical interpretation of the SEAP data are shown in Table 4. Following the existing SEAP guidelines, these strategies will be applied over a 10-year period (2020–2030). A cut-off is defined for action groups with applicability that is greater than 20% and a contribution above two percent to reducing final GWP emissions. Contrary to some European guidelines [53–55], this study will not consider any strategy associated with the strategic line of renewable technologies because of the low applicability of these strategies in the municipality of Donostia and the reduced impact on the final results proposed by SEAP. Regarding the final selected strategies, the applicability selection criterion has been maintained, but the selection criterion for the contribution to the reduced final GWP emissions has been changed to 0.4%. The strategies in this study are selected based on the percentage contribution to the total GWP reduction from the data reviewed (see Appendix B).

As shown in Table 4, for the 10 strategies to be evaluated by NEST in this study, SEAP has limited application to a particular building typology. For example, four of the ten strategies defined in Table 4 are limited to residential or commercial buildings. In addition, strategy 8 ("Acquisition of clean vehicles by the city") is limited to city vehicles. Therefore, although 10 strategies are proposed for evaluation at the municipal level in the study, only five of these strategies can be applied at the university campus level. Thus, universities should analyse different municipal policies for mobility on their campuses to achieve an optimal and coherent global mobility policy.

3.2.2. Results of Joint Refurbishment Scenario

A separate NEST model is developed for each strategy, and the reduction in the GWP emissions by the application of each strategy is shown in Table 5. In addition, three new scenarios are identified: In the first scenario (strategy 11), all of the energy efficiency strategies are applied together; in the second scenario (strategy 12), all of the mobility strategies are applied together; and in the third scenario (strategy 13), all of the strategies in Table 5 are applied together. The emissions reduction is calculated using the following values from Table 1; Table 2: GWP emissions from Donostia-San Sebastián of 6.96×10^8 and 9.68×10^6 kg CO_2-eq from the university campus.

Table 4. Applicability of SEAP strategies, between the city and the university campus.

Strategic Line	Acting Group	Strategy	Strategy to Integrate in NEST	Applicability	
				City of Donostia	University Campus
Energy Efficiency	Heating and cooling consumption reduction	Climate control regulation and insulation improvement of rehabilitation equipment	1-Improve the energy performance of existing equipment of tertiary buildings by 20%	YES	YES
		Public awareness campaigns	2-Reduce the energy demand of residential buildings by 25%	YES	NO
		Preparation of guides with saving measures in the tertiary sector	3-Reduce the energy demand of tertiary buildings by 20%	YES	YES
	Reduce lighting consumption	Lighting system regulation systems	4-Reduce the lighting consumption of tertiary buildings by 25%	YES	YES
		Improve efficiency of the home lighting system	5-Reduce the lighting consumption of residential buildings by 30%	YES	NO
		Renew street lighting	6-Reduce the lighting consumption of commercial buildings by 40%	YES	NO
	Reduce heating demand	Improve the efficiency of 20% of the windows	7-Reduce the heating energy demand of residential buildings by 36%	YES	NO
		Renewal of 5% of existing homes with high benefits			
Mobility	Increase biofuels	Acquire clean vehicles through the town hall	8-Adapt private car emissions	YES	NO
		Acquire clean vehicles by public bodies	9-Adapt the emissions of private cars and buses	YES	YES
		Empowerment of clean distribution vehicles			
	Reduce transport related consumption	Program to improve and boost pedestrian mobility	10-Adapt the current mobility model of NEST, mainly promoting pedestrians and bicycles and trying to replace the private vehicle as much as possible with the bus or/and the train	YES	YES
		Cycle mobility improvement and promotion program			
		Program to improve the competitiveness of public transport			
		Implement mobility management program			
		Implement education and communication program in sustainable mobility			
		Private vehicle and freight transport management program			

Table 5. Number of GWP emissions avoided in each strategy.

Strategy	GWP Emissions (kg CO_2-eq)		GWP Emissions (kg CO_2-eq/person)	
	City of Donostia	**University Campus**	**City of Donostia**	**University Campus**
1	6.93×10^8	9.29×10^6	3.71×10^3	8.40×10^2
2	6.89×10^8	9.68×10^6	3.69×10^3	8.75×10^2
3	6.80×10^8	8.57×10^6	3.65×10^3	7.74×10^2
4	6.89×10^8	9.12×10^6	3.69×10^3	8.25×10^2
5	6.92×10^8	9.68×10^6	3.71×10^3	8.75×10^2
6	6.72×10^8	9.68×10^6	3.60×10^3	8.75×10^2
7	6.79×10^8	9.68×10^6	3.64×10^3	8.75×10^2
8	6.93×10^8	9.68×10^6	3.71×10^3	8.75×10^2
9	6.88×10^8	9.56×10^6	3.69×10^3	8.64×10^2
10	5.53×10^8	8.53×10^6	2.96×10^3	7.71×10^2
11	6.19×10^8	7.62×10^6	3.31×10^3	6.89×10^2
12	5.42×10^8	8.40×10^6	2.90×10^3	7.59×10^2
13	4.64×10^8	6.35×10^6	2.49×10^3	5.73×10^2

The results of the analysis for the city of Donostia-San Sebastián show that, as shown by the SEAP plan for the city of Donostia-San Sebastián, strategy 10 for the city's mobility model results in the highest GWP emissions reduction of 20.6% (1.43×10^8 kg CO_2-eq) relative to the 2015 scenario. The implementation of efficiency strategies reduces GWP emissions by up to 11.2% (7.75×10^7 kg CO_2-eq). Within this strategic line, the following strategies stand out: "Renewable shop lighting" (6 strategy), "Improving the efficiency of residential buildings by replacing windows and energy-rehabilitating housing" (strategy 7) and "Developing guidelines with savings measures for the tertiary sector" (strategy 3), which reduce 2015 GWP emissions by 3.4% (2.37×10^7 kg CO_2-eq), 2.5% (1.73×10^7 kg CO_2-eq), and 2.2% (1.55×10^7 kg CO_2-eq), respectively. Finally, the application of all of the strategies considered in this study (strategy 13), would reduce GWP emissions by 33.3% (2.32×10^8 kg CO_2-eq) annually compared to the current scenario. The signatory cities to the 2017 Global Covenant of Mayors for Climate and Energy committed to reducing emissions in 2030 by 40% of those for the base year 2007. Considering that GWP emissions of the city of Donostia-San Sebastián were 9.92×10^8 kg CO_2-eq in 2007 [42], an evaluation of the scenarios proposed by this study shows that the city of Donostia-San Sebastián could meet its commitment by implementing this joint plan with the university. In turn, the city would meet the objective set by the European Commission [56], which defined an objective of reducing GWP emissions from reference year by a minimum of 40% by 2030.

As shown in Table 4, only five strategies for reducing GWP emissions from the university campus have been applied. In comparison to the results for the city, given that the impact of the buildings is 57% of the total GWP impact of the campus, the strategic line with the greatest amount of reduced GWP emissions is that of energy efficiency, at a reduction of up to 21.3% (2.06×10^6 kg CO_2-eq) of total campus emissions. Within this line, strategy 3 ("Preparation of guidelines with savings measures for the tertiary sector") stands out by reducing emissions by 11.5% (1.11×10^6 kg CO_2-eq). An improved mobility scenario can also significantly reduce total GWP campus emissions by up to 13.2% (1.28×10^6 kg CO_2-eq).

Table 5 shows a second measure that can be used to analyse the impact of GWP on each user in the city and the university campus. The effect of applying each strategy is similar to the total results. However, the impacts of the university campus are approximately 10 times below those of the city. There are two contributions to this difference. First, the level of energy efficiency of the different buildings of the campus is quite high, resulting in lower consumption than for older buildings in different city districts. Second, more users consume the same number of resources in the university than in the city. Therefore, the impact per person is lower for the university than for the city, where there is a lower density.

4. Discussion

Three considerations can be identified from the study and the obtained results: The city-campus relationship (1); the applicability of the strategic analysis; (2) and the integration of tools (3).

First, the single location of the campus within the urban network of the city of Donostia-San Sebastián results in a direct relationship between the two evaluated elements. At the same time, strategies or commitments of the city in terms of mobility or energy efficiency directly affect the two evaluation scales. However, university campuses normally develop their own strategies without considering the trends or commitments of the city in which they are located. This study is an attempt to reflect how strategies defined at the municipal level can be applied at the university campus level and how these actions environmentally impact the city. In this study, the small university campus contributed only 1.4% of the total GWP emissions of the city in 2015. The application of all of the strategies proposed in this study (scenario 13), reduces the total GWP emissions of the campus by 34.5% (with respect to 2015). This reduction is greater when the simulated reform scenario in NEST is in conjunction with the city and applying the municipal plans. Considering previous research carried out on the whole of the Basque Country University [38], where university campus reform scenarios based only on university policies were applied, (without taking into account municipal policies), the reduction was smaller.

The second consideration shows the need to constantly monitor the implementation and applicability of action strategies defined by documents such as SEAPs, which have a long perspective (10 years). The socio-economic or normative changes that occur over this timeframe can significantly alter any proposed scenario, reducing or eliminating the feasibility of application of a previously proposed strategy. The same happens with the university's plans that are linked to the Basque Government, following long-term European directives (2020, 2030, and 2050). Taking into account the socioeconomic changes that also affect the university, it can be periodically simulated, in a few weeks, what is the most sustainable option. However, from the comparison between the previous simulation carried out and the monitoring of the University consumption collected between 2015 and 2018, it shows that, although the simulation allows detecting the most sustainable reform scenario, in the case of the university the simulation data has suffered greater variations than in the city, so the university must take this in consideration, and corroborate the simulation data with constant monitoring.

The third consideration shows the potential of tools, such as NEST, to evaluate and define future scenarios. This study has shown that it is not always easy to obtain values for the energy consumption or GWP emissions that are in agreement with monitoring data. However, in this study, the highest difference between the monitored values and the NEST simulation results was 21% (mobility of the university campus), which could be viewed positively. Most significantly, the monitoring information for the city or the energy consumption of all of the buildings of the university campus were obtained by different public entities over a period of three years, whereas the modelling and calculation process was completed in three weeks for the city and one week for the campus. In addition to the rapidity of the simulations, this type of tool can be used to calibrate the input data to the actual application of each strategy. In this way, the evaluation model becomes a dynamic model that can be adapted to each moment, facilitating rapid decision-making on the most sustainable design solution. Lastly, these tools enable the potential of strategies to be analysed at either the municipal or local level, including for a given and smaller area of a municipality. In this way, those responsible for the university campus can estimate the reduction in GWP emissions of the campus from the application of municipal strategies or perform a parallel analysis of the effect of these strategies on the city. For this last consideration, different public actors of the city of Donostia-San Sebastián and the university campus, who currently work and choose strategies separately from each other, must work together in order to optimise resources, enabling the design, analysis, and quantification of the impact of each decision at different scales.

The evaluation of the proposed scenarios, wherein the strategy of the university is aligned with the municipal policies of the city of San Sebastian, can be used to achieve higher levels of sustainability. In this case, it has been possible to verify how a scenario of joint refurbishment between the city and the university, according to municipal sustainability plans (SEAP), allows cities to assure compliance with agreements at European level such as the Global Covenant of Mayors for Climate and Energy. Therefore, it is proven that the university can contribute to the environmental improvement of cities. The sustainability of a university can no longer be limited to improving a particular faculty building or the sustainability of the university as a whole but requires the establishment of strategies to develop synergies with the municipal environmental policies of the cities in which university campuses are located.

Author Contributions: Conceptualization, I.L., X.O., and C.M.; Investigation, I.L., X.O., and C.M.; Methodology, I.L. and X.O.; Data curation X.O.; Supervision, I.L. and C.M.; Writing original draft, I.L. and X.O.; Writing—review and editing: I.L., X.O., and C.M. All authors have read and agreed to the published version of the manuscript.

Funding: This research received no external funding.

Acknowledgments: The authors thank the UPV/EHU (Vicerrectorado de Innovación, Compromiso social y Acción cultural) for supporting the development of this type of work to achieve a more Sustainable University.

Conflicts of Interest: The authors declare no conflict of interest.

Appendix A

Table A1. Monitored data for the campus of Donostia-San Sebastián.

	ID	Faculty (PDI)	Students	Admin. Staff. (PAS)	Total Users	Electricity Consumption (kWh)	Natural Gas Consumption (kWh)	Diesel Consumption (kWh)
Gipuzkoa School of Engineering	B1	172	1335	30	1537	793,521	708,984	75
Faculty of Economics and Business	B2	79	1077	14	1170	196,750	527,404	0
Faculty of Education, Philosophy and Anthropology	B3	150	1377	21	1548	552,087	1,033,126	0
Faculty of Computer Science	B4	152	670	23	845	553,200	456,816	0
Faculty of Psychology	B5	136	1179	20	1335	605,183	862,678	0
Faculty of Chemistry	B6	184	691	26	901	685,031	697,161	0
Superior Technical School of Architecture	B7	99	807	14	920	293,733	622,594	0
Faculty of Low	B8	103	1124	31	1258	188,151	545,374	0
Teacher-Training College	B9	82	1087	13	1182	159,610	406,778	0
Carlos Santamaria Center	B10	0	1852	7	1859	1,302,316	524,763	0
Joxe Mari Korta Center	B11	3	4516	5	4524	1,957,695	624,214	0
Ignacio Mª Barriola Center	B12	0	1000	0	1000	216,114	646,357	0
Villa Julianategui (Vicerectorate of Campus)	B13	1	0	57	58	45,325	50,895	0
Multipurpose Training and Innovation Center (Elbira Zipitria Center)	B14	25	245	5	275	215,688	59,124	0

B.

Table 2. Complete table of data according to the SEAP, summarized in Table 3.

	Energy Saving (kwh)	Emissions Avoided (tCO$_2$-eq)	Percentage of Emissions Avoided with Respect to the Total (%)	Real Application	"Real" Emissions Avoided (tCO$_2$-eq)
Increase Performance Heating and Cooling Equipment					
1.1.1 Replace 17 boiler (condensation/low temperature) and heat pumps	1,029,420	214	0.087%	94%	201
1.1.2 Improve efficiency of 17 emitting systems	302,670	56	0.023%	53%	30
1.2.7 Replacement of existing boilers in 15,760 homes	21,568,943	3990	1.616%	6%	239
1.3.1 Improve boiler performance of 30%	10,875,000	2012	0.815%	10%	201
Heating and Cooling Consumption Reduction					
1.1.4 Regulation of air conditioning and improvement of insulation of rehabilitation equipment	6,131,692	993	0.402%	100%	993
1.1.8 Expand energy telemanagement systems	3,432,719	970	0.393%	35%	339
1.1.10 Good practices regarding heating and lighting control	119,723	30	0.012%	100%	30
1.1.11. Implement energy management software	1,837,862	517	0.209%	100%	517
1.1.14 Training of municipal technical staff	869,051	209	0.085%	100%	209
1.2.0 Citizen awareness campaigns	21,406,433	5326	2.157%	100%	5326
1.3.6 Energy management of 75% of buildings in the sector	43,345,000	11,220	4.544%	1%	112
1.3.8 Preparation of guides with saving measures for the tertiary sector	42,852,124	12,778	5.175%	25%	3194
Change the Energy Mix of the Generation System					
1.2.9 Centralized systems (DH) in residential areas of Asia	540,738	100	0.040%	25%	25
1.3.4 Install microcogeneration in the hotel and residential sector of the elderly (34 establishments). The average size of each installation of this type is 35 kWt and 16 kWe	3,848,891	410	0.166%	0%	0
1.3.9 Install centralized systems (DH) in the services sector (three systems are considered)	1,912,100	354	0.143%	0%	0
Reduce Lighting Consumption					
1.1.5 Improve the lighting installations of 68 buildings	879,134	334	0.135%	40%	134
1.1.15 Replacement mercury vapor lamps	37,454	14	0.006%	100%	14
1.1.16 Prevent light pollution, improving the energy efficiency of luminaires	1,530,000	568	0.230%	100%	568
1.1.17 Public lighting control system	20,000	7	0.003%	100%	7
1.1.18 Energy management software for public lighting	50,340	19	0.008%	100%	19
1.1.19 Implement innovative lighting technologies	39,850	15	0.006%	100%	15
1.1.20 Lighting system regulation systems	3,000,000	1113	0.451%	80%	890
1.2.6 Improve efficiency of the home lighting system (50,000 bulbs)	2,682,750	1020	0.413%	100%	1020
1.2.8 Renew electrical installations	3,900,722	1483	0.601%	20%	297
1.2.11 Environmental education in the school environment	2,378,493	627	0.254%	100%	627
1.3.2 Renew street lighting	45,675,000	17,361	7.031%	100%	17,361

Table 2. *Cont.*

	Energy Saving (kwh)	Emissions Avoided (tCO$_2$-eq)	Percentage of Emissions Avoided with Respect to the Total (%)	Real Application	"Real" Emissions Avoided (tCO$_2$-eq)
Reduce Heating Demand					
1.1.6 Improve the thermal envelope of buildings with characteristics that allow	1,888,188	349	0.141%	16%	56
1.2.1 Implement high-efficiency criteria in new urban developments	161,039	61	0.025%	100%	61
1.2.2 Energy certification A in 100% of new public housing	881,250	250	0.101%	50%	125
1.2.4 Improve the efficiency of 20% of the windows	15,988,662	2958	1.198%	100%	2958
1.2.13 Renovation of 5% of existing homes (3807) with high benefits	9,341,901	1731	0.701%	100%	1731
1.3.3 Certification A in 50% of the new tertiary buildings	3,314,000	614	0.249%	10%	61
Reduce Appliances Consumption					
1.2.5 Renew appliances with better energy efficiency in 25% of appliances	12,921,792	4168	1.688%	80%	3334
Photovoltaic					
2.1.1 Install PV of 149,985 m^2 (6.82 MWp) in public facilities	Non determin,	2585	1.047%	0.25%	6
2.3.1 Install PV on the sides of the tracks (65,127 m^2—2960 kWP)	Non determin,	1122	0.454%	0%	3
2.3.2 Install PV in the buckets of large surfaces (87,682 m^2—3980 kWP)	Non determin,	1511	0.612%	0%	4
2.3.3 Install PV in the car parks (2970 kWP)	Non determin,	1126	0.456%	0%	3
Aerogenerators					
2.1.2 Wind turbine installation. It is planned to install 7.2 MW	Non determin,	1887	0.764%	0%	0
Thermal Solar					
2.1.3 Ensure the efficient operation of solar thermal systems	Non determin,	40	0.016%	100%	40
2.1.4 Implement ST 2000 m^2 on roofs of public buildings	Non determin,	259	0.105%	0%	0
2.2.3 Install 13,000 m^2 of ST in the residential sectorl	Non determin,	1638	0.663%	0%	0
2.3.4 Install 10,000 m^2 of ST in the service sector	Non determin,	1297	0.525%	0%	4
Geotermia					
2.1.5 Incorporate nine units with a total of 630 kW of heat production with geothermal systems	Non determin,	684	0.277%	24%	164
2.3.6 Climate systems with geothermal support	Non determin,	1172	0.475%	0%	0
Biomass					
2.2.1 Forty installations with 3.2 MWt of biomass in the residential sector	Non determin,	754	0.305%	50%	377
Biogas					
2.2.2 Biogas exploitation biogas plant	Non determin,	5682	2.301%	0%	0

Table 2. *Cont.*

	Energy Saving (kwh)	Emissions Avoided (tCO$_2$-eq)	Percentage of Emissions Avoided with Respect to the Total (%)	Real Application	"Real" Emissions Avoided (tCO$_2$-eq)
Increase Biofuels					
3.1.1 Acquire clean vehicles by the town hall	4,371,335	1,182	0.479%	33%	390
3.1.2 Encourage the use of clean fuels in vehicles that provide public services	2,185,667	591	0.239%	100%	591
3.2.1 Acquire clean vehicles by the municipality	17,823,370	3563	1.443%	100%	3563
3.2.3 Promotion of clean distribution vehicles	4,798,600	1440	0.583%	100%	1440
Reduce Transportation Consumption					
3.1.3 Development of the sustainable mobility plan	337,634	99	0.040%	100%	99
3.2.4 Program to improve and boost pedestrian mobility	63,208,176	12,635	5.117%	100%	12,635
3.2.5 Program to improve and boost cycling mobility	63,208,176	12,635	5.117%	100%	12,635
3.2.6 Program to improve the competitiveness of public transportation	221,228,617	44,222	17.909%	100%	44,222
3.2.7 Private vehicle and freight transport management program	189,624,529	37,905	15.351%	100%	37,905
3.2.8 Implement mobility management program	63,208,176	12,635	5.117%	100%	12,635
3.2.9 Implement education and communication program in sustainable mobility	31,604,077	6317	2.558%	100%	6317
Electric Vehicle					
3.2.2 Promotion of the electric vehicle	28,151,007	4117	1.667%	20%	823

References

1. United Nations. *The World's Cities in 2016: Data Booklet*; Economic and Social Affair: New York, NY, USA, 2016.
2. Khazaei, M.; Razavian, M.T. Sustainable urban development (an innovative approach in the development of cities around the world). *Int. Res. J. Appl. Basic Sci.* **2013**, *4*, 1543–1547.
3. Swilling, M.; Hajer, M.; Baynes, T.; Bergesen, J.; Labbé, F.; Musango, J.K.; Ramaswami, A.; Robinson, B.; Salat, S.; Suh, S.; et al. The Weight of Cities Resource Requirements of Future Urbanization. Available online: https://europa.eu/capacity4dev/unep/documents/weight-cities-resource-requirements-future-urbanization (accessed on 18 December 2019).
4. UN Habitat. *World Cities Report 2016, Urbanization and Development: Emerging Futures, Key Findings and Messages*; UN Habitat: Nairobi, Kenya, 2016.
5. DESA 2018. Available online: https://www.un.org/development/desa/publications/2018-revision-of-world-urbanization-prospects.html (accessed on 18 December 2019).
6. IEA. *World Energy Outlook*; International Energy Agency: Paris, France, 2016; p. 28.
7. Muñoz, P.; Morales, P.; Letelier, V.; Muñoz, L.; Mora, D. Implications of life cycle energy assessment of a new school building, regarding the nearly zero energy buildings targets in EU: A case of study. *Sustain. Cities Soc.* **2017**, *32*, 142–152. [CrossRef]
8. Tronchin, L.; Fabbri, K. Energy performance building evaluation in Mediterranean countries: Comparison between software simulations and operating rating simulation. *Energy Build.* **2008**, *40*, 1176–1187. [CrossRef]
9. UNEP. *Building Day Brochure*; United Nations Environment Programme: Nairobi, Kenyan, 2015.
10. Dogruyol, K.; Aziz, Z.; Arayici, Y. Eye of Sustainable Planning: A Conceptual Heritage-Led Urban Regeneration Planning Framework. *Sustainability* **2018**, *10*, 1343. [CrossRef]
11. Catherine, C.-V.; Philippe, O. *Concevoir et Évaluer un Projet d'éco-Quartier: Avec le Référentiel INDI*; Editions du Moniteur: Paris, France, 2012.
12. Oliver-Solà, J.; Josa, A.; Arena, A.P.; Gabarrell, X.; Rieradevall, J. The GWP-Chart: An environmental tool for guiding urban planning processes. Application to concrete sidewalks. *Cities* **2011**, *28*, 245–250. [CrossRef]
13. Yigitcanlar, T.; Teriman, S. Rethinking sustainable urban development: Towards an integrated planning and development process. *Int. J. Environ. Sci. Technol.* **2015**, *12*, 341–352. [CrossRef]
14. Morais, P.; Camanho, A.S. Evaluation of performance of European cities with the aim to promote quality of life improvements. *Omega* **2011**, *39*, 398–409. [CrossRef]
15. Arundel, J.; Lowe, M.; Hooper, P.; Roberts, R.; Rozek, J.; Higgs, C.; Giles-Corti, B. *Creating Liveable Cities in Australia. Mapping Urban Policy Implementation and Evidence-Based National Liveability Indicators*; Centre for Urban Research RMIT University: Melbourne, Australia, 2017.
16. Beatriz, V.A.; Pilar, M.; David, R.G. Sustainable Urban Liveability: A Practical Proposal Based on a Composite Indicator. *Sustainability* **2019**, *11*, 86.
17. Qunxi, G.; Min, C.; Xianli, Z.; Zhigeng, J. Sustainable Urban Development System Measurement Based on Dissipative Structure Theory, the Grey Entropy Method and Coupling Theory: A Case Study in Chengdu, China. *Sustainability* **2019**, *11*, 293.
18. Pingtao, Y.; Weiwei, L.; Danning, Z. Assessment of City Sustainability Using MCDM with Interdependent Criteria Weight. *Sustainability* **2019**, *11*, 1632.
19. Wu, J. Landscape sustainability science: Ecosystem services and human well-being in changing landscapes. *Landsc. Ecol.* **2013**, *28*, 999–1023. [CrossRef]
20. Simón, M.; Cristina, G.; Antonio, H.; Vicente, A. Sustainability Assessment of Constructive Solutions for Urban Spain: A Multi-Objective Combinatorial Optimization Problem. *Sustainability* **2019**, *11*, 839.
21. Lee, W.L. A comprehensive review of metrics of building environmental assessment schemes. *Energy Build.* **2013**, *62*, 403–413. [CrossRef]
22. Reith, A.; Orova, M. Do green neighbourhood ratings cover sustainability? *Ecol. Indic.* **2015**, *48*, 660–672. [CrossRef]
23. Sharifi, A.; Murayama, A. A critical review of seven selected neighborhood sustainability assessment tools. *Environ. Impact Assess. Rev.* **2013**, *38*, 73–87. [CrossRef]
24. Braulio-Gonzalo, M.; Bovea, M.D.; Ruá, M.J. Sustainability on the urban scale: Proposal of a structure of indicators for the Spanish context. *Environ. Impact Assess. Rev.* **2015**, *53*, 16–30. [CrossRef]

25. Xabat, O.; Nekane, H.; Iñaki, P.; Jose Luis, I.; Lara, M.; Panagiotis, S. Automatised and georeferenced energy assessment of an Antwerp district based on cadastral data. *Energy Build.* **2018**, *173*, 176–194.

26. VITO. Dynamic Energy Atlas Tool. Available online: https://geoflex-solutions.eu/c/DEA%20-%20Dynamic%20Energy%20Atlas/ (accessed on 18 December 2019).

27. Walter, E.; Kämpf, J.H. A verification of CitySim results using the BESTEST and monitored consumption values. In Proceedings of the 2nd Building Simulation Applications Conference, Bolzano, Italy, 4–6 February 2015; pp. 215–222.

28. Yepez, G. *Construction d'un Outil D'évaluation Environnementale des Écoquartiers: Vers une Méthode Systémique de Mise en Oeuvre de la Ville Durable*; Université Bordeaux: Bourdeaux, France, 2011.

29. Xabat, O.; Maxime, P.; Lara, M.; Alexandre, E.; Iker, M. Sustainability assessment of three districts in the city of Donostia through the NEST simulation tool. *Nat. Res. Forum* **2016**, *40*, 156–168.

30. William, V.-C.; Xavier, P.-P.; Sergio, L.-M. Application of a Smart City Model to a Traditional University Campus with a Big Data Architecture: A Sustainable Smart Campus. *Sustainability* **2019**, *11*, 2857.

31. Ning, A.; Marc, K.; Cynthia, K.-B.; Thomas, L.T. Sustainability assessment of universities as small-scale urban systems: A comparative analysis using Fisher Information and Data Envelopment Analysis. *J. Clean. Prod.* **2019**, *212*, 1357–1367.

32. Alshuwaikhat, H.M.; Abubakar, I. An integrated approach to achieving campus sustainability: Assessment of the current campus environmental management practices. *J. Clean. Prod.* **2008**, *16*, 1777–1785. [CrossRef]

33. Disterheft, A.; da Silva Caeiro, S.S.F.; Ramos, M.R.; de Miranda Azeiteiro, U.M. Environmental Management Systems (EMS) implementation processes and practices in European higher education institutions top-down versus participatory approaches. *J. Clean. Prod.* **2012**, *31*, 80–90. [CrossRef]

34. Chen, S.; Lu, M.; Tan, H.; Luo, X.; Ge, J. Assessing sustainability on Chinese university campuses: Development of a campus sustainability evaluation system and its application with a case study. *J. Build. Eng.* **2019**, *24*, 100747.

35. Nikhat, P.; Avlokita, A. Assessment of sustainable development in technical higher education institutes of India. *J. Clean. Prod.* **2019**, *214*, 975–994.

36. Paola, M.; Federico, O.; Francesco, A.; Claudia, G. Environmental performance of universities: Proposal for implementing campus urban morphology as an evaluation parameter in Green Metric. *Sustain. Cities Soc.* **2018**, *42*, 226–239.

37. Paulo, J.R.; Lígia, M.C.P.; Nuno, G.; Helder, C.; Diogo, A. Sustainability Strategy in Higher Education Institutions: Lessons learned from a nine-year case study. *J. Clean. Prod.* **2019**, *222*, 300–309.

38. Iñigo, L.; Xabat, O.; Cristina, M. Environmental assessment of four Basque University campuses using the NEST tool. *Sustain. Cities Soc.* **2018**, *42*, 396–406.

39. Kottek, M.; Grieser, J.; Beck, C.; Rudolf, B.; Rubel, F. World Map of the Köppen-Geiger climate classification updated. *Meteorol. Z.* **2006**, *15*, 259–263. [CrossRef]

40. IES. IES Virtual Environment. MacroFlo User Guide. Integrated Environmental Solutions Limited, 2014. Available online: http://www.iesve.com/downloads/help/ve2014/Thermal/MacroFlo.pdf (accessed on 18 December 2019).

41. Informe anual de sostenibilidad. Donostia/San Sebastián. Observatorio de la Sostenibilidad, 2018. Available online: http://www.cristinaenea.eus/es/mnu/observatorio-de-la-sostenibilidad-informe-anual-de-sostenibilidad (accessed on 18 December 2019).

42. Plan de Acción para la Energía Sostenible (PAES-SEAP) del Municipio de Donostia—San Sebastián. Available online: https://www.donostia.eus/ataria/documents/8023875/8050877/Documento+PAES_cas.pdf/b9985321-6696-4ee6-884d-3de59a9851fe (accessed on 18 December 2019).

43. Plan de Acción Klima 2050 de Donostia—San Sebastián. Available online: https://www.donostia.eus/ataria/documents/8023875/8246263/Donostiako+Klima+2050+Ekintza+Plana_cas.pdf/d8c6f81c-1873-453d-b688-18b9d93f841b (accessed on 18 December 2019).

44. Udalsarea 2030. Red vasca de municipio hacia la sostenibilidad. Datos supramunicipales para el cálculo del inventariado GEI del municipio de Donostia-San Sebastián. Available online: http://www.udalsarea21.net/Usuarios/Acceso.aspx?IdMenu=657A0F24-A6D1-4E5A-9A4E-548A3D551DF0&Idioma=es-ES (accessed on 20 January 2020).

45. Plan Foral Gipuzkoa Energía. Diputación de Guipúzcoa. Dirección General de Medio Ambiente y Obras Hidráulicas. Available online: https://www.gipuzkoa.eus/documents/3767975/3808418/Plan+Foral+Gipuzkoa+Energ%C3%ADa.pdf/d162dce1-9eb6-48da-ada3-0dcff7854631 (accessed on 18 December 2019).
46. Plan de Movilidad Urbana Sostenible. Donostia/San Sebastián. 2008–2024. Ayuntamiento de Donostia—San Sebastián. Available online: http://www.donostiafutura.com/media/uploads/publicaciones/Plan_Movilidad__Urbana_Sostenible_2008_2024.pdf (accessed on 18 December 2019).
47. Agenda Local 21 Donostia—San Sebastián. Diagnóstico ambiental. Ayuntamiento de Donostia—San Sebastián, 2014. Available online: https://www.donostia.eus/ataria/documents/8023875/8050869/Des.pdf/bc9d6508-3963-4020-a874-9b397343c671 (accessed on 18 December 2019).
48. Registro de Certificados de Eficiencia Energética del País Vasco. Departamento de Desarrollo Económico e Infraestructuras. Gobierno Vasco. Available online: https://apps.euskadi.eus/y67paUtilidadSeccionWar/utilidadSeccionJSP/y67painicio.do?idDepartamento=51&idioma=es (accessed on 18 December 2019).
49. EHU-Aztarna Ecological and Social Footprint of the University of the Basque Country: How to Reduce Our Impact? University of the Basque Country. Available online: https://www.ehu.eus/documents/4736101/13145292/EHU-Aztarna.pdf/6cc5765d-e182-fd80-0cf8-4e9b866f375f (accessed on 18 December 2019).
50. Allard, I.; Olofsson, T.; Nair, G. Energy evaluation of residential buildings: Performance gap analysis incorporating uncertainties in the evaluation methods. *Build. Simul.* **2018**, *11*, 725–737. [CrossRef]
51. De Wilde, P. The gap between predicted and measured energy performance of buildings: A framework for investigation. *Autom. Constr.* **2014**, *41*, 40–49. [CrossRef]
52. Jensen, T.; Chappin, É.J.L. Reducing domestic heating demand: Managing the impact of behavior-changing feedback devices via marketing. *J. Environ. Manag.* **2017**, *197*, 642–655. [CrossRef] [PubMed]
53. European Commission. Directive 2012/27/EU of the European Parliament and of the Council of 25 October 2012 on energy efficiency, amending Directives 2009/125/EC and 2010/30/EU and repealing Directives 2004/8/EC and 2006/32/EC. *Off. J. Eur. Union* **2012**, *55*, 1–97.
54. European Commission. *EU 20-20-20 Objectives for 2020*; Energy Efficiency Plan: Brussels, Belgium, 2011.
55. Directorate-General for Research and Innovation (European Commission). *The Strategic Energy Technology (SET) Plan*; Joint Research Centre: Brussels, Belgium, 2018.
56. European Commission. A Policy Framework for Climate and Energy in the Period From 2020 to 2030. 2014. Available online: https://eur-lex.europa.eu/legal-content/EN/TXT/?uri=CELEX:52014DC0015 (accessed on 18 December 2019).

 *sustainability*

Article

Energy Management in PV Based Microgrids Designed for the Universidad Nacional de Colombia

Luis Fernando Grisales-Noreña [1], Carlos Andrés Ramos-Paja [2], Daniel Gonzalez-Montoya [3], Gerardo Alcalá [4] and Quetzalcoatl Hernandez-Escobedo [5,*]

[1] Departmento de Mecatrónica y Electromecánica, Instituto Tecnológico Metropolitano, Medellín 050015, Colombia; luisgrisales@itm.edu.co
[2] Facultad de Minas, Universidad Nacional de Colombia, Medellín 050041, Colombia; caramosp@unal.edu.co
[3] Departamento de Electrónica y Telecomunicaciones, Instituto Tecnológico Metropolitano, Medellín 050015, Colombia; danielgonzalez@itm.edu.co
[4] Centro de Investigación en Recursos Energéticos y Sustentables, Universidad Veracruzana, Coatzacoalcos, Veracruz 96535, Mexico; gacala@uv.mx
[5] Escuela Nacional de Estudios Superiores Juriquilla, UNAM, Queretaro 76230, Mexico
* Correspondence: qhernandez@unam.mx

Received: 13 December 2019; Accepted: 25 January 2020; Published: 7 February 2020

Abstract: Stand-alone Electrical microgrids (MGs) require power management strategies to extend the life-time of their devices and to guarantee the global power balance of non-critical loads such as lighting of small sections of an university campus or individual air conditioning systems. This paper proposes an energy management strategy (EMS) for an isolated DC microgrid formed by a photovoltaic system (PVS), an energy storage system (battery), and a noncritical load. This configuration enables the photovoltaic system to control the power generation and ensures that the storage element does not exceed the safe limits of the state of charge. To control the generation of the photovoltaic system, two operating modes based on the perturb and observe (P&O) algorithm are implemented. The first one performs a maximum power point tracking (MPPT) action, while the second one regulates the power generated by the PVS to match the load requirement (power demand tracking, PDT). The management strategy also considers different operating states for ensuring the battery safety: normal operation, overcharge (at the maximum state of charge), and bulk charge (at the minimum state of charge); in those states the disconnection/connection of both the battery and the load is also considered. The main contribution of this work is to design and test a control strategy for an EMS aimed at regulating a standalone microgrid based on a PV system and an energy storage device. This solution is validated using detailed MG circuital simulations, which includes the PV source model (single-diode model), lithium-ion battery model, constant power load model and the DC/DC converters equations; moreover, realistic power generation and demand from Universidad Nacional de Colombia, located at Medellín-Colombia, are considered. The results obtained demonstrate the effectiveness of the energy management strategy, and in this way, enable to extend the battery lifetime and reduce the costs associated to the maintenance and disconnection of the microgrid in educational buildings or other applications focused on this type of DC microgrid.

Keywords: DC microgrid; energy management system; photovoltaic (PV) system; energy storage system; constant power load; power generation control

---

1. Introduction

The integration of renewable energy sources (RES) into electrical power systems has been an active field of study for more than ten years, which is usually oriented to reduce the environmental impact of energy generation and to increase the coverage of electrical networks [1,2]. Microgrids (MGs)

have been used to integrate multiple RES, fossil fuels, energy storage systems (ESSs), electrical loads, and the electrical grid [3]; this by using AC or DC buses to exchange electrical power in two modes: stand-alone or connected to the grid [4]. As presented in [5,6], the integration of ESS into a microgrid mitigates the energy production intermittency of renewable sources caused by the night and weather conditions (cloudy operation, solar irradiance, temperature, wind speed, etc.); moreover, the ESS helps to control the power flows inside the MG to achieve the global power balance and maximize the benefits for the operator or proprietary. In addition, to manage the intermittency of renewable energy resources and load demand characteristics, an EMS is an essential part of the MGs for optimal use of the distributed energy resources in reliable and coordinated ways [5,6].

DC microgrids have advantages over AC MGs in terms of efficiency and connection simplicity; for example, as reported in [7], the DC MG avoids performing a frequency control for the main bus; which reduces the installation complexity. Moreover, DC MGs are formed by distributed generators (DGs), ESS, and electrical loads [8–11]. In particular, photovoltaic systems (PVSs) are commonly included in DC MGs as DGs due to the wide availability of solar energy [12]; and the integration of a PVS, an ESS, and loads is known as a stand-alone photovoltaic system (SPVS) [13]. Those SPVSs are used in multiple applications for attending non-critical loads, such as plug-in chargers for electrical vehicles, lighting systems, television sets, data centers (facility lighting or non-critical workstations used for log files inspection), air-conditioning, home applications, among others [14–17]. Many of the SPVS applications described above are present in educational buildings, e.g., schools, universities, among others [18]. In this sense, a correct energy management of the SPVS allows to reduce the operational cost and improve the quality of the electrical service of the educational building; thus improving the life conditions of the users in those spaces [19]. For example, in [20] a EMS for stand-alone DC microgrid based on a photovoltaic source, electrochemical storage, a supercapacitor, and a diesel generator was proposed. The proposed strategy allows to control the system to balance the power injection accounting the slow start-up characteristic of the diesel generator by means of a supercapacitor dynamics. A similar solution was published in [21], where an EMS is proposed for a stand-alone hybrid AC/DC microgrid formed by a PV system, a fuel cell as a secondary power source, and a battery and a supercapacitor as hybrid ESS. The proposed EMS allows to manage the system under different modes and SOC limits of the hybrid energy storage when all devices are connected to the DC bus. For the reasons discussed above, this paper is aimed to design an energy management strategy for PV based microgrids that consider the environmental conditions and power demand of the Universidad Nacional de Colombia, located at Medellín-Colombia.

SPVS devices must be integrated using a DC-bus interfaced with DC/DC converters, which enable the operation of each device in the corresponding safe and optimal operating condition [22]. For example, the DC/DC converter associated with the PVS is regulated using algorithms and/or control strategies aimed at tracking the maximum power point (MPP); hence, such algorithms are known as MPP tracking (MPPT) solutions. The ESS integration with the DC-bus is performed using bidirectional converters and control strategies [23,24] to regulate the DC-bus voltage and to guarantee the global power balance in the standalone MG when excess or shortage power exists. However, additional conditions can also be considered; for example, a limitation to the power generated by the PVS [25–27], which is applied when the ESS reaches the maximum state of charge (SOC) during low power demand, or a load disconnection when the ESS reaches the minimum SOC during high power demand [25,28]. Those protections are needed to prevent an accelerated reduction of the ESS lifetime [29–31], hence reducing costs.

Energy management systems (EMSs) used in SPVSs consider a load shedding strategy for preventing the ESS from violating the minimum SOC only when noncritical loads are connected to the MG [32–34]. However, this requires an oversizing of both the generators and the ESS when critical loads are present. The main challenge in EMS design concerns the limitation of the maximum SOC condition because traditional control techniques are unable to supply power to the load without using the ESS; therefore, ESS are constantly discharged and charged with small amounts of power.

This situation produces charge/discharge sub-cycles, which are integrated as full cycles that reduce the lifetime of the battery [35]. Therefore, this paper proposes a new EMS based on operating states for controlling a SPVS without violating both the maximum and minimum *SOC* of the ESS. For that purpose, the EMS considers three operating regions for the ESS: overcharge, normal operation, and bulk charge; and two operating modes for the PVS are also defined: MPPT and power demand tracking (PDT) [36]. The proposed EMS also includes a capacitor connected in parallel with the battery as backup ESS. The function of this capacitor is to enable the disconnection of the battery in both overcharge (high *SOC*) and bulk charge (low *SOC*) conditions, hence avoiding charge/discharge sub-cycles in the battery.

Therefore, the main paper contribution is the design and test of a control strategy for an EMS aimed at regulating a microgrid based on a PV system and ESS. The solution is also intended to avoid the sub-cycles problem present in power demand tracking operation, which causes an accelerated aging of the battery pack due to the integration of charge/discharge sub-cycles into full charge/discharge cycles. This is done by inserting an auxiliary capacitor into the ESS to support the PDT operation, which enables to disconnect the battery to avoid the sub-cycles problem. Moreover the EMS also considers all the states needed to protect battery, including the load disconnection when the power balance is not achievable due to both low SOC and low PV power production. This document is organized as follows: Section 2 presents the background of SPVSs based on DC microgrids. Then, Section 3 introduces the structure, control strategies, and energy management system used in the proposed solution. The results and simulations analysis are reported in Section 4, and the conclusions close the paper.

2. Background of SPVSs

Figure 1 shows the SPVS architecture adopted in this work, which is formed by a PVS, an ESS (lithium-ion battery in parallel with a capacitor), the controlled DC/DC converters, and the load [37,38]. Such microgrid structure is commonly used in several applications that require a global power balance [39–41].

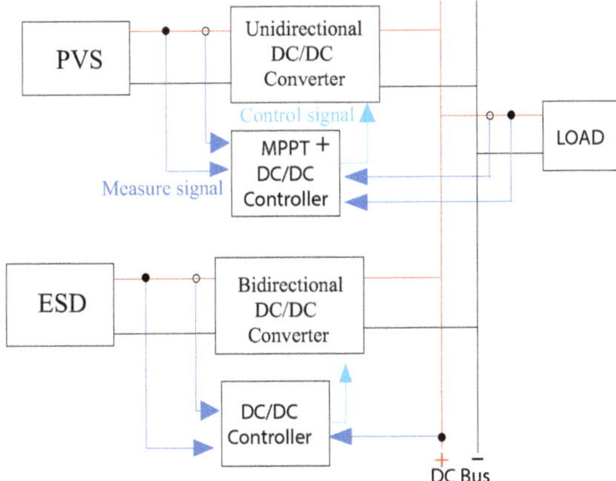

Figure 1. Conventional stand-alone DC microgrid.

Equation (1) formalizes the global power balance of the SPVS, where P_{ESD} represents the power supplied or stored by the external storage device (ESD), P_{PV} is the power generated by the PVS, and P_{Load} is the power demanded by the load. In this balance, the supplied power is positive and the consumed power is negative.

$$P_{ESD} + P_{PV} + P_{Load} = 0 \tag{1}$$

EMSs should consider the fact that batteries have two *SOC* limits: a maximum state of charge (SOC_{Max}), i.e., overcharge, and a minimum state of charge (SOC_{Min}), i.e., bulk charge. If the battery operates outside those limits, the ESS lifetime could be significantly reduced [42–44]. Therefore, EMSs should also integrate two operating modes: (1) constraining the power generated by the PVS so that it equals the power demanded by the load ($P_{PV} \approx P_{Load}$), which is needed when SOC_{Max} is achieved; and (2) disconnecting the load when the ESS reaches the SOC_{Min}. In the second mode, the PVS must be operated at the MPP until the battery reaches an acceptable *SOC* value to reconnect the load [36].

An example of this kind of EMSs is presented in [40], where a multi-loop and multi-segment adaptive droop-controller is used to manage the energy flow in a PV/battery hybrid system for a stand-alone DC MG. That solution controls the power generated by the PVS to regulate the charge/discharge battery power. However, since the battery is used to regulate the voltage in the DC-bus, such an action produces charge/discharge sub-cycles, which reduce the lifetime of the battery. Similarly, in [45] the authors proposed a double-layer hierarchical controller to ensure the global power balance of a DC MG. In such EMS, both the generators and the ESS can operate as voltage source converters or current source converters, which enable both types of devices to regulate the DC-bus voltage. This solution requires a virtual batteries disconnection when the SOC_{Max} is reached, which is performed by controlling the power generated by the PVS. Such a strategy may cause voltage instability due to the absence of a storage device because no device is capable of storing the small amount of excess power that could be produced by the PVS due to quantization or controller errors.

Other solutions have been focused on regulating the power generation of the PVS to prevent the batteries from operating outside the *SOC* limits. This is the case in [36], where PVS generation is constrained depending on battery *SOC* and grid availability. Similarly, in [46] is reported a real-time rule-based algorithm for the power management of an MG that considers a constraint to the power generated by the PVS. Different authors [47] have also proposed an EMS based on multiple MG operating states: MPPT mode and constant voltage (CV) mode. The CV mode is used when the battery *SOC* reaches the maximum limit; hence the PVS power generation is constrained to ensure a constant battery voltage. The same strategy has been applied by other authors to avoid an excessive *SOC* condition [36,40,45,46]. However, in those solutions the batteries are not disconnected; therefore, the small amount of excess power that could be produced by the PVS (due to quantization or controller errors) is stored or supplied by the battery, generating charge/discharge sub-cycles, which reduce the battery's lifetime.

The problem associated with the charge/discharge sub-cycles of the ESS is illustrated in Figure 2, which considers the control of PVS generation reported in [36]. That control strategy was implemented following the flowcharts reported in [36], but modifying the power profiles reported in that paper since those profiles do not illustrate the sub-cycles problem. Such solution has MPPT and PDT modes, where both modes are based on the Perturb and Observe algorithm. This simulation includes a single BP585 PV module with a constant irradiance of 1000 W/m^2, a constant power load of 40 W, and an initial battery *SOC* equal to 0.99995, which is near SOC_{Max}. The simulation considers constant power profiles for both the photovoltaic source and load to illustrate the sub-cycles problem in an easy way: those constant profiles enable to visualize the source of the sub-cycles problem without the action of other controller transitions. The results presented in Figure 2a show that the PVS operates in MPPT mode at the beginning of the simulation to provide the power required by the load, while charging the battery. The MPPT mode is active for 0.34 s; at that moment, the battery *SOC* reaches SOC_{Max}. From that moment on, the system operates in PDT mode to reduce the power generated by the PVS. In PDT mode, two different waveforms can be observed in the PVS power curves: (1) when the battery *SOC* is higher than SOC_{Max} (between 0.34 s and 0.63 s) and (2) when the *SOC* is almost equal to SOC_{Max} (from 0.63 s on). In the first part of this PDT operation, the control strategy uses the excess power from the battery to supply part of the load power, thus discharging the battery to respect the *SOC* limit. However, when the SOC is almost equal to SOC_{Max}, small errors of the PDT control produce small charge/discharge sub-cycles in the battery (observed in Figure 2b), which are integrated as full charge/discharge cycles that reduce the battery's lifetime. In this simulation, the PDT errors are caused by the discretization of the Perturb and Observe algorithm that is used to track the load power, which is an unavoidable condition. Similar errors will occur in other tracking solutions, such as incremental conductance, extremum seeking or linear and non-linear controllers.

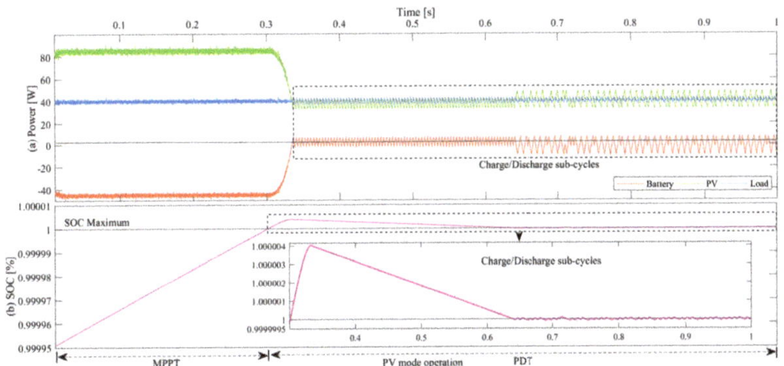

Figure 2. Photovoltaic systems (PVS) power generation control.

The following section proposes both a circuital structure and a control strategy to address this problem.

3. Circuital Structure and Control Strategy for the EMS

To solve the problem described in Section 2, this paper proposes an EMS for a SPVS that considers battery SOC limits, PVS power generation control, and the integration of a backup capacitor to avoid charge/discharge sub-cycles in the battery. The backup ESS (capacitor) is used to support the power balance when the main ESS (battery) is disconnected, which occurs when the battery SOC falls outside safe limits. Such procedure avoids charge/discharge sub-cycles in the battery and, hence, artificial battery degradation.

The EMS proposed in this work is based on three operating regions (OR), defined by the SOC limits, and two control modes for the PVS; a graphic representation of those regions is provided in Figure 3. Region 1 (OR1) covers the normal battery operation and PVS; hence, the PVS works in MPPT mode and the battery is connected to the EMS. The second operating region (OR2) occurs when the battery SOC reaches the maximum limit ($SOC \geq SOC_{Max}$), which forces the PVS to operate in PDT mode and disconnect the battery. OR2 uses the capacitor as backup ESD to ensure a correct power balance, thus eliminating the charge/discharge sub-cycles described in the previous section. The EMS remains in OR2 while the MPP of the PVS is higher than the power demanded by the load.

The third operating region (OR3) occurs when the battery SOC reaches the minimum limit ($SOC \leq SOC_{Min}$), which forces the battery disconnection and the use of the backup capacitor for ensuring the global power balance. Moreover, if the capacitor is discharged (hence the capacitor voltage reaches a safe minimum value), the load is disconnected to force the capacitor to charge until the capacitor voltage reaches the battery voltage, which enables the EMS to reconnect the battery. Finally, the load is reconnected when the battery has been charged to a safe SOC value.

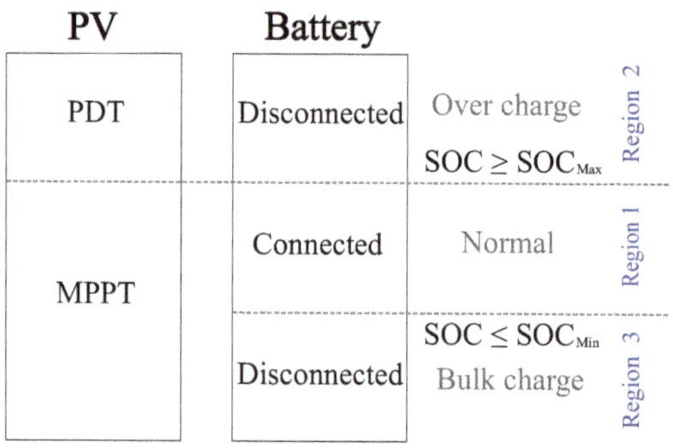

Figure 3. Proposed methodology.

3.1. Devices and Control Strategies

The components of the proposed EMS are depicted in Figure 4: PV panel, battery, capacitor, DC/DC converters, control strategies, and common DC-bus. The PVS is connected to the DC-bus using an unidirectional DC/DC boost converter, which is regulated using the sliding-mode controller (SMC) proposed in [48]. Such a SMC regulates the DC/DC converter to impose a desired PV voltage despite the environmental and DC-bus voltage conditions, hence the voltage of the PV source is always equal to the voltage reference (V_{PV-REF}); the design and stability proof of the SMC are reported in [48]. The reference of such a SMC is provided by the Perturb and Observe (P&O) algorithm proposed in [36], which has two operation modes: the MPPT mode tracks the optimal PV voltage that ensures the maximum power generation, while the PDT mode tracks the PV voltage to produce a given power reference. Therefore, the used strategy to control the power generation of the PVS does not require the implementation of forecasting methods, since the combined action of the P&O and SMC algorithms ensures the production of the maximum power possible (MPPT mode) or a lower power level to match the load demand (PDT mode) depending on the microgrid requirements. In addition, the proposed structure also has adjustable parameters to control the power generated by the PV sources, with one limitation: the maximum power that can be delivered is constrained by the environmental conditions, while the minimum power is controllable. This section of the power system is modeled using the electrical equations of both the PV panel and the unidirectional DC/DC converter given in Equations (2)–(4):

$$i_{pv} = i_{ph} - i_o \left[\exp \left(q \left(v_{pv} + i_{pv} R_s \right) / \left(N_c \eta k T_c \right) \right) - 1 \right] - \left(v_{pv} + i_{pv} R_s \right) / R_h \tag{2}$$

$$\frac{d v_{pv}}{dt} = \frac{i_{pv} - i_{L_{pv}}}{L_{pv}} \tag{3}$$

$$\frac{d i_{L_{pv}}}{dt} = \frac{v_{pv} - v_{dc} \left(1 - u_{pv} \right)}{L_{pv}} \tag{4}$$

Equation (2) describes the PV current i_{pv} and voltage v_{pv} imposed by the unidirectional DC/DC converter, the photo-induced current i_{ph}, which is almost proportional to the irradiance level, the inverse saturation current i_o, the series and shunt resistances R_s and R_h, the electron charge q, the Bolzmann constant k, the quality factor η, the number of cells N_c and the temperature in kelvins T_c. Equation (3) describes the dynamic behavior of the PV voltage, where C_{pv} is the capacitor in parallel with the PV source, L_{pv} is the inductance of the unidirectional DC/DC converter and $i_{L_{pv}}$ is the current of that inductor. Finally, Equation (4) describes the dynamic inductor current behavior, where v_{dc} corresponds to the bus voltage and u_{pv} is the activation signal of the converter MOSFET, which is generated by the SMC.

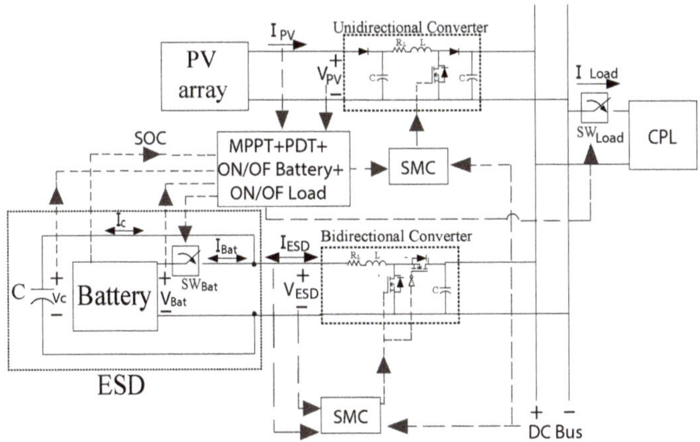

Figure 4. Structure and control of the proposed SPVS.

The backup capacitor C of the structure is used in OR2 and OR3 for supporting the power balance and, at the same time, enabling the control of the DC-bus voltage in both OR2 and OR3. The capacitor is always connected to the EMS, while the battery is disconnected in OR2 and OR3 using the switch SW_{Bat}. It must be noted that, independent of the size or the selected technology of the battery bank, the sub-cycles problems will remain; the quantization errors of the PDT control strategy will produce small charge/discharge sub-cycles, which are integrated as full charge/discharge cycles. The solution to avoid that problem is to disconnect the battery bank when the PV system must operate in PDT mode. A larger battery bank will suffer a small impact of the sub-cycles problem, since each sub-cycle will be divided into the batteries forming the bank, but the solution proposed in this paper is intended to remove the sub-cycles degradation independent of the battery size or selected technology.

The capacitor size is calculated using the average power values obtained when the system operates in PDT. Equation (5) calculates the average current in the capacitor, where P_c is the maximum allowable power for the capacitor, and V_{Bat} is the battery nominal voltage. Equation (6) can be used to calculate the capacity of the capacitor, where t represents the selected discharging time and ΔV is the maximum voltage drop allowed for those t seconds; both values are assigned by the MG designer according to the generator and load characteristics.

$$I_C = \frac{P_C}{V_{Bat}} \tag{5}$$

$$C = \frac{I_C}{\Delta V} * t \tag{6}$$

The hybrid ESS is connected to the DC-bus using the bi-directional converter depicted in Figure 4, which is controlled to regulate the DC-bus voltage. The control strategy adopted here was proposed in [49]; it consists in a SMC for the charger/discharger that ensures a stable DC-bus voltage in any operating condition: charging, discharging, or stand-by. This control strategy is active in all the operating regions by using the battery or capacitor as ESS, thus ensuring the global power balance of the system. This section of the model is represented by Equation (7), where v_{ESD} corresponds to the ESD voltage, L_{ESD} corresponds to the inductor of the bidirectional DC/DC converter and $i_{L_{ESD}}$ is the current of that inductor; finally, u_{ESD} is the activation signal of the MOSFETs generated by the SMC of this charger/discharger.

$$\frac{di_{L_{ESD}}}{dt} = \frac{v_{ESD} - v_{dc}\,(1 - u_{ESD})}{L_{ESD}} \tag{7}$$

To evaluate the proposed EMS, constant power load (CPL) [50] models were selected due to their widely use in DC MGs analyses [51–55]. That representation has the main characteristic of changing both the load current and voltage to keep the load consumption constant; therefore, when the bus voltage exhibited perturbations, the load current changed accordingly. This is different from the constant impedance load (CIL) representation, in which perturbations in the bus voltage causes a different load power consumption. In practical applications of DC Microgrids CPL is an accurate representation of the loads due to the presence of DC/DC converters interfacing the loads with the DC bus: the load current changes according to the perturbations in the bus voltage (load voltage) to consume the power requested by the load [51–53]. Therefore, CIL is not an accurate representation of the loads in DC Microgrids with commercial loads. In any case, CPL models can be used to represent loads interfaced with DC/DC converters having variable power consumptions. In that case the power requested by the load is just updated in the model, but the load keeps behaving as a CPL: the current changes according to the bus voltage perturbations to ensure the desired power consumption.

In addition, the load voltage was considered equal to the DC-bus voltage; hence, there was no need for an additional DC/DC converter for the load interface. Finally, since the load was disconnected in OR3, a switch SW_{Load} was included in the EMS circuit to disconnect and reconnect the load from the DC-bus.

3.2. Energy Management System Proposed to Control the SPVS

Figure 3 presents the operating regions defined around the technical operating limits established for the battery SOC. Such regions should consider the different operating modes of the PVS (MPPT and PDT) and the battery and load connection/disconnection depending on the requirements of each operating region. The following subsections explain the types and operating ranges of the battery SOC and the control strategies applied to the regions proposed in this work: overcharge, normal charge, and bulk charge. It also describes the requirements to move from one region into another and connect/disconnect the battery and the load so that voltage fluctuations are as small as possible. To provide a better explanation of the EMS, the three regions are described in the following order: Region 1 (normal operation); Region 2 (overcharge); and Region 3 (bulk charge).

3.2.1. Region 1 ($Normal : SOC_{Min} < SOC < SOC_{Max}$)

In this region, the PVS should always operate in MPPT in order to utilize its maximum power generation. Because the battery is responsible of the global power balance at all times, it should always be connected ($SW_{Bat} = 1$). Additionally, the load is always connected in this region ($SW_{Load} = 1$) because the PVS and the battery, combined, can produce sufficient power to supply the power demanded by the load. The restrictions and considerations to enter this operating region are detailed in the subsections describing Regions 2 and 3. To leave this operating region, the battery just needs to reach its SOC limits. The state diagram in Figure 5 enables a better interpretation of those interactions.

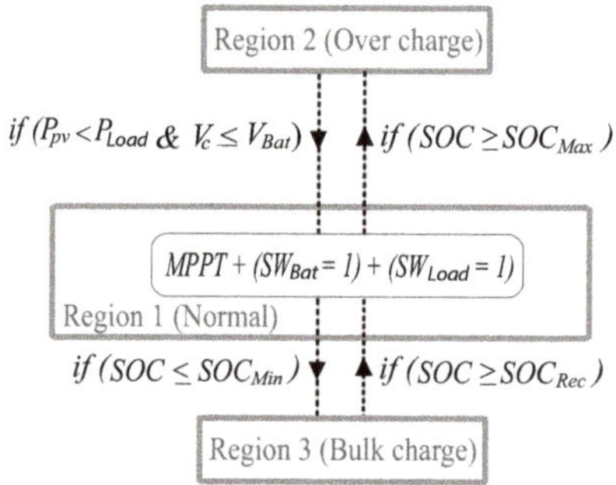

Figure 5. Flowchart of the control strategy in Region 1.

3.2.2. Region 2 ($Overcharge : SOC \geq SOC_{Max}$)

This operating region occurs when the power generated by the PVS is higher than that consumed by the load and the battery reaches SOC_{Max}. For that reason, the backup ESS (in this case, the capacitor) must replace the battery and take control of the power balance and the voltage at the DC-bus, keeping the load connected at all times ($SW_{Load} = 1$). It should be noted that, for the EMS to operate in this region, the battery SOC should be equal to or higher than the maximum limit ($SOC \geq SOC_{Max}$). In this OR, there are different operating scenarios depending on the values of the power generated by the PVS, the power demanded by the load, and the maximum and minimum voltage limits assigned to the capacitor, which are explained below and presented in the state diagram of Figure 6.

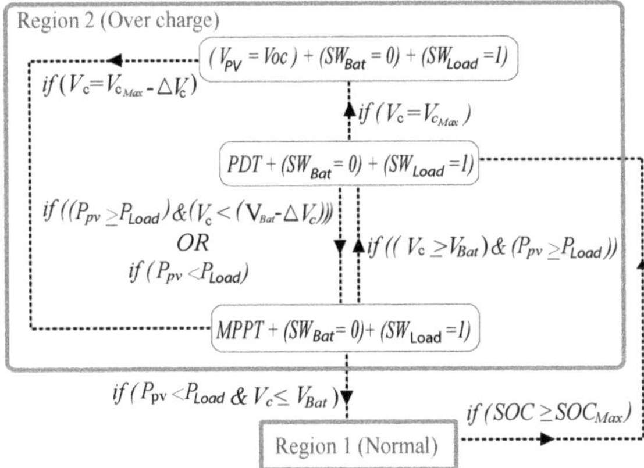

Figure 6. Flowchart of the control strategy in Region 2.

- ($P_{pv} \geq P_{Load}$): In this operating state, the battery must be disconnected ($SW_{Bat} = 0$) and, at the same time, the PVS should operate in PDT. As a result, the capacitor is in charge of supplying and storing the power required by the system to perform the PDT and to control the DC-bus voltage. In order to prevent the capacitor from fully discharging and the system from collapsing due to the lack of power to control the DC-bus, the system operates in PDT mode until the capacitor reaches a voltage below ($Vc < V_{Bat} - \Delta Vc$), being ΔVc the maximum allowable voltage drop for the capacitor. In the case that this limit is violated, the PVS should change to MPPT mode so that the capacitor is charged until it reaches the battery voltage (V_{Bat}), thus exploiting the excess power generated by the PVS. Note that, in the management system, the voltage of both the capacitor and the battery should always be sensed to perform this control action. Additionally, the capacitor should be protected against over voltage conditions. For that reason, if during the operation in PDT mode the capacitor reaches the maximum allowable voltage (Vc_{Max}), then the PVS is operated in open circuit voltage (Voc) until the capacitor reaches a safe voltage ($Vc = Vc_{Max} - \Delta Vc$). This is achieved because the capacitor is forced to supply the power required by the load, thus enabling the reduction of its voltage. When the capacitor reaches a safe voltage, the PVS starts to operate in MPPT to provide power. As a result, the operating mode can change to PDT if the power conditions are met ($P_{pv} \geq P_{Load}$); otherwise, the PVS continues operating in MPPT in Region 1 as long as the power generated by the panel is lower than that required by the load ($P_{pv} < P_{Load}$) and the capacitor's voltage is lower than or equal to that of the battery ($Vc \leq V_{Bat}$). This forces the capacitor to discharge before changing operating regions (from 2 to 1), thus reducing the voltage fluctuations at the DC-bus when the battery is connected.
- ($P_{pv} < P_{Load}$): In this state, the PVS power reaches a lower value than that of the load. This is due to the decrease in Solar Radiation (SR), the variation of T, or higher power demanded by the load; therefore, the PVS should increase the delivered power and start to operate in MPPT, maintaining

the battery disconnected ($SW_{Bat} = 0$). The system continues operating in Region 2 as long as the condition to change to OR1 (described in the previous paragraph) is not met. If that condition is not satisfied and the condition ($P_{pv} \geq P_{Load}$) remains, the capacitor is charged up to the battery nominal voltage and the PVS operates in PDT once again.

3.2.3. Region 3 ($Bulkcharge : SOC \leq SOC_{Min}$)

When the battery reaches the minimum allowable SOC, it should not supply power to the load. For that reason, the battery is disconnected when the operating state $SOC \leq SOC_{Min}$ is reached, while the capacitor is in charge of supplying the demanded power in such a way that the global power balance is satisfied and the voltage level at the DC-bus is controlled. The previous condition will be fulfilled as long as the capacitor does not reach the minimum voltage level established for its operation (Vc_{Min}). At the same time of the battery disconnection, a warning signal is sent to indicate that the system is operating in Region 3 and it might be de-energized. Such signal takes a value of one ($Warning = 1$) when the battery reaches its minimum SOC and continues that way until the SOC achieves a safe operating level, at which point it changes to zero ($Warning = 0$). That warning signal is adapted in the management system to alert users of the equipment that they should store their data because the system will stop supplying power for a period of time.

When the system is operating in Region 3, two operating conditions can be created: (1) The generated power is higher to or equal than the power demanded by the load ($P_{PV} \geq P_{Load}$); therefore, the system should charge the capacitor until its voltage reaches or exceeds the battery voltage in order to connect the latter ($SW_{Bat} = 1$), while charging the battery until it reaches a safe recovery level (SOC_{Rec}). Only when this condition is met, the warming is off ($Warning = 1$) and the system starts to operate in Region 1. (2) The generated power is lower than the power demanded by the load ($P_{PV} < P_{Load}$). If the capacitor reaches its minimum voltage in this condition, the load is disconnected, while the battery is kept disconnected until the capacitor reaches a voltage level equal or higher than the battery voltage. When this condition is satisfied, the battery is connected, and its charge cycle is started until its SOC is equal to or higher than SOC_{Rec}. Only when this condition is met, the load is reconnected, the warming signal is off and the system can start to operate in OR1. It is important to highlight that, by using the hysteresis band generated between SOC_{Min} and SOC_{Rec}, it is possible to avoid the connection and disconnection of the battery for short periods of time and the load connection and disconnection when the second operating condition is fulfilled. In an operating state of variable demand, this prevents fluctuations and continuous transients in the system. Additionally, in OR3 the PVS operates in MPPT at all times so that the maximum power is generated. The state diagram in Figure 7 is a graphic representation of this strategy.

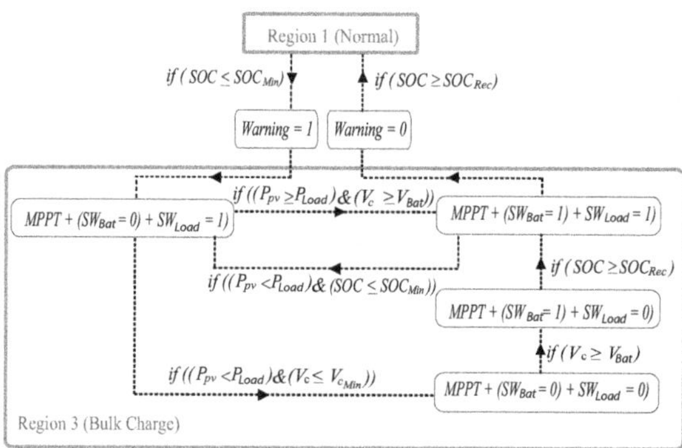

Figure 7. Flowchart of the control strategy in Region 3.

4. Simulation Results

This section describes the simulations that were conducted to make sure that each one of the regions and operating states in the EMS are satisfied and enable the SPVS to operate adequately. In order to simulate the components of the SPVS and the control strategies, it was employed the specialized software PSIM. The BP585 single mono-crystalline PV panel, formed by 36 cells in series, was selected. Table 1 presents the module characteristics in standard conditions (STC-irradiance, 1 kW/m^2; and cell temperature, 25 °C). To estimate the module parameters, the mathematical formulation proposed in [56] and the data in standard conditions were used. Changes in the irradiance and temperature are also considered by changing the short-circuit current of the PV source model, which is a common simulation practice as it is reported in [56]. It must be highlighted that the adopted PV model is also valid for representing modules currently available at the market, which exhibit characteristics similar to the ones of the PB585 module, e.g., M36PCS (85 W), P36PCS (85 W), among others. The nominal voltage of the lithium-ion battery (used to form the ESS) was 12V, with three cells in series, one set of cells in parallel, and a nominal capacity of 5.2 Ah. The limits $SOC_{Max} = 0.9$ and $SOC_{Min} = 0.1$ were defined in the management strategy to extend the lifetime of the device [57]. Additionally, for an adequate transition from Region 3 to Region 1, the EMS implemented a SOC_{Rec}, which was specified in each test scenario.

Table 1. Parameters of the BP585 panel.

Parameters (STC)	Value
Open circuit voltage (Voc, STC)	22.1 V
Short circuit current (Isc, STC)	5 A
MPP voltage (V_{MPP}, STC)	18.0 V
MPP current (I_{MPP}, STC)	4.70 A
Voc temperature constant (αVoc)	−0.088 V/°C
Isc temperature constant (αIsc)	0.047%/°C
Number of cells (Ns)	36
Estimated Parameters (STC)	
Iph (A)	5
Isat (nA)	2.808
Ideality factor of diode (η)	1.121
Rs (Ω)	0.227
Rp (kΩ)	2.029

The parameters of the different elements that formed the control strategies and the converters used to design this SPVS are described in the references cited in Section 3. For a better understanding and analysis of the results obtained by the proposed EMS, this section starts with the validation of each one of the operating regions considering the variation in photovoltaic generation and the power demanded by the load. For that purpose, it was necessary to use short simulation times and an analysis of the transients of each OR and their transitions, as well as the DC-bus voltage. Finally, a 24-h test scenario was defined (with load and solar irradiation variations sensed by a laboratory of the Universidad Nacional de Colombia, located in Medellín-Colombia) to evaluate the behavior of the SPVS in all the operating regions.

4.1. Simulation and Validation of the Proposed Operating Regions

This subsection analyzes each operating region in order to validate all the conditions and constrains considered in the proposed EMS. For that purpose, the EMS parameters were adjusted as follows.

The capacitor was calculated by analyzing the behavior of the PVS when PDT was applied in the presence of the battery and an irradiance of 1000 W/m^2; thus, obtaining the scenario of maximum power in which the capacitor will operate. After analyzing the results of this scenario (see Figure 2),

$P_c = 10$ W was defined so that the capacitor had sufficient power to assume the power excess generated in this operating mode. The maximum voltage drop of 1 V was also considered for a time of 1 s, which requires a 0.833 µF capacitor. Based on the capacitor calculation, the values $Vc_{Max} = 18$ V, $Vc_{Min} = 5$ V, and ($\Delta Vc = 1$ V) were assigned. It is considered a margin of 60% (upper limit) and 50% (lower limit) of the nominal voltage of the battery in order to establish the voltage levels in this EMS and offer an adequate power level in the capacitor. It is important to highlight that the capacitor voltage limits vary depending on the specific equipment and manufacturer. In addition, it was considered $\Delta V_{PV} = 0.2$ V as the voltage rate in the power control strategy of the PVS and assigned a value of $SOC_{Rec} = 0.10005$ as the recovery SOC of the battery. These values were selected in order to validate the operating states in a short period of time and analyze the transients generated by the fulfillment of the conditions as well as transitions from one OR to the next. Finally, each one of the proposed scenarios has an initial load, PVS power generation, and variable power demanded by the load so that all the states and regions in the EMS can take place. Note that all the test scenarios in this work use the same minimum and maximum SOC limits previously defined.

4.1.1. Region 1 ($Normal : SOC_{Min} < SOC < SOC_{Max}$)

This subsection describes different test cases in the same scenario, in which power generation and demand in the SPVS are changed to test the conditions established for the system operation in Region 1 (normal) and Region 2 (overcharge). Figure 8 presents this test scenario, describing the system dynamics under the proposed EMS. PVS power generation was divided into three levels. The first level was generated by a solar irradiance of 1000 W/m², which resulted in approximately 84 W produced by the PVS. At $t = 6.2$ s, a 40% reduction (400 W/m²) was applied, and the PV generated 50 W after that. Finally, another 40% reduction (400 W/m²) was applied at $t = 6.5$ s, which generated a total drop of 80% in the maximum irradiation from the start of the test scenario until the time under analysis. Thus, the final power delivered by the PVS was approximately 16 W. Additionally, this scenario considered the variation in the power demanded by the load, which was 20 W at $t = 0$ s, 90 W at $t = 2$ s, and, finally, 30 W at $t = 3$ s. The time interval between seconds 4 and 5 presents a load increase with positive slope from 30 W to 50 W, which is maintained until $t = 6$ s. Finally, the load drops to 30 W from $t = 6$ s to the end of the period under analysis. The previous scenario of PVS generation and power demand by the load is defined in such way that transitions between ORs 1 and 2 occur and the PVS operates in PDT and MPPT. The test simulation was divided into 5 time intervals for a correct interpretation of the test scenario as well as the PVS behavior based on the different variations. The intervals represent a scenario where power generation and demand change, one exceeding the other.

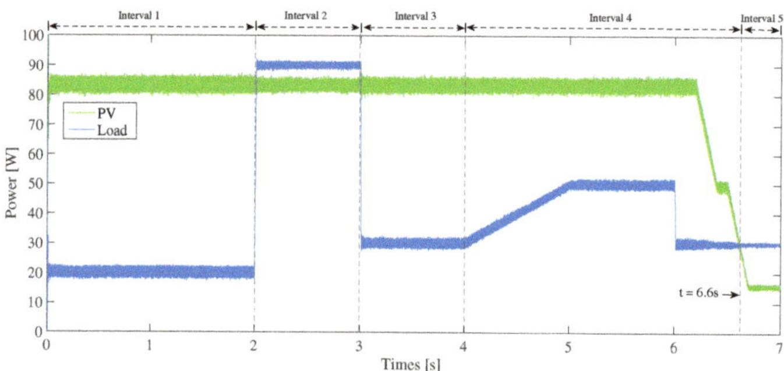

Figure 8. Power demanded and generated by the SPVS: test system in Operating Region (OR)1 and OR2.

To validate the operation of the proposed EMS in OR1, the battery SOC started at 0.5 so that, during the entire simulation time, the EMS stayed within the ranges assigned to operate in this region. Figure 9 presents the behavior of the system when load and demand variations were applied. Remarkably, the PVS operated in MPPT at all times, keeping the battery and load connected throughout the simulation, while the battery was in charge of supplying and storing the excess power generated by the PVS so that a global power balance could be achieved. The subfigure that presents the voltage assigned to the PVS by the DC/DC converter shows the way the EMS modified that voltage as the irradiation on the panel varied (fell), always staying in MPPT mode. This validates the correct operation of the external control strategy applied to obtain the maximum power from the PVS. Moreover, it was proved that the management strategy selected to control the ESS could be used to control the DC-bus voltage, always keeping the voltage near its nominal value ($V_{DC} = 48$ V), and the power demanded by the load is supplied and absorbed without affecting the MG operation. In this test system, the power flow supplied by the ESD is provided by the battery, since the switch of the ESD is closed (please refer to Figure 4 to check the switch). The voltage peaks in DC bus are associated with strong variations in the demand, which are mitigated by the control strategy. This test scenario does not consider the transition to other ORs because that is validated in ORs 2 and 3.

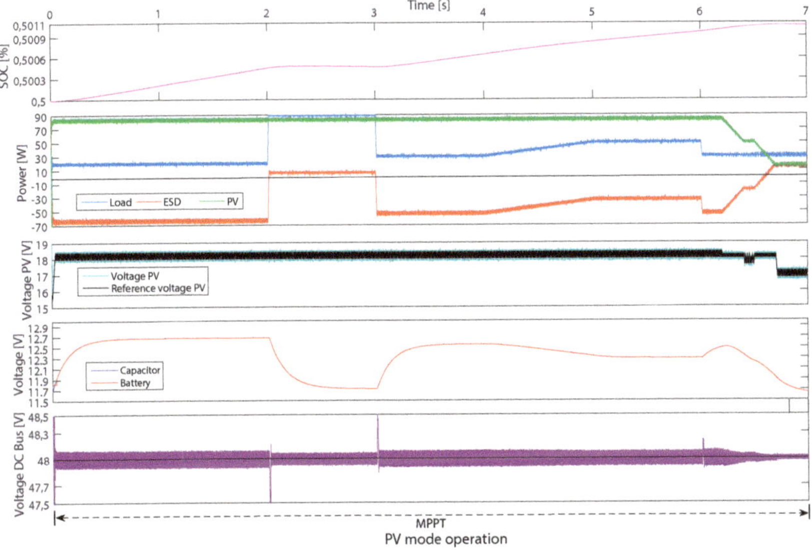

Figure 9. Energy Management System: Test system in OR 1.

4.1.2. Region 2 ($Overcharge : SOC \geq SOC_{Max}$)

In order to validate the conditions proposed for OR2, the same test system presented in Figure 8 was used. The validation of the correct operation of the EMS is presented in Figure 10, which details the control actions performed by the EMS in different operating and demand scenarios of OR2 divided into time intervals based on variations in load and PVS generation. The waveforms of battery SOC and voltage demonstrate that, during most of the simulation time, the system operated with $SOC \geq SOC_{Max}$ and the battery was disconnected, while the capacitor was in charge of supplying and storing the excess power required by the system to achieve the global power balance. As a result, the sub-cycles of the battery associated with the operation in PDT in this OR could be eliminated. In Interval 1, the generated power was higher than the demand of the system and, because the SOC level was lower than the maximum established limit, 0.9 ($SOC = 0.89995$), the system started operating in OR1. As the simulation progressed, the battery was charged until it reached the maximum allowable limit for the SOC at $t = 0.36$ s. When this happened, the operation of the system changed to OR2, the battery was disconnected, and the PVS system was operated in PDT. This can be observed in the graph that illustrates the voltage of the battery and the voltage assigned by the converter to the PVS (PV voltage). This operating mode continued until Interval 2. The capacitor voltage shows that this device was in charge of supplying and storing the excess power associated with the PDT during Interval 1, when the system operated in that mode. In Interval 2, the power state in the SPVS changed: the demanded power was higher than the power generated by the PVS. For that reason, the photovoltaic system changed to operate in MPPT at $t = 2$ s, reducing the PV voltage so that the power required by the load was delivered. Because the power generated by the PVS was not sufficient to cover the load demand, the capacitor should have compensated for that deficit. During this entire interval the battery was disconnected because the system did not meet the conditions to change operating regions.

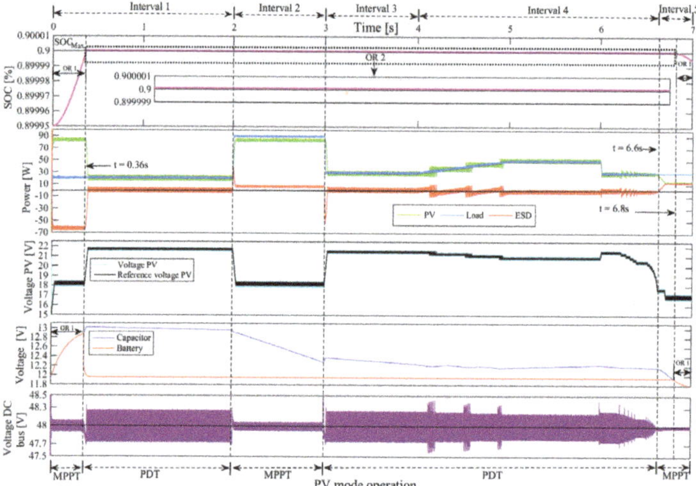

Figure 10. Energy Management System: test system in OR2.

In Interval 3, the load conditions led the system to operate in PDT once again, increasing the voltage of the PVS until it reached an adequate power level. Because the capacitor should have delivered the power difference in this operating mode, its voltage started to drop (as can be seen from the voltage waveform), keeping the battery disconnected. Interval 4 presents different load levels, which were correctly managed by the EMS. In the simulation, the 4 s $\leq t \leq 5$ s interval exhibited a step-shaped load increase during which the PDT never lost control over the generation and immediately adapted to such an increase. This produces voltage dropped in V_{pv} and, as a result, the power generated by the PVS rose in a controlled manner. In that interval, the DC-bus presented higher voltage fluctuations than in other intervals in the test scenario, to which the capacitor and the management strategy responded adequately. After $t = 5$ s, the load was constant ($P_{Load} = 50$ W) until $t = 6$ s, while the system was operating in PDT with a disconnected battery without any problem. At $t = 6$ s, the demanded power dropped to 30 W, because the PDT conditions were maintained, and the system continued operating in this OR without any problem. Additionally, Interval 4 presented a drop in SR, which had an impact on the power generated by the PVS and resulted in an immediate reduction in power produced by the PV panel. During the first power reduction, which occurred in the 6.2 s $\leq t \leq 6.4$ s interval, the power generation stayed above the power demanded by the load; for that reason the PVS operateed in PDT mode at all times. Note that, although the reduction in power generated by the PVS was important, the EMS was not affected at all, correctly responding by reducing the reference voltage of the PVS. Nevertheless, charge and discharge power fluctuations were produced in the capacitor, although smaller than those generated during the step-shaped increase in the load in the 4 s $\leq t \leq 5$ s interval. Finally, a second drop in SR took place between Intervals

4 and 5, 6.5 s $\leq t \leq$ 6.7 s, which caused the power generated by the PVS to drop below the power demanded by load at t = 6.6 s. The PVS then changed from PDT to MPPT mode. The EMS for OR2 maintained the system in that OR until the capacitor reached the battery voltage at t = 6.8 s, satisfying the condition for battery reconnection in Interval 5. For that reason, the SPVS started to operate in OR1, while the battery *SOC* noticeably fell since t = 6.8 s until the end of the simulation. Remarkably, in the entire test scenario, the operating voltage of the DC-bus was maintained near its nominal voltage (V_{DC} = 48 V), with a maximum variation of 0.72%.

The test scenario in Figure 10 does not show the charge/discharge action of the capacitor when the maximum and minimum limits are reached. For that reason, Figures 11 and 12 present the behavior of the system when forced into these two operating states in order to demonstrate the adequate operation of the system and the satisfactory performance of the proposed EMS. Figure 11 illustrates a scenario where the capacitor reaches Vc_{Max} when the PVS is operating in PDT. Thus, we assigned an initial voltage (Vc = 18.1), $SOC \geq SOC_{MAX}$, and the power demanded by the load (40 W). These conditions forced the system to discharge the capacitor to a safe voltage level (Vc = 17 V in this case). For that purpose, the EMS operated the system in Voc until the capacitor voltage reached the safe level established by the control strategy, thus generating P_{PV} = 0 W and forcing the capacitor to supply all the power required by the load. When the capacitor reached the safe voltage level, the system switched modes to operate in MPPT at t = 0.388 s until it reached the required power level to continue operating in PDT at t = 0.397 s, which can be seen in the graph of PVS, load, and capacitor power. Figure 11 presents the voltage levels assigned by the control strategy (reference) and the converter to the PVS, as well as the stabilization of the capacitor voltage in the levels assigned by the EMS. Note that if (while the capacitor is being discharged) the system changed the load conditions and it was necessary to switch to OR1, this would only occur if the capacitor voltage equals that of the battery. This was guaranteed by the condition for the transition from OR 2 to OR1 in Figure 6.

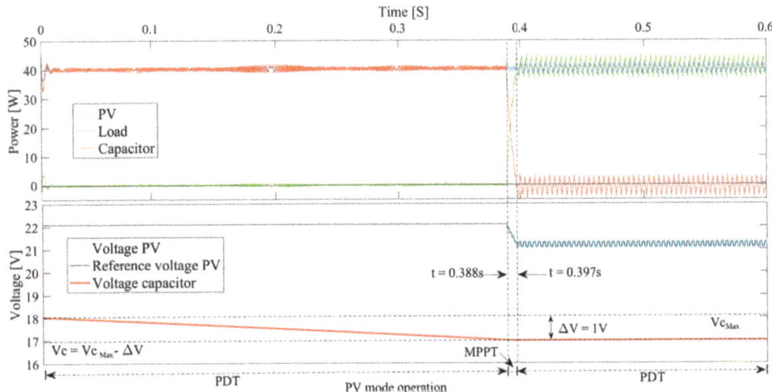

Figure 11. Control strategy to reduce the power generation in the PVS in order to discharge the capacitor when $Vc \geq Vc_{Max}$ in PDT mode.

Figure 12 illustrates the test scenario in which the capacitor reaches the minimum voltage limit allowable in OR1, with the PVS operating in PDT. In this scenario, the load power was 40 W and SOC = 0.91, thus generating V_{Bat} = 11.94 V. In order to violate the minimum voltage limit ($Vc < V_{Bat} - \Delta Vc$), an initial voltage of 10.92 V was assigned to the capacitor so that the system charged it. Due to the previous conditions, the power graph of the devices in Figure 12 shows that the system, when the simulation started, immediately operateed in MPPT until the capacitor reached the minimum allowable voltage level to operate in PDT in OR2 at t = 0.242 s, when it d to PDT $Vc \geq V_{Bat}$. Importantly, the power to charge the capacitor was extracted from the excess power of PVS generation. Additionally, the figure of the voltage levels in this test scenario presented different voltage signals that can be used to interpret the way the system worked when the capacitor was charged in OR2. At the start of the simulation, the capacitor voltage was under the allowable level, i.e., a volt below that of the battery voltage $(\Delta Vc = 1 \text{ V})$. For that reason, the voltage assigned to the PVS by the control strategy resulted in the system operating in MPPT until the capacitor voltage equals that of the battery at the moment of the disconnection. After that point, the system operated in PDT again until the load and generation conditions in the SPVS changed.

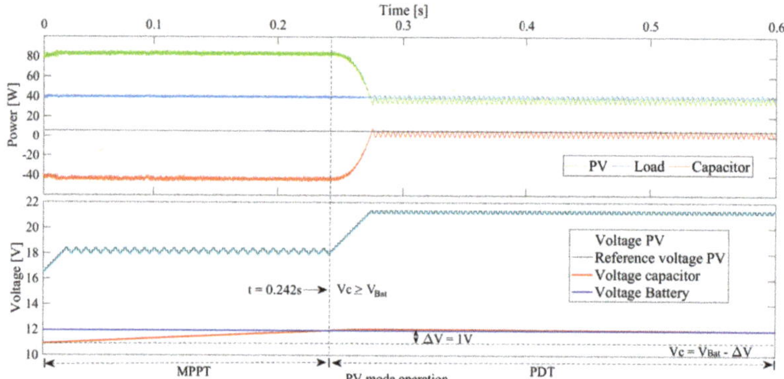

Figure 12. Control strategy to increase the power generation in the PVS in order to charge the capacitor when $Vc = V_{Bat} - \Delta Vc$ in PDT mode.

4.1.3. Region 3 $(Bulkcharge : SOC \leq SOC_{Min})$

In order to validate the adequate system operation in OR3, this subsection proposes two test scenarios to study the conditions that can exist when the SPVS works in that region. Figure 13 shows the test scenario that represents the operating condition in which the battery reached the SOC_{min} and the power of the PVS changed to $P_{PV} \geq P_{Load}$. Such a scenario considered an initial SOC of $SOC = 0.100025$ and $SOC_{Rec} = 0.10005$, and the system started to operate in OR1. Additionally, the power generated by the PVS ($P_{PV} = 84$ W) was constant and the power demanded by the load was variable. More specifically, the load started with a demand of 120 W and, as a result, the system changed from OR1 to OR3 at $t = 0.104s$. When the SPVS changed to OR3, the warning signal took a value of 1, which indicated the users that the system may have been close to a complete load disconnection. After $t = 0.2$ s, the power demanded by the load decreased ($P_{Load} = 60$ W) until $t = 1.6$ s. Due to that drop in the load, the power state of the system changed ($P_{PV} \geq P_{Load}$), and as a result, the capacitor started to charge. When the condition $Vc \geq V_{Bat}$ was satisfied at $t = 0.806$ s, the EMS reconnected the battery and started the charging process. Importantly, only when the SOC reached the recovery level at $t = 1.446$ s the system started to operate in OR1. At that point, the warning signal took a value of zero, which indicated there was no risk of load disconnection anymore. Finally, in OR1 there was a load increase ($P_{Load} = 60$) W, but the system continued operating without any problem and storing the generated excess power in the battery. Note that, during the entire test scenario, the system operated in MPPT to take advantage of the power generated by the PVS and escape the bulk charge condition.

In this simulation the EMS maintained the voltage level of the DC-bus approximately equal to the nominal voltage ($V_{DC} = 48$) V, presenting maximum variations of 4.5% and 0.2% on average. Remarkably, the strongest voltage fluctuations were caused by variations in the power demanded by the load, under which the sliding-mode control strategy used for the batteries offered an excellent performance, as can be seen in Figure 14.

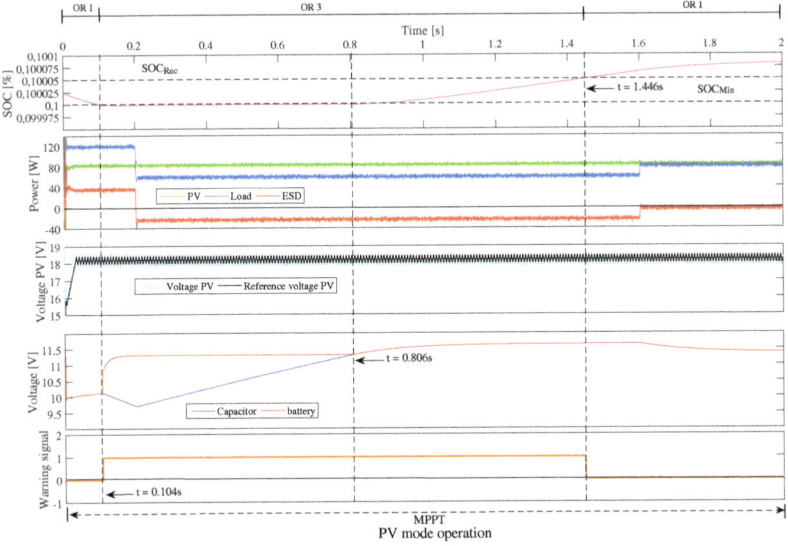

Figure 13. Test system in OR3 (operating condition ($P_{PV} \geq P_{Load}$)).

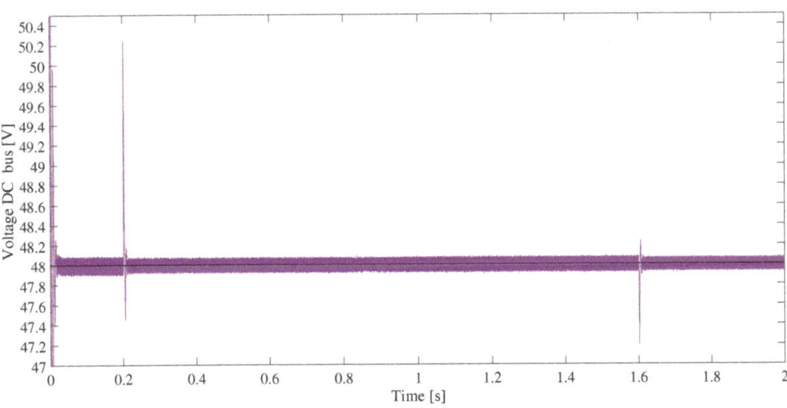

Figure 14. Voltage of the DC-bus test system in OR3 (operating condition ($P_{PV} \geq P_{Load}$)).

Figure 15 presents a test scenario to validate the operating conditions when the load is disconnected. Such a scenario considers an initial SOC of 0.100025 and $SOC_{Rec} = 0.10005$. As in the previous scenario, the power generated by the PVS ($P_{PV} = 84$) W was considered to be constant and the power demand by the load, variable. The systems started its operation in OR1, with a power demand of $P_{Load} = 120$ W; for that reason, it changed to OR3 at $t = 0.036$ s. Immediately after the SPVS changed operating regions, the battery was disconnected and the warning signal took a value of 1, indicating load disconnection risk. At the same time, the capacitor started its discharge process, which was accelerated at $t = 0.4$ s due to an increase in the power demanded by the load ($P_{Load} = 140$ W). At $t = '0.507$ s, the capacitor reached its minimum voltage limit ($Vc \leq Vc_{Min}$); for that reason, the load was disconnected and the charging process of the capacitor was started. When the capacitor reached the battery voltage at $t = 1.021$ s, the latter was reconnected and its charging process was started until it reaches the SOC_{Rec}. When the recovery SOC was reached at $t = 1.28$ s, the load was reconnected with a value of $P_{Load} = '140$ W, thus forcing the system to operate in OR1.

At that time the warning signal took a value of zero, indicating the end of the disconnection risk. During the remaining simulation time, the SPVS operated in OR1 while the battery was supplying the missing power to the load. In addition, the excess power produced by the reduction in the power demanded by the load (P_{Load} = 80 W) at t = 1.6 s was stored. In this test scenario (as in the previous one), the PVS operated in MPPT all the time, correctly following the reference voltage assigned by the control strategy of the DC/DC converter.

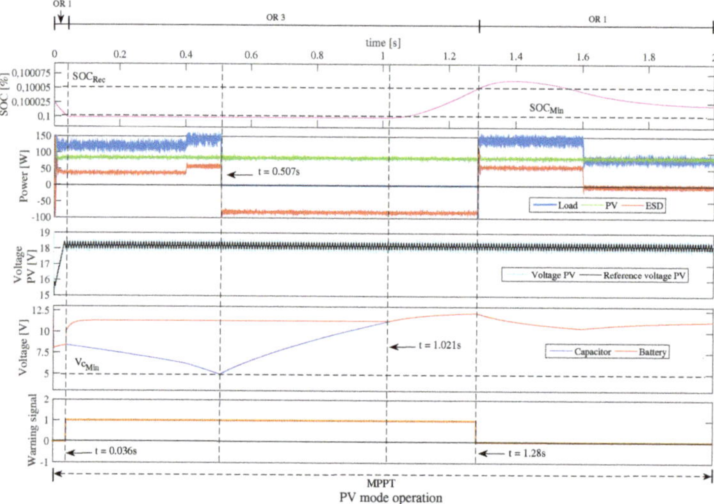

Figure 15. Test system in OR3 (with disconnected load).

Regardless of the load and battery disconnection, the capacitor maintained the voltage level in the DC-bus constant. The bus exhibited high voltage variations when the load was connected and disconnected, which were correctly mitigated by the adopted control strategy. The capacitor dealt with the excess power associated to the control of the DC-bus when the load was disconnected (t = 0.507 s), and the battery was responsible for managing such power when the load was reconnected (t = 1.285 s). This situation can be seen in Figure 16.

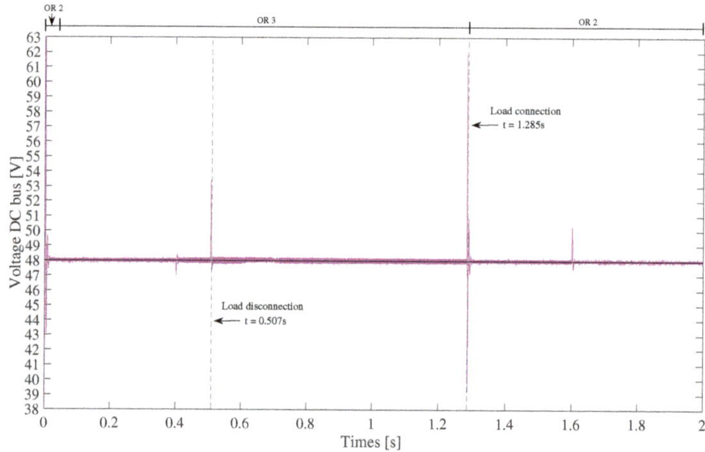

Figure 16. Voltage of the DC-bus test system in OR3 (with disconnected load).

4.2. 24-h Test Scenario

After the correct operation of the SPVS was validated in each one of the operating regions defined in the proposed EMS, a 24-h test scenario was defined using the irradiation values reported by NASA for the University Nacional de Colombia located in Medellín, Colombia, [58] and the same PVS and operating conditions as in the previous power generation tests. Moreover, such scenario uses a power curve that represents the typical behavior of a residential building in the location under study [59]. The power levels were adjusted to observe the different operating regions proposed in this work; Figure 17 presents those values. Because the simulation times in this scenario were longer than in the previous subsection, the EMS parameters were adjusted: for the calculation of the capacitor, $P_c = 10$ W was considered, as well as a maximum voltage drop of 1 V, and a discharge time $t = 60$ s. Those values result in a 49.9 F (almost 50 F) capacitor. In addition, a $Vc_{Max} = 18$ V, $Vc_{Min} = 5$ V, and $(\Delta Vc = 1)$ V were defined. For the PVS voltage rate, the control strategy considered $\delta V_{PV} = 1$ V. Finally, $SOC_{Rec} = 0.2$ was established to create an adequate hysteresis band (10% of the SOC) and to avoid multiple transitions between OR1 and OR3.

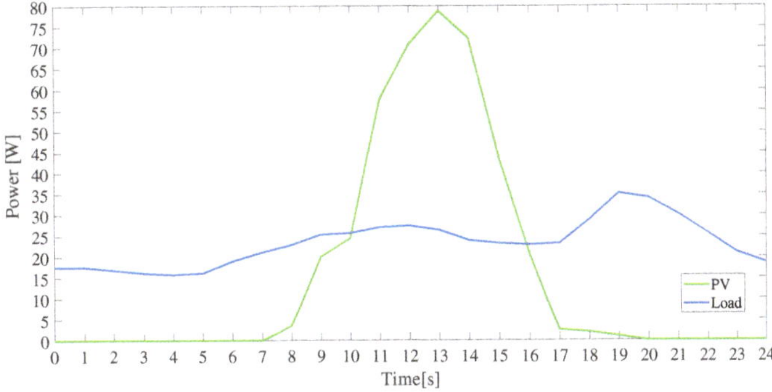

Figure 17. PV generation and power demanded by the load in a 24-h scenario.

Figure 18 presents the results obtained when the EMS was applied in the power and demand scenario described in Figure 17. The analysis begins at hour 6, when the system started with $SOC = SOC_{Min}$ and, for that reason, it entered in OR3, disconnecting the battery and producing the load disconnection. Because the solar irradiance was not available, power was not generated at that time. Nevertheless, the PVS operated in MPPT mode waiting for an increase in irradiance. The previous operating condition was maintained until hour 7, when the PVS started to generate power due to an increase in the irradiance. First, it charged the capacitor and then connected the battery when $Vc \geq V_{bat}$ at $t = 7.67$ h. The voltage waveforms of both the battery and the capacitor show that the battery charging process started after that point. At $t = 8.64$ h the battery reached $SOC_{Rec} = 0.2$; therefore, the load was connected and the SPVS moved into OR1, keeping the PVS operating in MPPT. Noticeably, in this state, the excess power and power requirements were managed by the battery without any inconvenience. At $t = 12.10$ h the battery reached $SOC_{Max} = 0.9$; for that reason, the system moved to OR2, which caused the PVS to switch modes to PDT and the immediate disconnection of the battery, while the capacitor was in charge of supplying and absorbing the excess power associated with PDT mode. At $t = 12.60$ h the capacitor reached the maximum voltage limit allowable in OR2; as a result, the EMS set the PVS to operate in open-circuit voltage so that the capacitor was forced to supply the power required by the load and reach an adequate operating voltage level using the ΔVc previously established. This can be seen in the sub-figures of both battery and capacitor voltage and the power of the elements in the MG. In the 13.07 h $\leq t \leq$ 13.93 h interval, it is observed that the EMS did not allow the capacitor to violate the voltage drop limit established above, $(\Delta Vc = 1)$ V, forcing the capacitor to charge up to battery voltage every time that limit was reached. This was achieved using the transition of the PVS from PDT to MPPT, which led to an increase in the power generation in the PVS, and the excess power was injected into the capacitor to charge it (see Power sub-figure). In the 13.93 h $\leq t \leq$ 16.09 h interval, the system operated in PDT, while the capacitor was responsible for guaranteeing the global power balance operating in accordance with the conditions defined for OR2. Due to the reduction in generation after hour 13, the system met the conditions to change to OR1 at $t = 16.10$ h; and in that instant the capacitor reached the voltage level of the battery, and the battery was reconnected. From that moment on, the PVS operated in MPPT, while the battery was responsible of supplying the excess power needed to serve the load due to the reduction in PVS generation. The battery SOC started to decrease until $t = 18.53$ h, when it reached the minimum allowable limit; consequently, the system changed to OR3. Due to the amount of power demanded by the load at that time, the capacitor was taken to the minimum allowable voltage limit and, as a result, the load was disconnected. After that instant, the capacitor started to charge by absorbing the small amount of energy generated by the panel, reaching the voltage of the battery at $t = 19.23$ h (when the latter is connected). When the battery was reconnected, the power generated by the PVS was limited by the low irradiance, and the battery did not reach the SOC_{Rec}; for that reason the load remained disconnected until hour 6. The bottom sub-figure in Figure 18 presented the warning signal of the system, which was activated every time the system entered OR3 and deactivated when the recovery SOC was reached, it forced the system to enter in OR1. The simulation demonstrated the correct operation of the warning signal in the scenario under analysis.

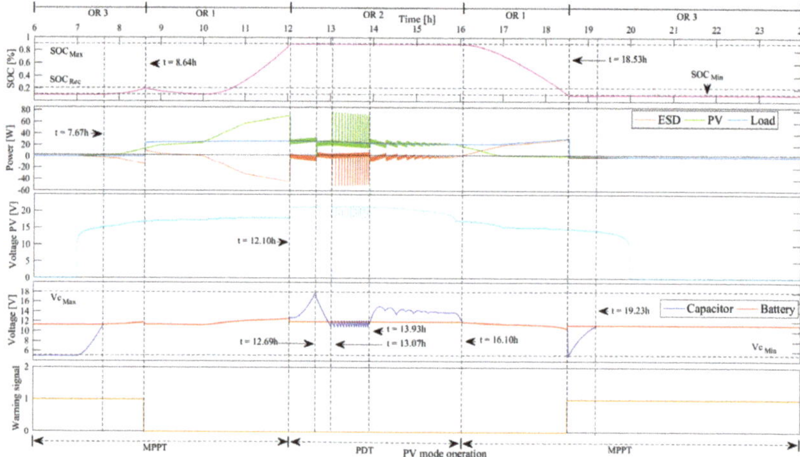

Figure 18. 24-h test scenario.

5. Conclusions

This work proposes an energy management strategy applied to a stand-alone DC microgrid formed by a PVS, a battery, a not-critical DC load, and a capacitor as a backup storage element. Such EMS manages the connection and disconnection of the battery and the load, as well as the generation of the photovoltaic system and the ESD charge/discharge process, in order to guarantee the power balance and the operation of the system within allowable technical limits, thus increasing the lifetime of the devices that form the MG. The validation of the proposed EMS demonstrated that, by implementing a backup energy storage element (in this case a capacitor) and PVS generation control, load disconnections and inappropriate use of the main storage device can be reduced. Furthermore, using the auxiliary capacitor, the charge and discharge sub-cycles produced by the PDT control strategy can be eliminated. Those advantages extend the battery lifetime and reduce the costs associated with the maintenance and disconnection of the microgrid. In addition, the installation of SPVSs with the proposed EMS, into education buildings, allow to reduce the energy purchasing from the utility grid, the dependency of fossil fuels to reduce the environmental impact, and also reduce the operational complexity of the system in comparison with systems based on other renewable resources, e.g., wind generation, small scale hydroelectric, among others. The main limitation of the proposed solution is the total load disconnection; hence as future work, the fragmentation of the load or the integration of multiple critical and non-critical loads into the microgrid will be considered. Those considerations allow the implementation of more complex load shedding strategies, multiple generation and storage systems integration to improve the energy storage capacity of the MGs. Moreover, microgrids could be connected to electrical grid where possible, thus enabling a cheaper energy dispatch that enhances the financial impact of the microgrid and improves the quality of the service provided to final users.

Author Contributions: Conceptualization, L.F.G.-N., C.A.R.-P., D.G.-M., G.A. and Q.H.-E.; methodology, L.F.G.-N., C.A.R.-P., D.G.-M., G.A. and Q.H.-E.; formal analysis, L.F.G.-N., C.A.R.-P., D.G.-M., G.A. and Q.H.-E.; investigation, L.F.G.-N., C.A.R.-P., D.G.-M., G.A. and Q.H.-E.; resources, L.F.G.-N., C.A.R.-P., D.G.-M., G.A. and Q.H.-E.; writing—original draft preparation, L.F.G.-N., C.A.R.-P., D.G.-M., G.A. and Q.H.-E. All authors have read and agreed to the published version of the manuscript.

Funding: This research was funded by the Instituto Tecnológico Metropolitano, Universidad Nacional de Colombia, and Colciencias under the project "Estrategia de transformación del sector energético Colombiano en el horizonte de 2030—Energética 2030"—"Generación distribuida de energía eléctrica en Colombia a partir de energía solar y eólica" (Code: 58838, Hermes: 38945).

Acknowledgments: Authors want to acknowledge to the research groups "Materiales avanzados y energía" from Instituto Tecnologico Metropolitano, "Mecánica Eléctrica" from Universidad Veracruzana and "GAUNAL" from the Universidad Nacional de Colombia.

Conflicts of Interest: The authors declare no conflict of interest.

Reference

1. Rahbar, K.; Chai, C.C.; Zhang, R. Energy Cooperation Optimization in Microgrids with Renewable Energy Integration. *IEEE Trans. Smart Grid* **2018**, *9*, 1482–1493. [CrossRef]
2. Justo, J.J.; Mwasilu, F.; Lee, J.; Jung, J.W. AC-microgrids versus DC-microgrids with distributed energy resources: A review. *Renew. Sustain. Energy Rev.* **2013**, *24*, 387–405. [CrossRef]
3. Parhizi, S.; Lotfi, H.; Khodaei, A.; Bahramirad, S. State of the Art in Research on Microgrids: A Review. *IEEE Access* **2015**, *3*, 890–925. [CrossRef]
4. Guerrero, J.M.; Chandorkar, M.; Lee, T.L.; Loh, P.C. Advanced Control Architectures for Intelligent Microgrids—Part I: Decentralized and Hierarchical Control. *IEEE Trans. Ind. Electron.* **2013**, *60*, 1254–1262. [CrossRef]
5. Jing, W. Battery-supercapacitor hybrid energy storage system in standalone DC microgrids: A review. *IET Renew. Power Gener.* **2017**, *11*, 461–469. [CrossRef]
6. Zia, M.F.; Elbouchikhi, E.; Benbouzid, M. Microgrids energy management systems: A critical review on methods, solutions, and prospects. *Appl. Energy* **2018**, *222*, 1033–1055. [CrossRef]
7. Schonbergerschonberger, J.; Duke, R.; Round, S. DC-Bus Signaling: A Distributed Control Strategy for a Hybrid Renewable Nanogrid. *IEEE Trans. Ind. Electron.* **2006**, *53*, 1453–1460. [CrossRef]
8. Gee, A.M.; Robinson, F.V.P.; Dunn, R.W. Analysis of Battery Lifetime Extension in a Small-Scale Wind-Energy System Using Supercapacitors. *IEEE Trans. Energy Convers.* **2013**, *28*, 24–33. [CrossRef]
9. Ismail, M.; Moghavvemi, M.; Mahlia, T. Design of an optimized photovoltaic and microturbine hybrid power system for a remote small community: Case study of Palestine. *Energy Convers. Manag.* **2013**, *75*, 271–281. [CrossRef]
10. Augustine, S.; Mishra, M.K.; Lakshminarasamma, N. Adaptive Droop Control Strategy for Load Sharing and Circulating Current Minimization in Low-Voltage Standalone DC Microgrid. *IEEE Trans. Sustain. Energy* **2015**, *6*, 132–141. [CrossRef]
11. Dizqah, A.M.; Maheri, A.; Busawon, K.; Kamjoo, A. A Multivariable Optimal Energy Management Strategy for Standalone DC Microgrids. *IEEE Trans. Power Syst.* **2015**, *30*, 2278–2287. [CrossRef]

12. López, J.; Seleme, S.; Donoso, P.; Morais, L.; Cortizo, P.; Severo, M. Digital control strategy for a buck converter operating as a battery charger for stand-alone photovoltaic systems. *Solar Energy* **2016**, *140*, 171–187. [CrossRef]

13. Abu Eldahab, Y.E.; Saad, N.H.; Zekry, A. Enhancing the design of battery charging controllers for photovoltaic systems. *Renew. Sustain. Energy Rev.* **2016**, *58*, 646–655. [CrossRef]

14. Jeon, J.-Y.; Kim, J.S.; Choe, G.-Y.; Lee, B.-K.; Hur, J.; Jin, H.-C. Design guideline of DC distribution systems for home appliances: Issues and solution. In Proceedings of the 2011 IEEE International Electric Machines & Drives Conference (IEMDC), Niagara Falls, ON, Canada, 15–18 May 2011; pp. 657–662. [CrossRef]

15. Weixing, L.; Xiaoming, M.; Yuebin, Z.; Marnay, C. On voltage standards for DC home microgrids energized by distributed sources. In Proceedings of the 7th International Power Electronics and Motion Control Conference, Harbin, China, 2–5 June 2012; Volume 3, pp. 2282–2286. [CrossRef]

16. Shimomachi, K.; Hara, R.; Kita, H. Comparison between DC and AC microgrid systems considering ratio of DC load. In Proceedings of the 2015 IEEE PES Asia-Pacific Power and Energy Engineering Conference (APPEEC), Brisbane, Australia, 31 October 2015; Volume 2016, pp. 1–4. [CrossRef]

17. Whaite, S.; Grainger, B.; Kwasinski, A. Power Quality in DC Power Distribution Systems and Microgrids. *Energies* **2015**, *8*, 4378–4399. [CrossRef]

18. Diaz-Acevedo, J.A.; Grisales-Noreña, L.F.; Escobar, A. A method for estimating electricity consumption patterns of buildings to implement Energy Management Systems. *J. Build. Eng.* **2019**, *25*, 100774. [CrossRef]

19. Anastasi, G.; Corucci, F.; Marcelloni, F. An intelligent system for electrical energy management in buildings. In Proceedings of the 2011 11th International Conference on Intelligent Systems Design and Applications, Cordoba, Spain, 22 November 2011; pp. 702–707. [CrossRef]

20. Yin, C.; Wu, H.; Sechilariu, M.; Locment, F. Power Management Strategy for an Autonomous DC Microgrid. *Appl. Sci.* **2018**, *8*, 2202. [CrossRef]

21. Sharma, R.K.; Mishra, S. Dynamic Power Management and Control of a PV PEM Fuel-Cell-Based Standalone ac/dc Microgrid Using Hybrid Energy Storage. *IEEE Trans. Ind. Appl.* **2018**, *54*, 526–538. [CrossRef]

22. Ma Lei.; Sun Yaojie.; Lin Yandan.; Bai Zhifeng.; Tong Liqin.; Song Jieqiong. A high performance MPPT control method. In Proceedings of the 2011 International Conference on Materials for Renewable Energy & Environment, Shanghai, China, 22 May 2011; Volume 1, pp. 195–199. [CrossRef]

23. Ayad, M.; Becherif, M.; Henni, A. Vehicle hybridization with fuel cell, supercapacitors and batteries by sliding mode control. *Renew. Energy* **2011**, *36*, 2627–2634. [CrossRef]

24. Ramos-Paja, C.A.; Bastidas-Rodríguez, J.D.; González, D.; Acevedo, S.; Peláez-Restrepo, J. Design and control of a buck-boost charger-discharger for DC-bus regulation in microgrids. *Energies* **2017**, *10*, 1847. [CrossRef]

25. Lu, X.; Wan, J. Modeling and Control of the Distributed Power Converters in a Standalone DC Microgrid. *Energies* **2016**, *9*, 217. [CrossRef]

26. Kumar, M.; Srivastava, S.C.; Singh, S.N. Control Strategies of a DC Microgrid for Grid Connected and Islanded Operations. *IEEE Trans. Smart Grid* **2015**, *6*, 1588–1601. [CrossRef]

27. Sechilariu, M.; Wang, B.C.; Locment, F.; Jouglet, A. DC microgrid power flow optimization by multi-layer supervision control. Design and experimental validation. *Energy Convers. Manag.* **2014**, *82*, 1–10. [CrossRef]

28. Kwasinski, A.; Onwuchekwa, C.N. Dynamic Behavior and Stabilization of DC Microgrids With Instantaneous Constant-Power Loads. *IEEE Trans. Power Electron.* **2011**, *26*, 822–834. [CrossRef]

29. Linden, D. Handbook of batteries. *Fuel Energy Abstr.* **1995**, *36*, 265.

30. Belvedere, B.; Bianchi, M.; Borghetti, A.; Nucci, C.a.; Paolone, M.; Peretto, A. A Microcontroller-Based Power Management System for Standalone Microgrids With Hybrid Power Supply. *IEEE Trans. Sustain. Energy* **2012**, *3*, 422–431. [CrossRef]

31. Chen, C.; Duan, S.; Cai, T.; Liu, B.; Hu, G. Optimal allocation and economic analysis of energy storage system in microgrids. *IEEE Trans. Power Electron.* **2011**, *26*, 2762–2773. [CrossRef]

32. Anand, S.; Fernandes, B.G.; Guerrero, J. Distributed Control to Ensure Proportional Load Sharing and Improve Voltage Regulation in Low-Voltage DC Microgrids. *IEEE Trans. Power Electron.* **2013**, *28*, 1900–1913. [CrossRef]

33. Matrawy, K.; Mahrous, A.F.; Youssef, M. Energy management and parametric optimization of an integrated PV solar house. *Energy Convers. Manag.* **2015**, *96*, 377–383. [CrossRef]

34. Olatomiwa, L.; Mekhilef, S.; Ismail, M.; Moghavvemi, M. Energy management strategies in hybrid renewable energy systems: A review. *Renew. Sustain. Energy Rev.* **2016**, *62*, 821–835. [CrossRef]

35. Li, J.; Gee, A.M.; Zhang, M.; Yuan, W. Analysis of battery lifetime extension in a SMES-battery hybrid energy storage system using a novel battery lifetime model. *Energy* **2015**, *86*, 175–185. [CrossRef]

36. Baochao Wang.; Sechilariu, M.; Locment, F. Intelligent DC microgrid with smart grid communications: Control strategy consideration and design. In Proceedings of the 2013 IEEE Power & Energy Society General Meeting,Vancouver, BC, Canada, 25 July 2013; Volume 3. [CrossRef]

37. Mojallizadeh, M.R.; Badamchizadeh, M.A. Adaptive Passivity-Based Control of a Photovoltaic / Battery Hybrid Power Source via Algebraic Parameter Identification *IEEE J. Photovoltaics* **2016**, *6*, 532–539. [CrossRef]

38. Adly, M.; Strunz, K. Irradiance-adaptive PV Module Integrated Converter for High Efficiency and Power Quality in Standalone and DC Microgrid Applications. *IEEE Trans. Ind. Electron.* **2017**, *65*, 436–446. [CrossRef]

39. Mbodji, A.K.; Ndiaye, M.L.; Ndiaye, M.; Ndiaye, P.A. Operation optimal dynamics of a hybrid electrical system: Multi-agent approach. *Procedia Comput. Sci.* **2014**, *36*, 454–461. [CrossRef]

40. Mahmood, H.; Michaelson, D.; Jin Jiang. A Power Management Strategy for PV/Battery Hybrid Systems in Islanded Microgrids. *IEEE J. Emerg. Sel. Top. Power Electron.* **2014**, *2*, 870–882. [CrossRef]

41. Gaurav, S.; Birla, C.; Lamba, A.; Umashankar, S.; Ganesan, S. Energy Management of PV—Battery Based Microgrid System. *Procedia Technol.* **2015**, *21*, 103–111. [CrossRef]

42. Jenkins, D.; Fletcher, J.; Kane, D. Lifetime prediction and sizing of lead–acid batteries for microgeneration storage applications. *IET Renew. Power Gener.* **2008**, *2*, 191–200. [CrossRef]

43. Zhao, B.; Zhang, X.; Chen, J.; Wang, C.; Guo, L. Operation optimization of standalone microgrids considering lifetime characteristics of battery energy storage system. *IEEE Trans. Sustain. Energy* **2013**, *4*, 934–943. [CrossRef]

44. Oliveira, T.R.; Wodson, W.; Gonçalves, A.; Donoso-garcia, P.F. Distributed Secondary Level Control for Energy Storage Management in DC Microgrids. *IEEE Trans. Smart Grid* **2016**, *8*, 2597–2607. [CrossRef]

45. Dragicevic, T.; Guerrero, J.M.; Vasquez, J.C.; Skrlec, D. Supervisory Control of an Adaptive-Droop Regulated DC Microgrid With Battery Management Capability. *IEEE Trans. Power Electron.* **2014**, *29*, 695–706. [CrossRef]

46. Locment, F.; Sechilariu, M. Modeling and Simulation of DC Microgrids for Electric Vehicle Charging Stations. *Energies* **2015**, *8*, 4335–4356. [CrossRef]

47. Han, Y.; Chen, W.; Li, Q. Energy Management Strategy Based on Multiple Operating States for a Photovoltaic/Fuel Cell/Energy Storage DC Microgrid. *Energies* **2017**, *10*, 136. [CrossRef]

48. Montoya, D.G.; Ramos-Paja, C.A.; Giral, R. Improved Design of Sliding-Mode Controllers Based on the Requirements of MPPT Techniques. *IEEE Trans. Power Electron.* **2016**, *31*, 235–247. [CrossRef]

49. Serna-Garcés, S.I.; Montoya, D.G.; Ramos-Paja, C.A. Sliding-mode control of a charger/discharger DC/DC converter for DC-bus regulation in renewable power systems. *Energies* **2016**, *9*, 245. [CrossRef]

50. Emadi, A.; Khaligh, A.; Rivetta, C.H.; Williamson, G.A. Constant power loads and negative impedance instability in automotive systems: Definition, modeling, stability, and control of power electronic converters and motor drives. *IEEE Trans. Veh. Technol.* **2006**, *55*, 1112–1125. [CrossRef]

51. Huddy, S.R.; Skufca, J.D. Amplitude Death Solutions for Stabilization of DC Microgrids with Instantaneous Constant-Power Loads. *IEEE Trans. Power Electron.* **2013**, *28*, 247–253. [CrossRef]

52. Herrera, L.; Zhang, W.; Wang, J. Stability Analysis and Controller Design of DC Microgrids with Constant Power Loads. *IEEE Trans. Smart Grid* **2015**, *8*. [CrossRef]

53. Guo, L.; Zhang, S.; Li, X.; Li, Y.W.; Wang, C.; Feng, Y. Stability Analysis and Damping Enhancement Based on Frequency-Dependent Virtual Impedance for DC Microgrids. *IEEE J. Emerg. Sel. Top. Power Electron.* **2017**, *5*, 338–350. [CrossRef]

54. Microgrids, V.s.c.b.D.C.; Gao, F.; Bozhko, S.; Costabeber, A.; Patel, C.; Wheeler, P.; Member, S.; Hill, C.I.; Asher, G. Comparative stability analysis of droop control approaches in voltage-source-converter-based DC microgrids. *IEEE Trans. Power Electron.* **2017**, *32*, 2395–2415.

55. Zadeh, M.K.; Gavagsaz-Ghoachani, R.; Martin, J.P.; Pierfederici, S.; Nahid-Mobarakeh, B.; Molinas, M. Discrete-Time Tool for Stability Analysis of DC Power Electronics-Based Cascaded Systems. *IEEE Trans. Power Electron.* **2017**, *32*, 652–667. [CrossRef]

56. Bastidas-Rodriguez, J.; Petrone, G.; Ramos-Paja, C.; Spagnuolo, G. A genetic algorithm for identifying the single diode model parameters of a photovoltaic panel. *Math. Comput. Simul.* **2017**, *131*, 38–54. [CrossRef]

Sustainability **2020**, *12*, 1219

57. Li, J.; Danzer, M.A. Optimal charge control strategies for stationary photovoltaic battery systems. *J. Power Sour.* **2014**, *258*, 365–373. [CrossRef]

58. NASA. *NASA Surface Meteorology and Solar Energy: HOMER Data*; NASA: Washington, DC, USA, 2018.

59. CREG: Comisión de Regulación de Energía y Gas. *Propuesta para Remunerar la Generación, Distribución y Comercialización de Energía Eléctrica en las ZNI*; CREG: Comisión de Regulación de Energía y Gas: Bogota, Colombia, 2014.

Article

Benchmarking Energy Use at University of Almeria (Spain)

Mehdi Chihib [1], Esther Salmerón-Manzano [2] and Francisco Manzano-Agugliaro [1],*

[1] Department of Engineering, CEIA3, University of Almeria, 04120 Almeria, Spain; chihib.mehdi@gmail.com
[2] Faculty of Law, Universidad Internacional de La Rioja (UNIR), Av. de la Paz, 137, 26006 Logroño, Spain; esther.salmeron@unir.net
*   Correspondence: fmanzano@ual.es

Received: 20 December 2019; Accepted: 5 February 2020; Published: 12 February 2020

Abstract: Several factors impact the energy use of university campus buildings. This study aims to benchmark the energy use in universities with Mediterranean climates. The University of Almeria campus was used as a case study, and different types of buildings were analyzed. The second goal was to model the electricity consumption and determinate which parameter correlate strongly with energy use. Macro-scale energy consumption data during a period of seven years were gathered alongside cross-sectional buildings information. Eight years of daily outdoor temperature data were recorded and stored for every half hour. This dataset was eventually used to calculate heating and cooling degree-days. The weather factor was recognized as the variable with the greatest impact on campus energy consumption, and as the coefficient indicated a strong correlation, a linear regression model was established to forecast future energy use. A threshold of 8 GWh has been estimated as the energy consumption limit to be achieved despite the growth of the university. Finally, it is based on the results to inform the recommendations for decision making in order to act effectively to optimize and achieve a return on investment.

Keywords: benchmark; campus energy consumption; heating and cooling degree-days; energy model; occupancy rate

---

1. Introduction

Fuel constraints are a relevant issue in both industrialized and developing countries and are related to energy prices and accessibility of energy services [1]. Public buildings such as universities, schools. and hospitals are challenged to manage the exponential growth of their energy demand and transform their buildings into energy efficient ones. The design of buildings should logically be adapted to the lowest energy consumption levels, but in most cases, it is necessary to focus on existing buildings [2]. Therefore, the reduction of both energy consumption and CO_2 emissions from buildings is one of society's main targets today [3]. In Spain, there is a climatic classification according to the technical code of the building that contemplates these issues, which has been mandatory since 2006 [4]. Acting as models for communities, universities are supposed to provide innovative solutions through research in order to support the sustainability and reduce the carbon footprint [5]. One of the key operating aspect for universities is related to enhance students and teachers comfort levels, which may have a significant effect on their performance [6]. Visual, acoustic, and thermal comforts should not be considered as luxuries but rather as basic standard for schools [7]. However, maintaining indoor quality will eventually lead to a significant growth of electricity consumption; therefore, transforming university locals into energy efficient ones is a necessity. To ensure that these locals have optimal energy performance, researchers and professionals have developed management systems such as energy benchmarking and energy audit [8].

The energy benchmarking technique allows us to compare the energy consumption of buildings by dividing the key performance metric by gross floor area [9]; this index is usually expressed in (kBtu/ft^2/yr or kWh/m^2/yr), and it is labeled as Energy Use Intensity (EUI) or Energy Intensity (EI). This gives the opportunity to the portfolio manager to track the key performance metric overtime [8]. EUI is expressed as energy per square meter per year. It is calculated by dividing the total energy consumed by the building in one year by the total gross floor area of the building. The main benefit of using EUI is that the performance of a building can be compared with similar buildings across the country. EUI can vary significantly depending on building type; therefore, it is necessary to calculate it in buildings used in which there is no data so far. Energy audit is a tool that allows building owners and managers to determine which energy efficiency measures meet their sustainability goals and their investment return criteria [10]. The energy efficiency directive (201/27/EU) requires the auditing of the energy performance of old schools to assess them and propose future retrofitting if necessary [11]. In Italy, over 28% of schools are energy inefficient [12]. A previous experiment executed an energy management program in a high school located in Dubai, UAE [13], and its results show that energy performance can be basically improved by 35%. Many evaluation programs for green schools have been designed to assess managers towards sustainable solutions, like the program whole-school approaches, this initiative integrated different elements of school life such as governance, pedagogical methods, curriculum, resource management, school operations and grounds [14]. In the case of the University in Spain, particular studies have been carried out for the Universidad Politecnica de Valencia (UPV) in order to predict electricity consumption patterns in buildings [15] or the use of algorithms using demand and generation forecasts and costs of the available resources, so the benefit obtained in a whole year is five times higher, with a percentage of participation in demand response programs (DRPs), which is accepted as 60.27% or higher [16]. At this same university, with the use of energy efficiency measures (EEMs), in three different types of buildings (a research building (Building 8G), a teaching and staff building (School of Telecom Engineering building 4P), and a greenhouse building 8I-8J), the savings representing about 10% of total annual energy consumption [17].

HVAC and lighting systems have drastically changed in the last decade. Today, they incorporate sophisticated sensors and computer networking programs to monitor and adjust building systems and energy usage. These new technologies are called building automation systems, and they control, monitor, and collect data on the buildings performance technology [18,19]. University campuses serve different functions by providing spaces such as teaching rooms, academic offices, laboratories, restaurants, and sport facilities. This research outlines the classification of categories by their ECs and EUIs. The building category that influences substantially the overall EC of the University by 47% even though it covers only 27% of its total GFA. This category is the science and research category, and it is also the most energy intense by an average EUI 119.5 kWh. Similar results were reported by a study that was ran to support the ASHRAE standard 100. It has determined the EUI median for 18 major categories by climate zone in the USA, according to CBECS 2003. The national median of the laboratory has the highest energy intensity on a university campus (98 kWh/m^2). Our case study provides an opportunity to treat a diverse dataset of buildings. A study carried out in Australia reported that laboratory energy intensity was the highest among other categories, and it was three times higher than non-laboratory buildings [20]. In addition, another study divided laboratories into different classes of science, applied science, and intervention, and the results show that the HVAC and electric appliances load, as well as the long operating hours, are the main reasons behind the high energy consumption of this category [21].

The quantity of energy used in universities can change from a country to another, as a recent study in Taiwan has demonstrated that gross domestic product (GDP) of the country has a positive correlation with the energy consumption [22]. Furthermore, a study carried out by Catherine and Byrne et al., (2014) summarized the major factors that significantly impact university buildings' energy use are as follows: occupancy rate, HVAC load and artificial lighting, number of computers and electric equipment, and weather conditions [23]. The influence of these various parameters on the energy use and their correlated relationship to each other define the stochastic nature of the EC. This paper focused on two parameters—weather and the size of the active network inside the campus. The choice of those variables was made based on an energy survey that was conducted inside the campus and the analysis of energy consumption patterns over the last eight years. Unlike many previous studies that focused on modeling the occupant behavior and its influence on the EC, this study tested the impact of the network on a yearly basis. We gathered the number of occupants active inside the university, including the number of students, professors, and administrative staff, and since this parameter varies during the academic year, we had to line up the two other parameters in order to have them all on the same sequence. Then, we tested the correlation with both variables (the number of occupants and the sum of CDD and HDD) with macro-scale energy data. On the other hand, energy benchmarking seeks to give a reference value by defining reliable indicators, we split the buildings portfolio by the following six categories: research, administration office, teaching and seminary room, library, sport facility, and restaurants. This will allow us to benchmark within the same category and identify the benchmark value of each category. These values could be used in the future to define a national baseline for universities or in the Mediterranean region. As the majority of studies have proven that outdoor conditions are the main variable that influences energy use, in this study, we will test the correlation of the total EC with the size of network from one side and with the sum total of the HDD and CDD during each academic year. Moreover, identifying the variable that has the strongest correlation is primarily in order to take suitable actions to achieve better energy management.

2. Materials and Methods

The University of Almeria is a Spanish Public University located in the south coast of Spain, with the coordinates of latitude 36°49′45″N and longitude −2°24′16″E, see Appendix A. The university campus spreads on a surface of 17 hectares and has 33 buildings (see Figure 1). In the 2018–2019 academic year, the university offered 38 different degree programs, with 883 lecturers and 13,547 students.

Figure 1. The University of Almeria ground plan.

Almeria is a coastal city located on the southern region of Spain, the climate is particularly arid and semi-continental, with relatively dry warm winters with an average temperature at 16°C (60 °F) and hot summer with an average temperature of 28°C (80 °F). The most daily sunshine hours are scored in July and the wettest month is January with an average of 30mm of rain and an annual average percentage of humidity of 61.0% [24,25].

The dataset used in this analysis consists of (1) daily outdoor temperature scored every half hour during the last eight years, (2) Total energy consumption on a monthly basis and gross floor area data from 2011–2018 (Table 1), (3) campus buildings' energy consumption data during the last three years 2016–2018 (Table 2), (4) the average EUI within each category (Table 3), (5) Building energy performance classification of all the buildings by category (Table 4), the number of students, professors, and administration staff per academic year (Table 5). Figure 2 outlines the methodology flow chart, starting from defining objectives to collecting data to developing results.

Table 1. Monthly evolution in energy consumption of the university campus per year.

Campus Monthly EC (kWh)								
Month/Year	2011	2012	2013	2014	2015	2016	2017	2018
January	722,623	717,867	706,880	708,499	765,785	702,918	773,539	770,454
February	689,088	746,940	672,895	657,712	737,049	693,787	652,036	725,147
March	729,218	685,622	681,843	689,979	728,747	666,108	706,072	716,676
April	573,686	571,444	630,393	597,792	640,762	644,299	580,052	680,127
May	700,648	687,262	644,661	667,437	717,944	672,034	713,215	730,265
June	762,909	765,368	656,218	696,597	798,527	761,444	936,273	740,233
July	737,705	724,290	707,900	703,602	877,181	776,462	801,608	704,690
August	550,274	527,979	487,813	513,183	597,359	545,341	617,208	608,966
September	760,301	6933	731,827	753,425	785,407	837,467	821,574	645,724
October	693,375	654,832	727,983	710,872	721,047	742,409	780,345	699,640
November	640,273	632,938	647,683	671,644	623,533	685,437	705,548	757,993
December	621,586	629,833	627,630	683,565	616,653	645,593	691,188	872,744

Table 2. Space category, energy consumption, and gross floor area.

Space Category	Building	EC (kWh/year)			GFA (m^2)
		2016	2017	2018	
Administration Office	1	226,192	220,042	239,366	5880
	2	329,354	331,988	320,119	11,430
	18	64,759	63,880	62,840	2620
	19	196,189	208,161	188,432	8290
	20	64,553	56,955	56,261	2450
	21	104,633	129,335	122,294	4605
	8	141,530	143,240	144,591	3994
Teaching and Seminary Room	3	300,566	330,608	327,658	5585
	5	120,802	137,864	142,172	5611
	15	107,047	119,255	123,319	4118
	12	137,493	161,042	152,614	6016
	4	13,273	13,938	16,132	12
	23	156,424	168,703	151,046	6,605
	27	296,197	362,304	369,768	8,618
	11	68,441	59,409	51,887	3,089
Research Building	9	176,596	178,167	156,995	5487
	30	812,983	809,544	842,943	4301
	28	388,289	384,574	361,428	4828
	29	734,370	788,962	796,318	4975
	16	156,650	186,889	150,348	2100
	31	294,533	246,824	199,478	1072
	24	280,959	465,018	478,491	3089
	13	735,213	767,650	691,523	12,341
Library Building	26	905,166	19,215	947,826	16,194
	32	2311	213,344	202,611	2026
Sports Facilities	10	257,182	155,856	306,779	5548
	7	89,013	76,623	78,963	3280
	17	49,892	38,184	32,967	547
Restaurant Buildings	33	42,169	41,811	61,249	1190
	6	43,690	52,910	62,919	1280

Table 3. Average energy use index per year and average energy consumption of each space category.

Building Category	Average EUI (kWh/m^2 ·Year)	EC (kWh)
Research	119.50	3,694,915
Library	82.67	1,169,721
Sport facilities	47.30	361,820
Restaurant	41.11	101,583
Teaching and seminary	28.99	1,295,988
Administration Office	28.78	1,38,239
Others	-	28,007
Public Lighting	-	416,812

Table 4. Building energy performance classification of all the buildings by category.

Building Category	Energy Performance Classification							
	Poor Practice <===> Best Practice							
Research	31	30	29	24	16	28	13	9
Library	32	26	-	-	-	-	-	-
Sport facilities	17	10	7	-	-	-	-	-
Restaurant	6	33	-	-	-	-	-	-
Teaching and seminary	3	4	27	15	12	23	5	11
Administration Office	1	8	2	21	18	20	19	-

Table 5. Inputs and outputs used for the correlation analysis and regression model.

Academic Year	CDD & HDD (°C)	N of Occupants	N of Staff		EC (kWh)
			N of Professors	N of Administrative Staff	
2011–2012	1456.30	15,062	475	806	8,142,307
2012–2013	1225.90	14,978	476	698	7,799,209
2013–2014	1091.20	15,234	477	732	7,969,924
2014–2015	1449.40	15,295	475	752	8,682,860
2015–2016	1220.40	15,417	468	791	8,209,033
2016–2017	1398.80	15,392	464	780	8,690,909
2017–2018	1417.70	15,680	468	809	8,675,213
2018–2019	1664.20	15,166	482	883	8,453,842

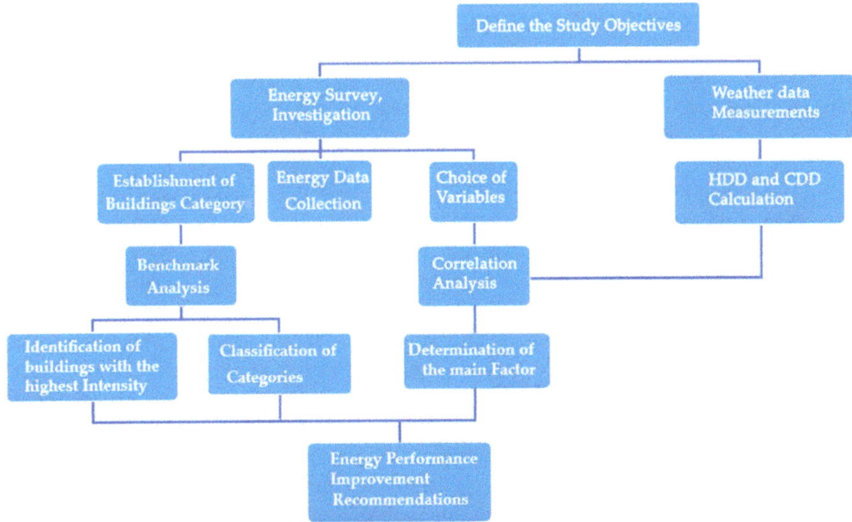

Figure 2. Methodology top-down chart.

Correlation Approach

There is a lack of data for most of the electric components and the physical characteristics of the university buildings (building materials, building geometric sizes). Thus, the given data set of observations gives us the opportunity to establish a statistical approach that allows us to measure the relationship between two variables by defining there correlation coefficient, which will provide us with a straightforward interpretation of the two variables on the overall electricity consumption on a yearly basis. This method relies on historic values of overall energy consumption and background knowledge of the input variables that influence perception. In this case study, we define the first explanatory variable occupancy rate as the total number of students, professors, and administration staff for the academic year. If we get a weak correlation, we proceed by dividing the number of occupants into two groups, students and staff (professors and administration staff), and then test them separately. The second explanatory variable is the weather explanatory variable, defined as the sum of the heating and cooling degree-days during one academic year; its unit is in degrees Celsius.

Heating and cooling degree-days (HDD and CDD) are defined as the differences between the average daily outdoor temperature T(o/d) and corresponding base temperatures Tb [26]. The base temperature for heating and cooling is different from place to place. It also depends on the type of building (household, administration, hospital). In this case study, the CDD temperature base is $(T(b, CDD) = 28\ °C)$ and the HDD temperature base is $(T(b, HDD) = 14\ °C)$. These assumptions are based on a survey conducted inside the campus [24,25]. Note that the temperature base values indicate the outside temperature, and there is usually a minimum of two degrees of difference between the inside and outside temperatures. We sum up both variables on a sequence of every academic year so that it can be lined up with the quantity of interest (EC).

$$CDD = T(o/d) - T(b, CDD) \quad \text{if} \quad T(b, CDD) > T(o/d) \quad \text{then} \quad CDD = 0 \tag{1}$$

$$HDD = T(b, HDD) - T(o/d) \quad \text{if} \quad T(b, HDD) < T(o/d) \quad \text{then} \quad HDD = 0 \tag{2}$$

The extensive weather data set will be used in this paper to develop and validate statistical models. The complex nature of EC inside buildings and the lack of the data of most components of buildings drive us to a black box model that relies on a simple input and output system [27]. The statistical model of the linear regression is set up according to the Formula (3):

$$Ei = b \times (Xi) + a \tag{3}$$

where Ei is the annual energy consumption corresponding to the academic year i, the input Xi is the explanatory variable, b is the slope, a is the y-intercept [28].

3. Results

3.1. Benchmark Analysis

Table 3 contains values of the electricity performance metrics of all the categories in the campus, and they are cited in Table 3 from the highest to lowest intensity. Average EUIs were calculated by first calculating the average in the three years of each building. Then, we sum up the EUIs within the category, and we dived them by the number of facilities of each category. The section of others is excluded from the benchmark study because it includes three buildings (14,22,25) that represent, respectively, a warehouse, a parking garage, and a nursery. These buildings have a weak EC, do not have an impact on the campus EC, and do not fit into any of the major categories. The EUI averages represent benchmark values of the cited categories in the Mediterranean climate.

Figures 3 and 4 summarize as percentages the total EC by categories and the sum total GFA by categories. The research and science category have the biggest share by 47% of the EC, even though it accounts for only 27% of the gross floor area (GFA), followed by the teaching and seminary category that accounts 17% of the EC and 26% of the GFA, the library category that accounts 15% of the EC and 7% of the GFA, the administration office category that accounts 15% of the EC and 27% of the GFA, by the sport facilities category that account 5% of the EC and 6% of the GFA, and finally the restaurant category that accounts 1% and 2% of the GFA. These distribution shows that there is no direct relation relationship between EC and GFA because each category has its own characteristics.

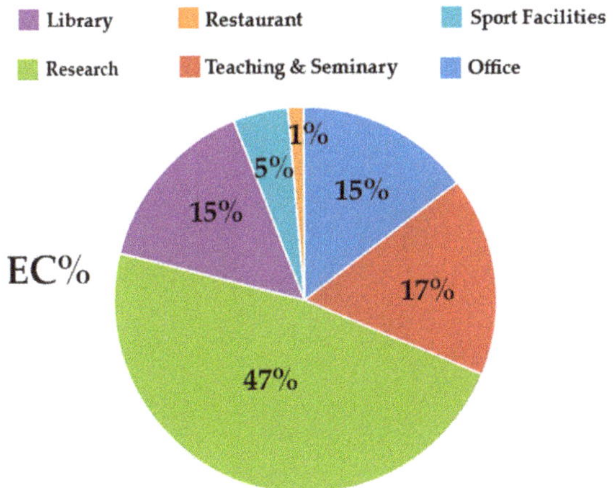

Figure 3. Energy consumption (EC) proportion of all the categories.

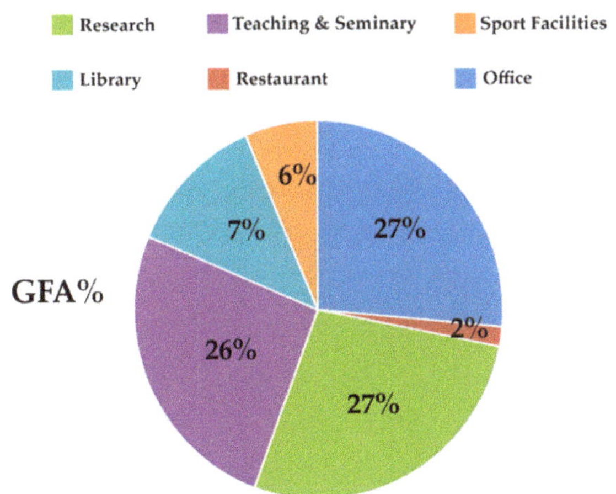

Figure 4. Gross floor area (GFA) proportion of all the categories.

Figure 5 summarizes the energy consumption evolution by building category from 2016 to 2018. The evolution of the total sum campus EC had a minimum value in 2016, then had peaked in 2017, and had medium value in 2018. Figure 5 reveals that all the categories followed the overall trend, except restaurant and sport facilities, where both categories account combined 6% of the total EC. Research building EUI varies from 32.5–230.3 kWh/m^2. Furthermore, its facilities include spaces like academic offices, computer rooms, and laboratories, and those spaces are characterized by a longer period of operation and a large number of computers, laboratory freezers, and other electric equipment. However, the majority of research buildings have a value superior to 80 kWh/m^2; the highest intensity value—230.3 kWh/ m^2—was scored by the solar energy center building (31). One of the reasons behind this high consumption is that a lot of research takes place in the center, and the researchers and students working on solar chemistry and water detoxification use several compressors with high energy consumption that cannot be powered small solar field installed on the roof of the building. The technology of information and communication center building (30) has the second highest value 188.3 kWh/m^2, and the lowest value was scored by the engineering school building (9).

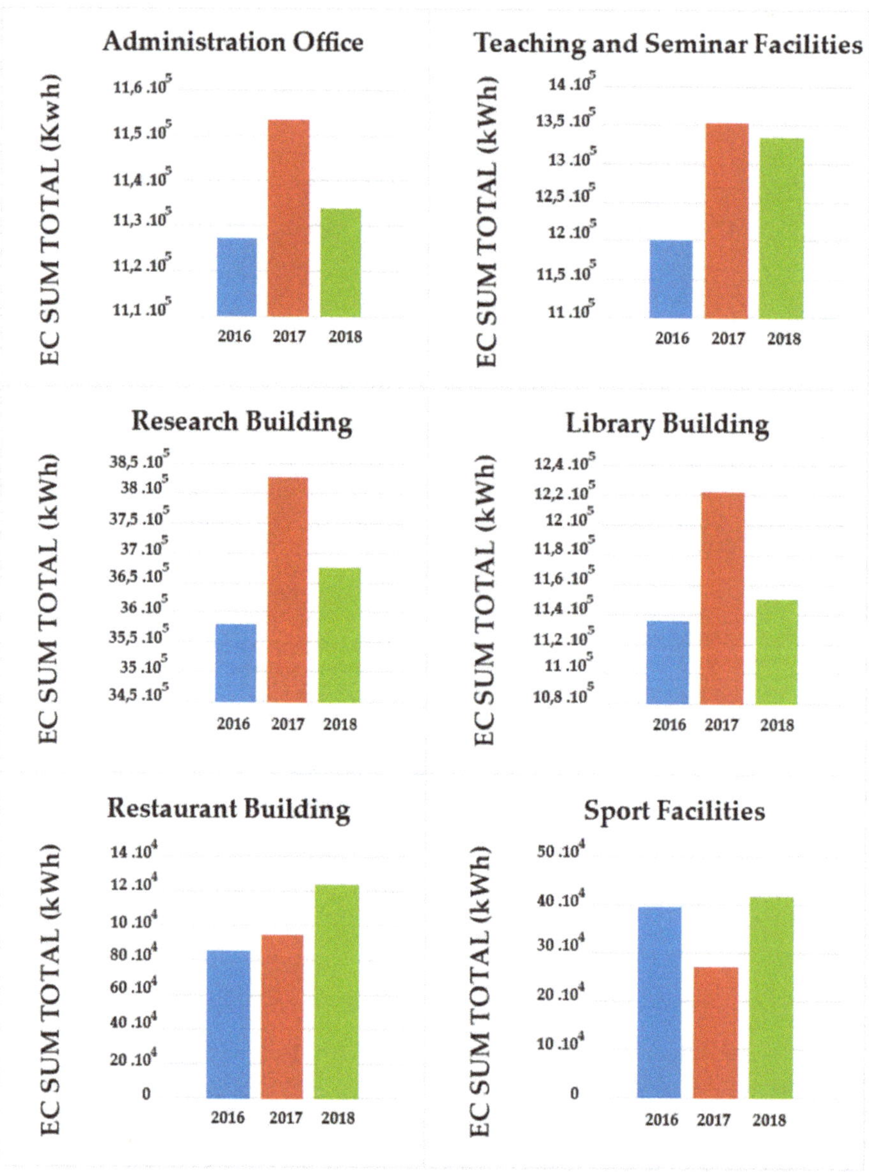

Figure 5. Energy consumption evolution by building category.

Library building EUI varies from 62.4–105.3 kWh/m^2. This category is the second most energy intensive. Their locals include spaces like reading rooms, computer rooms, and common spaces and are characterized by a centralized air conditioning system, longer operating time, a high number of occupants (especially during the exam period), and a substantial number of computers and laptops.

Sport facilities are the third biggest consumer of energy by GFA, and their EUI varies from 28.02–69.9 kWh/m^2. It contains spaces like swimming pool, a covered multitask hall, and gym rooms.

Restaurant buildings had the fourth highest energy intensity, and their EUI varies from 35.13–41.34 kWh/m^2. They are characterized by longer operating time and different electric equipment.

Teaching and seminary rooms had the fifth highest energy intensity, and their EUI varies from 13.59–59.2 kWh/m^2. This category includes spaces like regular classrooms, computer rooms, and room theaters, and it is characterized by a high number of occupants. Four buildings out of eight have similar EUI values, which are close to average energy use index of this category. The lowest value is scored by building (4), a seminary building that has a low operating frequency, while the highest value is scored by building (3), which is an exception in this category because of its infrastructure that includes a water pumping system to get rid of the used water for the whole campus.

Administration offices are the least energy intensive category, and their EUI varies from 23.3–32.5 kWh/m^2. They include offices, meeting rooms, and common spaces and are characterized by a low number of occupant and shorter operating time.

Figure 6 outlines the scatter plot of the average EUI during last three years. In the function of the average EC, we can observe that only five buildings (31,30,29,24,32) have an EUI superior to the EUI median of universities (MU) in the Mediterranean climate given by the 2003 CBECS data [9]. Twenty-five buildings from our portfolio have an energy intensity inferior than the M.U, and 19 out of those 25 buildings are three time less intense than the k–12 schools in hot and humid zones [9]. In addition, the scatter plot demonstrates that four buildings (30,29,13,26) have an EC greater than 7.0 × 10^5 kWh. These facilities belong to the research and library category. However, two of those big consumers have EUI values bellow the MU, and the category that has shown more harmony in their sample of buildings is the restaurant category, which include two properties that has almost identical values in their EC, GFA, and EUI.

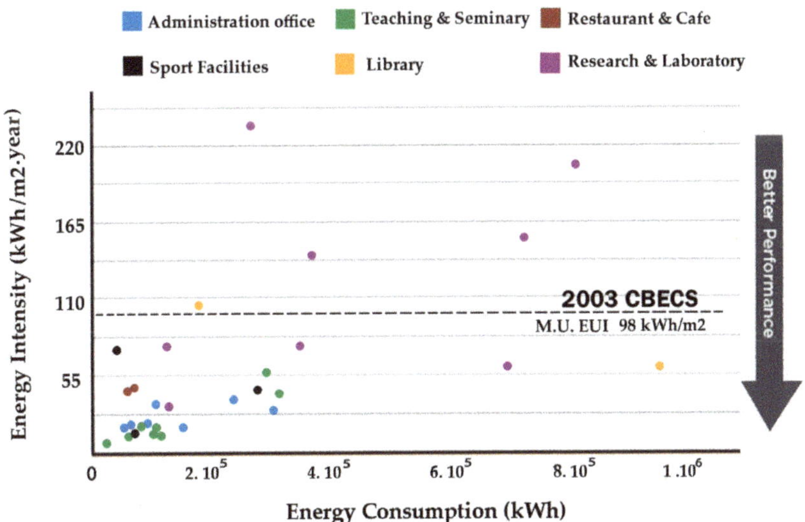

Figure 6. Scatter plot of energy use intensity (EUI) as a function of energy consumption (EC) by buildings category.

Table 4 outlines the classification of energy performance of all the buildings within each category. This classification is a useful tool for managers because it reports insights into which of the building should be prioritized in term of investments to achieve efficiency. Managers an divide the portfolio of every category into three groups (poor practice, usual practice, and best practice).

3.2. Case Study

Figure 7 summarizes the campus yearly energy consumption from 2011 to 2018. The overall energy consumption has been varying in the range of 7.93–8.8 GWh, and it has a known growth of 9% since 2011. However, the patterns do not represent a clear trend through the calendar years; this is one of the reasons we decided to proceed with the academic years since we have the data of the campus monthly EC (Table 1).

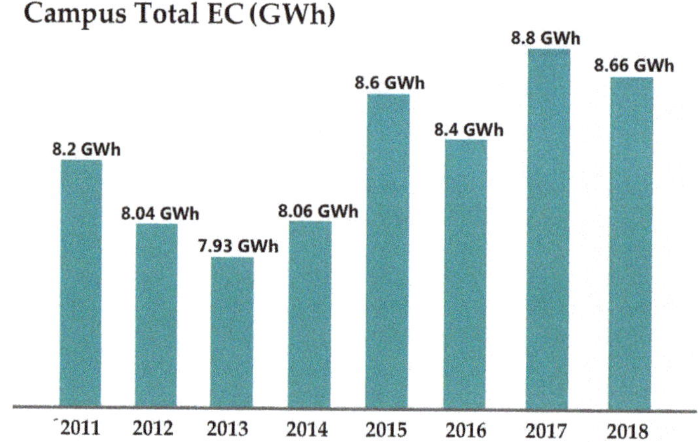

Figure 7. Campus yearly consumption 2011–2018.

Campus energy consumption data on a monthly basis (Table 1) shows that the evolution over the years is generally the same, which means that it plunges and peaks in the same period of the year. The peaks usually happen during the months of May and June. In this period, campus buildings make substantial use of HVAC systems, and some buildings like the library start operating for a longer period because of the exam period. The down trend starts in October after the weather begins to be cooler, and in the beginning of July, it plunges to hit the lowest values in August—during this period, the campus is practically empty, and the majority of the university buildings are non-operational because of the summer vacation.

3.3. Correlation Analysis and Regression Model

Data presented in Table 5 sums up the inputs and the outputs used for the correlation analysis and regression model which corresponds to weather parameters (CDD and HDD), the number of occupants (the number of students, professors, and administration staff), and the EC of the campus.

Figure 8 outlines the scatter plots of EC sum total of the campus in function of the following variables: CDD and HDD, number of occupants, number of students, and number of staff. In the case of the correlation of outdoor temperature with EC, EC (Kwh) = 1204.8 CDD&HCC (°C) + $7 \cdot 10^{6}$, its correlation coefficient of 0.72 indicates a positive correlation. According to the scatter plot of the number of occupants, its correlation coefficient of 0.62, which also indicates a positive correlation (EC (Kwh) = 1073.9.7 N—$8 \cdot 10^{6}$). However, the remaining scatter plots represented a weak correlation, especially for the number of students. In addition, the number of occupants is the main factor behind the excessive consumption.

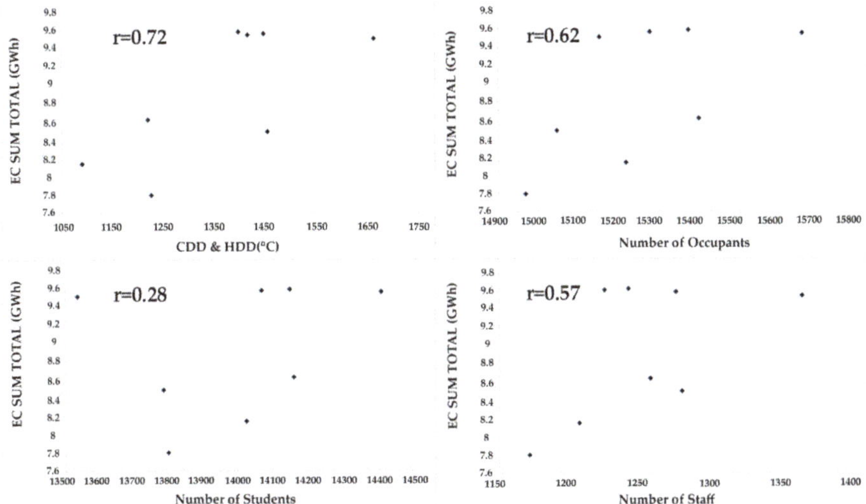

Figure 8. Scatter plots with their correspondent coefficient of correlation.

Figure 9 outlines the linear regression model of the total EC as a function of CDD and HDD. Both variables are statistically related because the correlation coefficient is 0.719. Figure 10 shows the energy thresholds of good practice of the University of Almeria.

$$Lc = \frac{\text{Min1(EC)1} + \text{Min2(EC)} + \text{Min3(EC)}}{3}$$

$$Hc = \frac{\text{Med1(EC)1} + \text{Med2(EC)} + \text{Med3(EC)}}{3}$$

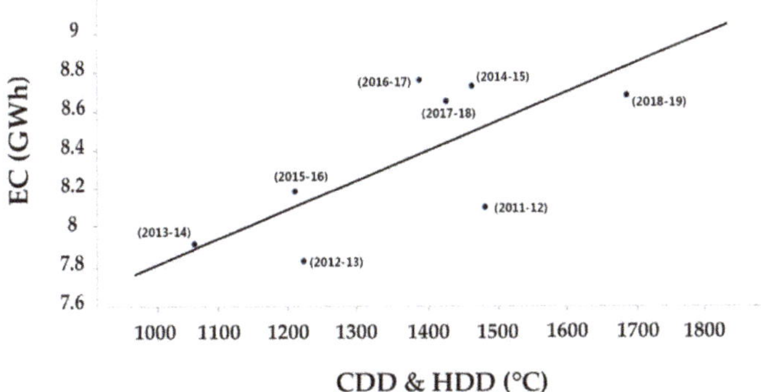

Figure 9. Linear regression model of EC in function of CDD and HDD.

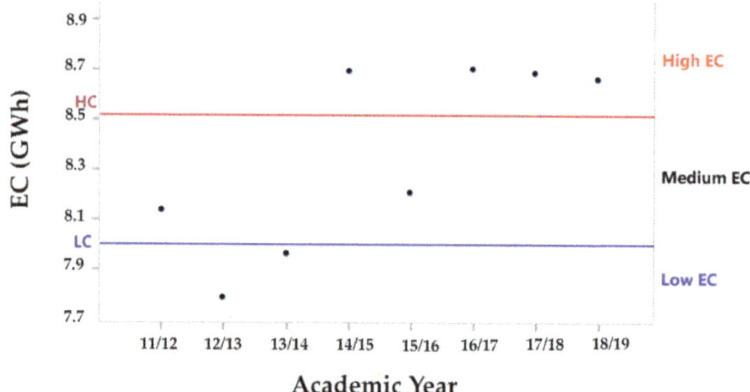

Figure 10. Scatter plot of the total EC in function of academic year

4. Discussion

4.1. Benchmark Analysis

Dividing campus facilities into categories allows us to compare buildings based on their utility and to identify the category that has the lowest and largest energy consumption by GFA. The category that scored the highest EUI average in this case study is the research and science category by an average metric score of 119.5 kWh/m^2, which is inferior to the laboratory intensity median given by CBECS 2003 data by Oak Ridge National Lab and the Department of Energy, which is 226 kWh/m^2 for the Mediterranean climate [9]. In another comparative study, which was conducted at the regional scale in the state of California (Mediterranean climate), they showed that some laboratories scored the highest energy intensity with a value of 909.5 kWh/m^2, and it is four times bigger than the state average energy intensity [29–31]. The results in our study have proven that this category can provide a wide range of energy intensity, which makes it worthy of more in-depth study. The observations indicate that the longer operation hours inside the facilities of the research and science category, its heavy plug load materials (like ultra-low freezers and incubators), and other laboratory equipment are the reason behind the high intensity. In addition, the high number of computers used are the reason behind the high intensity. The second most intense category is the library category, which accounts for only 7% of the total GFA but contributes to 15% in of total EC. The buildings in this category peak during the months of the preparation of exams, especially in the summer session, when their EC becomes three times higher, unlike the teaching and seminary rooms, which have a slight increase during the same period when the teaching days are relatively lower. Our portfolio have an EUI inferior than 50 kWh/m^2 (Figure 6), which is the equivalent of one third of the k–12 schools in hot and humid zones. This brings back the question of which is more energy intense—schools or universities—and how much can the weather parameter contribute to the increase of electricity consumption. Thus, there are several parameters other than the weather to take into consideration, such as occupancy rate, number of COM, plug load, and operating time, that are responsible for the EC gap between different categories. On the other hand, simulation techniques represent one of the efficient alternatives to evaluate the energy performance of a building regardless of its utility. This method was used to develop a benchmark analysis based on models of equipment and system performance, which proved that plug loads and HVAC are some of the biggest influencers of high energy consumption in laboratory buildings.

4.2. Correlation Analysis

Despite the complexity of EC in university campuses, we were able to demonstrate that outdoor temperature and number of occupants positively correlate with the overall energy use, which confirmed our choice of variables. Nevertheless, another study that developed a simulation of the building occupants' decision-making and information communication process found that the network size has no significant impact on the EC [32]. Still, that result needs to be confirmed in non-residential building, especially in cases like schools and universities, where the number of occupants changes substantially over the year. On the other hand, a study that was conducted on 10 universities in the US and confirmed that EC correlates highly with outdoor temperature [33]. Weather variations can easily change cooling and heating use by 20–30% [34]. The estimation of the occupancy rate for each building remains challenging, especially in this case study because of the irregular patterns of the student movement inside the campus, which is not only related to the classes or other scheduled activities. Nonetheless, some studies used CO_2 measuring, relative humidity, and acoustic sensors to estimate occupancy [35–37]; however, those techniques are hard to implement in our case of study because of the several components that the buildings incorporate. On the other hand, many studies [38,39] have focused on the behavior of occupants rather than the size of active occupants. In order to evaluate the energy saving potential, one study developed an occupancy model of individuals moving in and out in offices [40], while another study found an alternative to analyze occupancy patterns using physical-statistical approaches to improve energy demand forecasting [35,41]. Still, identifying how occupant's behavior influences EC is complex because of the stochastic nature of individual actions [37].

The five buildings (31,30,29,24,32) from our portfolio scored a higher energy intensity than the university in a Mediterranean climate median. Installing solar panels on the roof as a backup is highly recommended since the campus is located on southern coast of Spain, where the yearly sum of global irradiation is over 1900 kWh/m^2 [41]. Therefore, efforts and investments have to prioritize these facilities because they are driving the high consumptions and there is a considerable gain potential to achieve from their high energy intensity.

If a model with the two explicative variables (N and CDD&HCC) is established and a multiple regression is conducted, Equation (4) is obtained.

$$EC \text{ (Kwh)} = 1050.375608 \text{ N} + 1172.322578 \text{ CDD\&HCC (°C)} - 9320518.243 \qquad (4)$$

with $R^2 = 0.83$. If the model obtained is plotted (see Figure 11), the range of expected energy consumption can be found according to the parameters of number of occupants (N) and CDD and HCC in °C. Therefore, a threshold of 8 GWh has been estimated as the energy consumption limit to be achieved for 15,000 persons and for a CDD and HCC of 1350 °C, both factors being the average of the last eight years.

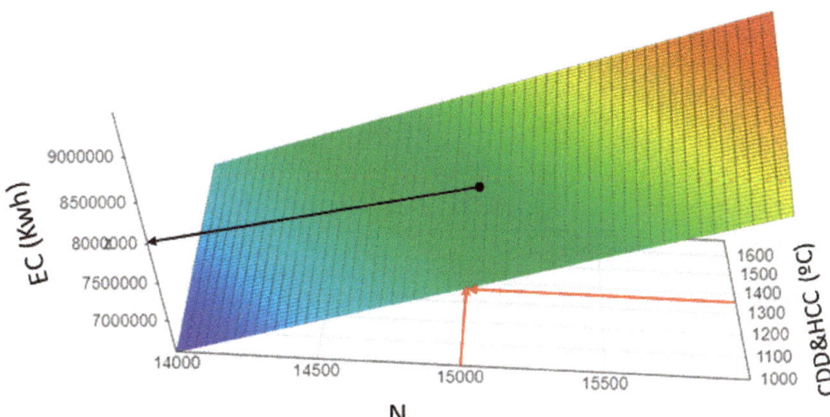

Figure 11. Energy consumption (EC) model obtained for the explicative variables: N (number of occupants) and CDD and HCC (°C).

5. Conclusions

Gathering high resolution outdoor temperature data during the last eight years with half an hour time frame was important for this case of study because it provided the opportunity to calculate cooling and heating degree-days. The weather factor is the most significant variable in this case study, which means that the university administration will achieve better results in term of reducing end user costs by investing in the efficiency of the HVAC system and then improving the thermal performance of the campus buildings. It has been found that research buildings consume four times more energy than teaching or administration buildings. In addition, behavioral changing programs are recommended, especially in cases like ours, where the properties are open to the public and the managers are challenged to lead the users toward sustainable actions—for instance, starting a program that aims to share information about the evolution and the gains of EC with the students and staff which could be useful to raise awareness about the continuous increase of energy use inside the campus. A similar experiment was executed in dormitory residences, and the results show that the group of residents who received real-time data feedback were more effective in energy conservation gains. The buildings from our portfolio that scored a higher energy intensity should consider installing solar panels on their roofs. Additionally, setting up systems like occupancy sensors for automatic lighting will increase efficiency. Nonetheless, energy conservation measures should not affect the indoor quality; for this reason, we must be able to reduce EC while retaining indoor quality.

Author Contributions: M.C. dealt with literature review, analyzed the data and wrote the draft. E.S.-M. and F.M.-A. Research idea and article writing. They share the structure and aims of the manuscript, paper drafting, editing and review. All authors have read and approved the final manuscript.

Funding: This research did not receive any external funding.

Conflicts of Interest: Declare no conflict of interest.

Abbreviations

Acronym	Meaning
GFA	Gross floor area
EUI	Energy use index
EC	Energy consumption
HVAC	Heating ventilation air conditioning
HDD	Heating degree-days
CDD	Cooling degree-days
GDP	Gross domestic product
KPM	Key performance metrics
ASHRAE	American society of heating, refrigeration and air conditioning engineers
CBECS	Commercial energy consumption survey
RECS	Renewable energy certificates

Appendix A

The outdoor temperature data was given by the weather station of the Almeria Airport, Latitude 36.846911° N, Longitude −2.356989° E), which is located 4 km away from the University of Almeria, at the same height above the sea level. The dataset does not figure in this paper due to its extensive format, which reached 140,160 tags. The energy consumption reports of the university campus and the number of students, professors, and administration staff were given by the university infrastructure department and are available to the public.

References

1. Pino-Mejías, R.; Pérez-Fargallo, A.; Rubio-Bellido, C.; Pulido-Arcas, J.A. Artificial neural networks and linear regression prediction models for social housing allocation: Fuel Poverty Potential Risk Index. *Energy* **2018**, *164*, 627–641. [CrossRef]

2. Forcael, E.; Nope, A.; García-Alvarado, R.; Bobadilla, A.; Rubio-Bellido, C. Architectural and Management Strategies for the Design, Construction and Operation of Energy Efficient and Intelligent Primary Care Centers in Chile. *Sustainability* **2019**, *11*, 464. [CrossRef]

3. Bienvenido-Huertas, D.; Rubio-Bellido, C.; Pérez-Ordóñez, J.L.; Martínez-Abella, F. Estimating adaptive setpoint temperatures using weather stations. *Energies* **2019**, *12*, 1197. [CrossRef]

4. Sánchez-García, D.; Rubio-Bellido, C.; Tristancho, M.; Marrero, M. A comparative study on energy demand through the adaptive thermal comfort approach considering climate change in office buildings of Spain. *Build. Simul.* **2020**, *13*, 51–63. [CrossRef]

5. Wright, T.S.; Wilton, H. Facilities management directors' conceptualizations of sustainability in higher education. *J. Clean. Prod.* **2012**, *31*, 118–125. [CrossRef]

6. Hodgson, M.; Brodt, W.; Henderson, D.; Loftness, V.; Rosenfeld, A.; Woods, J.; Wright, R. Needs and opportunities for improving the health, safety, and productivity of medical research facilities. *Environ. Health Perspect.* **2000**, *108* (Suppl. 6), 1003–1008. [CrossRef]

7. Rastegar, M.; Fotuhi-Firuzabad, M.; Zareipour, H. Home energy management incorporating operational priority of appliances. *Int. J. Electr. Power Energy Syst.* **2016**, *74*, 286–292. [CrossRef]

8. *ANSI/ASHRAE/IES Standard 100-2015, Energy Efficiency in Existing Buildings*; ANSI/IES/ASHRAE: Atlanta, GA, USA, 2015.

9. *ANSI/ASHRAE/IES Standard 55-2015, Developing Energy performance Targets for ASHRAE Standard*; ANSI/IES/ASHRAE: Oak Ridge, TN, USA, 2015.

10. The Energy Efficiency Directive. (2012/27/EU). Available online: https://ec.europe.eu/energy (accessed on 22 December 2019).

11. Krarti, M. *Energy Audit of Building Systems: An Engineering Approach*; CRC Press: Boca Raton, FL, USA, 2016.

12. Lara, R.A.; Pernigotto, G.; Cappelletti, F.; Romagnoni, P.; Gasparella, A. Energy audit of schools by means of cluster analysis. *Energy Build.* **2015**, *95*, 160–171. [CrossRef]

13. AlFaris, F.; Juaidi, A.; Manzano-Agugliaro, F. Improvement of efficiency through an energy management program as a sustainable practice in schools. *J. Clean. Prod.* **2016**, *135*, 794–805. [CrossRef]

14. Henderson, K.; Tilbury, D. *Whole School Approaches to Sustainability: An International Review of Sustainable School Programs*; Macquarie University: Sydney, NSW, Australia, 2004.

15. Serrano-Guerrero, X.; Escrivá-Escrivá, G.; Roldán-Blay, C. Statistical methodology to assess changes in the electrical consumption profile of buildings. *Energy Build.* **2018**, *164*, 99–108. [CrossRef]

16. Roldán-Blay, C.; Escrivá-Escrivá, G.; Roldán-Porta, C. Improving the benefits of demand response participation in facilities with distributed energy resources. *Energy* **2019**, *169*, 710–718. [CrossRef]

17. Escrivá-Escrivá, G.; Roldán-Blay, C.; Roldán-Porta, C.; Serrano-Guerrero, X. Occasional Energy Reviews from an External Expert Help to Reduce Building Energy Consumption at a Reduced Cost. *Energies* **2019**, *12*, 2929. [CrossRef]

18. Wang, K.; Wang, Y.; Sun, Y.; Guo, S.; Wu, J. Green industrial Internet of Things architecture: An energy-efficient perspective. *IEEE Commun. Mag.* **2016**, *54*, 48–54. [CrossRef]

19. Mathew, P.A.; Dunn, L.N.; Sohn, M.D.; Mercado, A.; Custudio, C.; Walter, T. Big-data for building energy performance: Lessons from assembling a very large national database of building energy use. *Appl. Energy* **2015**, *140*, 85–93. [CrossRef]

20. Khoshbakht, M.; Gou, Z.; Dupre, K. Energy use characteristics and benchmarking for higher education buildings. *Energy Build.* **2018**, *164*, 61–76. [CrossRef]

21. Mills, E.; Bell, G.; Sartor, D.; Chen, A.; Avery, D.; Siminovitch, M.; Greenberg, S.; Marton, G.; de Almeida, A.; Lock, L.E. *Energy Efficiency in California Laboratory-Type Facilities*; LBNL Report #39061; Ernest Orlando Lawrence Berkeley National Laboratory: Berkeley, CA, USA, 1996. Available online: https://www.aivc.org/sites/default/files/airbase_11236.pdf (accessed on 20 December 2019).

22. Chen, Y.-T. The Factors Affecting Electricity Consumption and the Consumption Characteristics in the Residential Sector—A Case Example of Taiwan. *Sustainability* **2017**, *9*, 1484. [CrossRef]

23. Pickering, C.; Byrne, J. The benefits of publishing systematic quantitative literature reviews for PhD candidates and other early-career researchers. *High. Educ. Res. Dev.* **2014**, *33*, 534–548. [CrossRef]

24. Aemet. Spanish State Meteorological Agency (Agencia Estatal de Meteorología en España). Available online: http://www.aemet.es/ (accessed on 20 December 2019).

25. Manzano-Agugliaro, F.; Montoya, F.G.; Sabio-Ortega, A.; García-Cruz, A. Review of bioclimatic architecture strategies for achieving thermal comfort. *Renew. Sustain. Energy Rev.* **2015**, *49*, 736–755. [CrossRef]

26. Kipping, A.; Trømborg, E. Modeling aggregate hourly energy consumption in a regional building stock. *Energies* **2017**, *11*, 78. [CrossRef]

27. Cheng, V.; Steemers, K. Modelling domestic energy consumption at district scale: A tool to support national and local energy policies. *Environ. Model. Softw.* **2011**, *26*, 1186–1198. [CrossRef]

28. Fumo, N.; Biswas, M.R. Regression analysis for prediction of residential energy consumption. *Renew. Sustain. Energy Rev.* **2015**, *47*, 332–343. [CrossRef]

29. Huizenga, C.; Liere, W.; Bauman, F.; Arens, E. *Development of Low-Cost Monitoring Protocols for Evaluating Energy Use in Laboratory Buildings*; Center for Environmental Design Research; University of California: Berkeley, CA, USA, 1996.

30. Federspiel, C.; Zhang, Q.; Arens, E. Model-based benchmarking with application to laboratory buildings. *Energy Build.* **2002**, *34*, 203–214. [CrossRef]

31. Chen, J.; Taylor, J.E.; Wei, H.H. Modeling building occupant network energy consumption decision-making: The interplay between network structure and conservation. *Energy Build.* **2012**, *47*, 515–524. [CrossRef]

32. Jafary, M.; Wright, M.; Shephard, L.; Gomez, J.; Nair, R.U. Understanding Campus Energy Consumption–People, Buildings and Technology. In Proceedings of the 2016 IEEE Green Technologies Conference (GreenTech), Kansas City, MO, USA, 6–8 April 2016; pp. 68–72.

33. Cox, R.A.; Drews, M.; Rode, C.; Nielsen, S.B. Simple future weather files for estimating heating and cooling demand. *Build. Environ.* **2015**, *83*, 104–114. [CrossRef]

34. Erickson, V.L.; Carreira-Perpiñán, M.Á.; Cerpa, A.E. OBSERVE: Occupancy-based system for efficient reduction of HVAC energy. In Proceedings of the 10th ACM/IEEE International Conference on information processing in sensor networks, Chicago, IL, USA, 12–14 April 2011; pp. 258–269.

35. Khan, A.; Nicholson, J.; Mellor, S.; Jackson, D.; Ladha, K.; Ladha, C. Occupancy monitoring using environmental & context sensors and a hierarchical analysis framework. In Proceedings of the BuildSys 2014: 1st ACM International Conference on Embedded Systems for Energy-Efficient Buildings, Memphis, TN, USA, 4–6 November 2014; pp. 90–99. [CrossRef]

36. Chen, J.; Ahn, C. Assessing occupants' energy load variation through existing wireless network infrastructure in commercial and educational buildings. *Energy Build.* **2014**, *82*, 540–549. [CrossRef]

37. Masoso, O.T.; Grobler, L.J. The dark side of occupants' behaviour on building energy use. *Energy Build.* **2010**, *42*, 173–177. [CrossRef]

38. Pilkington, B.; Roach, R.; Perkins, J. Relative benefits of technology and occupant behaviour in moving towards a more energy efficient, sustainable housing paradigm. *Energy Policy* **2011**, *39*, 4962–4970. [CrossRef]

39. Chang, W.K.; Hong, T. Statistical analysis and modeling of occupancy patterns in open-plan offices using measured lighting-switch data. *Build. Simul.* **2013**, *6*, 23–32. [CrossRef]

40. Lü, X.; Lu, T.; Kibert, C.J.; Viljanen, M. Modeling and forecasting energy consumption for heterogeneous buildings using a physical–statistical approach. *Appl. Energy* **2015**, *144*, 261–275. [CrossRef]

41. Montoya, F.G.; Aguilera, M.J.; Manzano-Agugliaro, F. Renewable energy production in Spain: A review. Renewable and Sustainable Energy Reviews, 33, 509–531. *Renew. Sustain. Energy Rev.* **2014**, *33*, 509–531. [CrossRef]

Article

Sustainable Thermal Energy Generation at Universities by Using Loquat Seeds as Biofuel

Miguel-Angel Perea-Moreno [1], Francisco Manzano-Agugliaro [2],
Quetzalcoatl Hernandez-Escobedo [3] and Alberto-Jesus Perea-Moreno [1,*]

[1] Departamento de Física Aplicada, ceiA3, Campus de Rabanales, Universidad de Córdoba, 14071 Córdoba, Spain; k82pemom@uco.es
[2] Department of Engineering, ceiA3, University of Almeria, 04120 Almeria, Spain; fmanzano@ual.es
[3] Escuela Nacional de Estudios Superiores Juriquilla, Universidad Nacional Autonoma de México, Queretaro 76230, Mexico; qhernandez@unam.mx
*   Correspondence: aperea@uco.es; Tel.: +34-957-212633

Received: 31 December 2019; Accepted: 6 March 2020; Published: 9 March 2020

Abstract: Global energy consumption has increased the emission of greenhouse gases (GHG), these being the main cause of global warming. Within renewable energies, bioenergy has undergone a great development in recent years. This is due to its carbon neutral balance and the fact that bioenergy can be obtained from a range of biomass resources, including residues from forestry, agricultural or livestock industries, the rapid rotation of forest plantations, the development of energy crops, organic matter from urban solid waste, and other sources of organic waste from agro-food industries. Processing factories that use loquats to make products such as liqueurs and jams generate large amounts of waste mainly in the form of skin and stones or seeds. These wastes are disposed of and sent to landfills without making environmentally sustainable use of them. The University of Almeria Sports Centre is made up of indoor spaces in which different sports can be practiced: sports centre pavilion (central court and two lateral courts), rocodrome, fitness room, cycle inner room, and indoor swimming pool. At present, the indoor swimming pool of the University of Almeria (UAL) has two fuel oil boilers, with a nominal power of 267 kW. The main objective of this study is to propose an energetic analysis to determine, on the one hand, the energetic properties of the loquat seed and, on the other hand, to evaluate its suitability to be used as a solid biofuel to feed the boilers of the heated swimming pool of the University of Almeria (Spain), highlighting the significant energy and environmental savings obtained. Results show that the higher calorific value of loquat seed (17.205 MJ/kg), is like other industrial wastes such as wheat straw, or pistachio shell, which demonstrates the energy potential of this residual biomass. In addition, the change of the fuel oil boiler to a biomass (loquat seed) boiler in the UAL's indoor swimming pool means a reduction of 147,973.8 kg of CO_2 in emissions into the atmosphere and an annual saving of 35,739.5 €, which means a saving of 72.78% with respect to the previous fuel oil installation. A sensitivity analysis shows that fuel cost of base case is the variable with the most sensitivity changing the initial cost and net present value (NPV).

Keywords: loquat seed; sustainability; renewable energy; universities; biomass boiler

---

1. Introduction

Social and economic development, together with an increase in human welfare, has led to an increase in energy consumption. Energy services in societies become essential to satisfy the demands of light, heating, transport, communication, etc., as well as for the manufacture of goods. Fossil fuels have been the main source of energy since 1850, and since then there has been a growing demand worldwide [1]. This has led to a progressive increase in the levels of carbon dioxide in the atmosphere, reaching a record 415.70 ppm in May 2019 [2]. Such a concentration of CO_2 in the Earth's atmosphere had not been reached for more than three million years, when the global sea level was a few meters higher and Antarctica was partially covered with forests.

The provision of energy services to a world population in continuous growth has increased the emission of greenhouse gases (GHG), these being the main cause of global warming. In December 2015, the Paris Climate Change Conference, also known as COP21, was held in Paris. It established the world's first binding climate agreement, through which the 195 signatory countries establish a global action plan to keep global warming below 2 °C between 2020 and 2030, and below 1.5 °C if possible [3].

In order to tackle global warming, two types of strategies have been developed: those for mitigating climate change and those for adapting to it. Mitigation strategies aim to reduce greenhouse gas emissions in such a way that the point of no return is not reached, while adaptation strategies aim to reduce the consequences of an already apparent climate change and address its impact.

Climate change mitigation measures that can be taken to reduce pollutant emissions include increased energy efficiency, greater use of green energy within the global energy mix, electrification of industrial processes, promotion of more sustainable transport (electric mobility, cycling) or higher carbon taxes as well as a market for CO_2 emissions [4]. Therefore, it can be said that climate change is a global problem demanding a shift from the current energy model [5].

Measures to adapt to climate change that can be taken in order to reduce its effects include the construction of adapted housing and workplaces, reforestation, adaptation of crops to new agronomic variables, development of emergency plans to deal with possible natural disasters, or research into possible effects on human health [6].

Renewable energies have demonstrated their potential in mitigating climate change, but they also have other benefits. Such forms of energy favour local employment and thus contribute to economic and social development, facilitate access to energy, increase energy resilience, and reduce pollution in cities, and therefore the effects on human health [7]. The concept of renewable energy encompasses heterogeneous categories of technologies. Some renewable energy sources can provide electricity, others supply thermal or mechanical energy, and others can provide biofuels to meet various energy demands. Some renewable energy technologies can be adopted at the point of consumption (centralized) in rural and urban environments, while others are implemented mainly in large supply networks (decentralized). Although more and more technically advanced renewable energy technologies have been adopted on a medium scale, others are at a less advanced stage and have a more incipient commercial presence or supply specialised market niches.

Within renewable energies, bioenergy has undergone a great development in recent years. This is due to its carbon neutral balance and the fact that bioenergy can be obtained from a range of biomass resources, including residues from forestry, agricultural, or livestock industries, the rapid rotation of forest plantations, the development of energy crops, organic matter from urban solid waste, and other sources of organic waste from agro-food industries [8]. These residues can be burnt directly to produce electricity or heat or can be transformed by using physicochemical processes to generate gaseous, liquid or solid fuels. Bioenergy technologies are very diverse, and their degree of technical sophistication differs considerably. Some already marketed are small or large boilers, district heating systems, or the production of ethanol from sugar and starch. Bioenergy technologies have therefore applications in both, centralized and decentralized contexts, and their most widespread application is conventional use of biomass in industrialized countries for heating and power generation [9]. Bioenergy production is often constant or controllable. Bioenergy projects generally depend on locally and regionally available fuel, although recently there seem to be indications that solid biomass and liquid biofuels are increasingly present in international trade. Within centralized context and small-scale systems, biomass boilers are an emerging technology with numerous economic and environmental advantages [10].

Traditionally biomass boilers have been powered by pellets composed mainly of chips and sawdust from the wood industry [11]. Due to the increasing trend in the prices of biomass from the forest and wood industry, it has been necessary to investigate other sources of bioenergy, which allow lower costs while reducing the pressure on the agroforestry sector. Many investigations have shown the potential of certain fruit stones and nutshells in the generation of thermal energy at industrial and residential levels [12,13]. It is therefore necessary to find alternative biofuels as enacted by the European standard EN 14961-1.

Eriobotrya japonica, commonly called Japanese loquat, or simply loquat is a perennial fruit tree of the Rosaceae family, originating in south-eastern China, where it is known as "pi ba". It was introduced in Japan, where it was naturalized and has been cultivated for more than a thousand years. It was also naturalized in India, the Mediterranean Basin, Canary Islands, Pakistan, Argentina and many other areas. Today, Japanese loquat cultivation has spread throughout the world, both for its ornamental value and for its prized fruits. The loquat is cultivated mainly in China, Japan, India, Pakistan, Mediterranean countries (Spain, Portugal, Turkey, Italy, Greece, Israel), United States (California and Florida), Brazil, Venezuela, and Australia. According to data provided by the Food and Agriculture Organization (FAO), China, in addition to be its country of origin, is the world's largest producer. In the last ten years China has doubled its area of cultivation and production, reaching 118,270 ha. and 453,600 Tn per year [14]. Spain is the world's second largest producer and exporter of loquat with an annual production of around 30,000 tn. Andalusia has around 1,100 hectares of loquat spread between Granada (815 hectares) and Malaga (275 hectares) [15]. The consumption of this tropical fruit has gradually increased over the last two decades and a good performance of demand is forecast for the coming years. Figure 1 shows the loquat production in the main countries in the world.

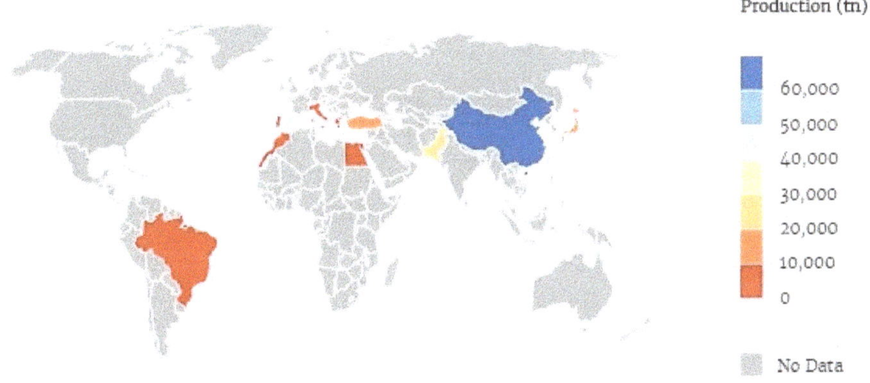

Figure 1. Loquat production in the main countries [16].

Processing factories that use loquats to make products such as liqueurs and jams generate large amounts of waste mainly in the form of skin and stones or seeds. Inside the loquat contains between 3 and 7 large brown seeds representing between 15% and 18% of the total weight of the fruit [17]. These wastes are disposed of and sent to landfills without making environmentally sustainable use of them.

The water in heated swimming pools needs an external supply of energy in order to maintain thermal comfort, as the natural tendency of the water will be to equalize the temperature of its environment. If the temperature of your environment is lower, the temperature of the water will decrease depending on the conditions of the environment such as: temperature of the air, of the walls and floor of the pool, etc. As a first consideration, we must bear in mind that the energy consumption of this type of installation can be higher than 700 MWh per year, depending on the climate of the place where it is located, the conditions and seasonality of the use, and the demand required by the systems included. Therefore, a small improvement in energy efficiency for proper operation translates into a big saving in global terms in CO_2 emissions to the atmosphere and the use of scarce energy resources.

It should be noted that not only is the demand very high, but in most cases, there is a priori ignorance of the amount of energy that will demand a heated pool [18–20].

The main objective of this study is to determine, on the one hand, the energetic properties of the loquat seed and, on the other hand, to evaluate its suitability to be used as a solid biofuel to feed the boilers of the heated swimming pool of the University of Almeria (Spain), highlighting the significant energy, economical, and environmental savings obtained, contributing to reduce greenhouses gases.

2. Case Study

As a case study has been taken a university sport centre located in the University of Almería (UAL), Southern Spain (Figure 2).

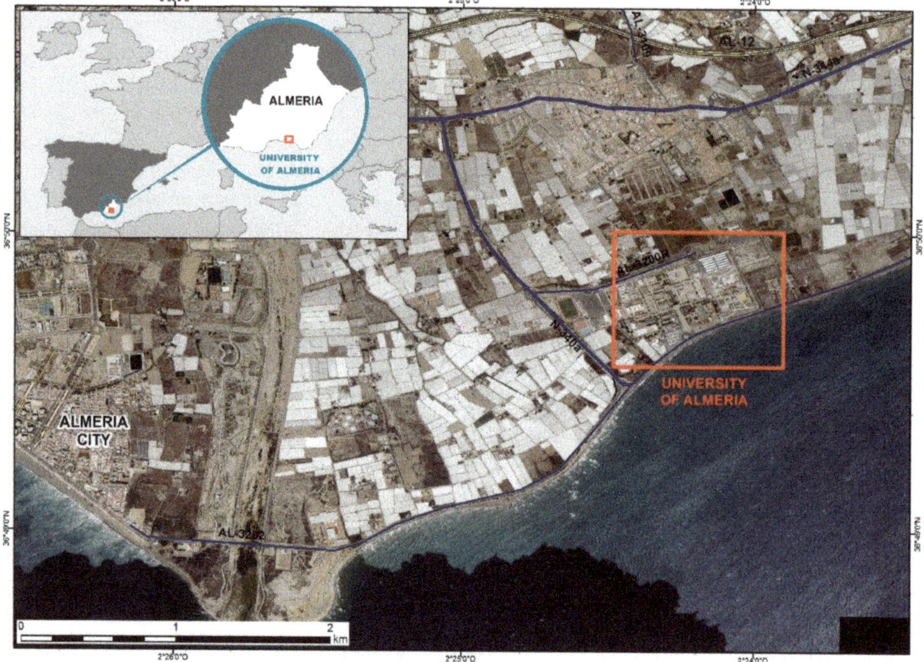

Figure 2. Location of University of Almeria.

The UAL Sports Centre is made up of indoor spaces in which different sports can be practised: sports centre pavilion (central court and two lateral courts), rocodrome, fitness room, cycle inner room and indoor swimming pool.

The water volume of the heated swimming pool is 494.15 m^3, has a compensation vessel of 30 m^3 and has an internal dimension of the water sheet of 12.5 × 25.10 m. Figure 3 shows the indoor swimming pool of the UAL Sports Centre.

Figure 3. Indoor swimming pool of the University of Almeria.

Firstly, an energy audit of the installation was carried out to establish a technical basis for this study. In addition, all the necessary data was collected to study the possibility of replacing the existing boiler with a biomass boiler using loquat seed as biofuel.

2.1. Meteorological Data

Almeria has a Mediterranean climate, with mild winters, hot summers, and little rain. The average annual temperature is 17.9 °C and the average rainfall is 228 mm. Table 1 shows the most important meteorological data of Almeria.

Table 1. Meteorological conditions in Almeria.

Parameters	Values
Longitude	2°24′23.3″ W
Latitude	36°49′40.9″ N
Altitude	16 m
Average maximum annual temperature	23.1 °C
Average minimum annual temperature	14.3 °C
Average annual temperature	17.9 °C
Winter dry temperature	8 °C
Summer dry temperature	31 °C

2.2. Description of Existing Thermal Facilities

When planning the heating of an indoor swimming pool, some fundamental differences must be considered compared to a residential building heating system: firstly, there is a high level of evaporation on the premises, and secondly, the comfort conditions for bathers are different.

According to current regulations, the temperature and relative humidity of the room must be adequate to protect the health of the users (RITE, Complementary Technical Instruction) [21]. As for the temperature of the ambient air, the water temperature, and the environmental humidity, the following were taken as comfort conditions:

- Setpoint water temperature: 28 °C
- Maximum water temperature: 29 °C
- Air temperature: 29 °C
- Relative Humidity: 60%.
- Daily renewal of pool water: 2.5%.

At present, the indoor swimming pool of the University of Almeria has two fuel oil boilers, with a nominal power of 267 kW. One of them is a backup in case of failure.

The current installation has the following auxiliary elements:

- Two tanks of 2000 litres.
- Main hydraulic circuit of impulsion and return of steel pipe of 1′.
- Boiler room with dimensions of 40 m^2.
- Fuel oil tank of 3000 litres.
- The facility of the indoor swimming pool is presented in Figure 4.

Figure 4. Boiler room of the indoor swimming pool.

This installation has an annual consumption of 52,239 liters, as Table 2 shows.

Table 2. Annual consumption of the indoor swimming pool boiler for the year 2018.

Month	Fuel Consumption (L)
January	7582
February	8370
March	8110
April	6434
May	4473
June	2186
July	1018
September	671
October	3477
November	3926
December	5992
Total	52,239

In Spain, Royal Decree 742/2013 [22] classifies swimming pools with respect to public access and classifies them into public and private use swimming pools. In relation to the temperature of the air and of the bath water, heated or covered swimming pools are defined as those in which the enclosure where the glasses are located is closed, has a fixed structure, and the water is kept at a more or less hot temperature.

Figure 5 shows the operating diagram of the fuel oil boiler.

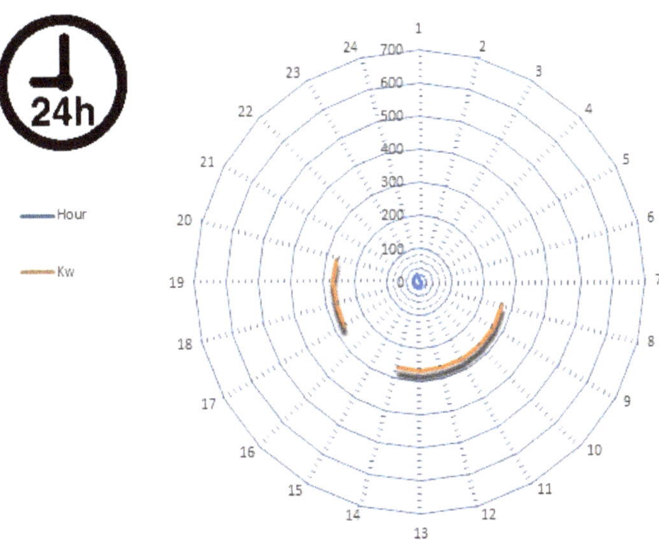

Figure 5. Operating diagram of the fuel boiler..

3. Materials and Methods

Regarding the use as biofuel of loquat seeds from agro-food industries, 2000 g of such seeds were collected for their subsequent energy and chemical composition analysis (Figure 6).

Figure 6. Eriobotrya japonica, commonly called Japanese loquat.

UNE-EN 14961-1 standard "Solid biofuels–Specifications and fuel classes—Part 1: General requirements", were used to assign the biomass quality parameters. This standard has been developed by the Spanish Association for Standardisation and Certification (AENOR).

Table 3 shows the standards and the measuring equipment used.

Table 3. Standards and the measuring equipment used in this study.

Parameter	Unit	Standards	Equipment	Standard Deviation (SD)
Higher heating value	MJ/kg	EN 14918	Calorimeter Parr 6300	0.02
Total sulphur	%	EN 15289	Analyzer LECO TruSpec S 630-100-700	0.002
Total hydrogen	%	EN 15104	Analyzer LECO TruSpec CHN 620-100-400	0.03
Total chlorine	mg/kg	EN 15289	Titrator Mettler Toledo G20	6.73
Total carbon	%	EN 15104	Analyzer LECO TruSpec CHN 620-100-400	0.12
Total nitrogen	%	EN 15104	Analyzer LECO TruSpec CHN 620-100-400	0.009
Ash	%	EN 14775	Muffle Furnace NABERTHERM LVT 15/11	0.02
Moisture	%	EN 14774-1	Drying Oven Memmert UFE 700	According to EN14774-1

3.1. Humidity

The moisture content of the biomass is the ratio of the mass of water contained per kilogram of dry matter. An excess of humidity in the biomass, leads to [23]:

-       A large amount of volatile elements, which offer a loss in energy efficiency.
-       A low calorific value, which would call into question the expectations regarding the replacement
        of other fuels.
-       Ashes in large quantities. This can cause equipment cleaning problems.
-       Boilers would suffer continuous problems, affecting the durability of their life.
-       Have much more storage volume for biomass.

In order to avoid these problems, a solar greenhouse dryer system for loquat seed improvement as biofuel is proposed.

The solar greenhouse dryer built in this study was the feature on conventional greenhouse type tunnel with north-south facing (Figure 7). Figure 8 shows the dimensions of this greenhouse dryer.

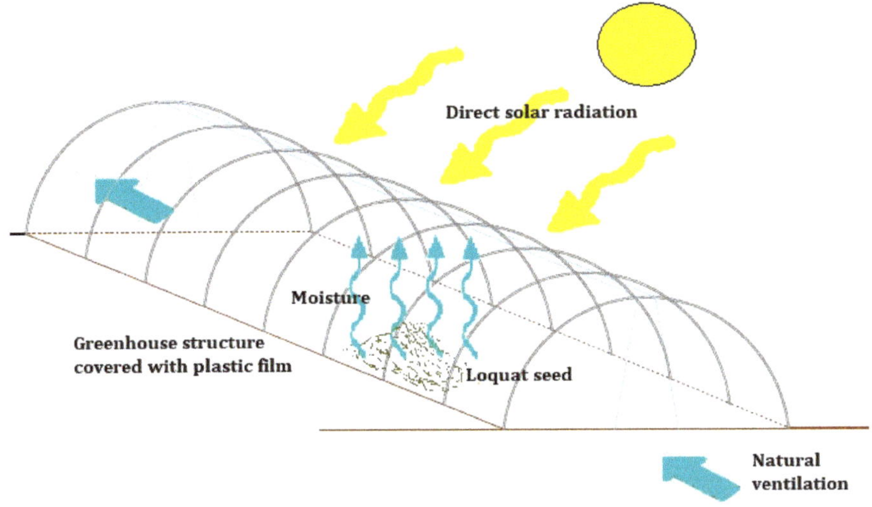

Figure 7. Scheme of the greenhouse dryer.

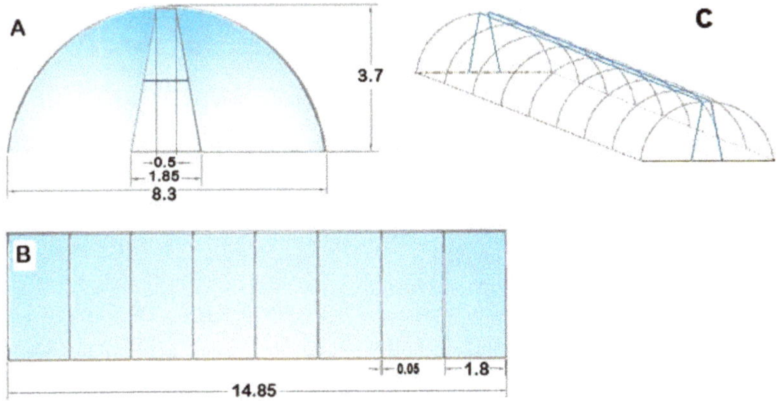

Figure 8. Dimensions of the greenhouse dryer proposed.

The drying of the loquat seeds in the solar dryer started on 1 October 2019 and was completed on the 17th of the following month. The moisture content was measured using Extech hygrometer M0210.

Before introducing the loquat seeds into the solar dryer to evaluate its operation, the air temperature and relative humidity levels were recorded during a week at different times of the day with an Extech RHT20 recorder, in order to know precisely their magnitude and the conditions under which the seeds would be exposed.

In order to show solar monthly resource and air temperature, Figure 9 is presented.

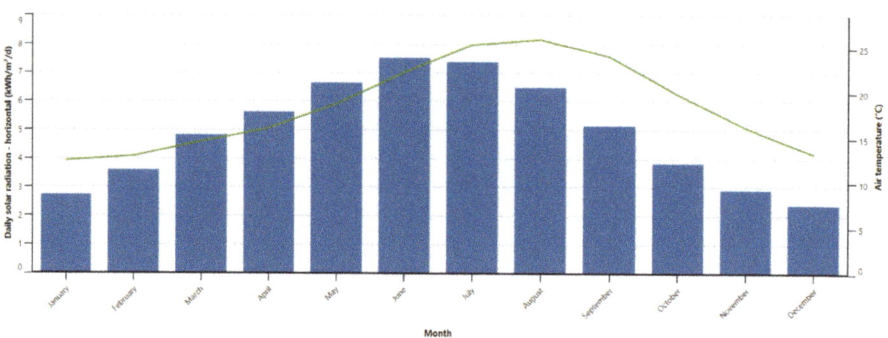

Figure 9. Relationship between monthly solar radiation and air temperature in Almeria.

As seen in Figure 9, in Almería the lowest solar radiation is January and December with 4 kWh/m^2/d, the months with the highest solar resource are July and August with more than 8 kWh/m^2/d.

3.2. Elemental Composition

The elemental analysis technique consists of determining the content of hydrogen, nitrogen, carbon and sulphur present in samples of an organic and inorganic nature, both solid and liquid. The UNE-EN 15104 standard has been consider in this analysis and the analyzer LECO TruSpec CHN 620-100-400 was used.

3.3. Ash Content

The determination of ashes will be based on UNE EN 14775 standard. Ashes are the residue of air incineration of coal and come from the inorganic compounds initially present in carbonaceous substances and associated mineral materials.

The ash content is expressed as the percentage ratio between the mass of the residue after incineration and the mass of the original sample.

The percentage of ash in biofuels is an important parameter to consider in boilers, since it can produce corrosion and accelerated erosion of the metal that makes up each of the elements of the boiler.

3.4. Chlorine and Sulfur Content

Titrator Mettler Toledo G20 was used to obtain the chlorine content and Analyzer LECO TruSpec S 630-100-700 to obtain sulfur content and the standard UNE EN 15289 was followed.

Riedl et al. [24] describe the corrosion mechanism in the presence of chlorine as "active oxidation" and assign to it the accelerated corrosion rate observed in boiler tubes. The authors describe the enrichment of alkaline metal chlorides on the surface of pipes by a condensation process. They also suggest that the chlorides react with SO_2 and SO_3 in the gases to form sulphates with the subsequent generation of chlorine gas.

An important study on this subject has been carried out in Denmark where accelerated corrosion tests were conducted on boilers that were using fuels such as straw and cereals with ash levels in the range of 5%–7%, chlorine levels between 0.3% and 0.5%, and sulfur content less than 0.2%, on a dry basis [25].

3.5. Higher Heating Value (HHV) and Lower Heating Value (LHV)

The higher heating value (HHV) is defined assuming that all elements of the combustion (fuel and air) are taken to 0 °C and the products (combustion gases) are also taken to 0 °C after the combustion, so the water vapor will be completely condensed.

In the calculation of the LHV, it is assumed that the water vapor contained in the flue gas does not condense and therefore there is no additional heat input due to the condensation of the water vapor. Only the heat of oxidation of the fuel is available. While determining the HHV experimentally in a calorimeter, the LHV is obtained by a calculation from the HHV [26–28].

$$\mathrm{LHV}\left(\frac{kJ}{kg}\right) = \mathrm{HHV}\left(\frac{kJ}{kg}\right) - 212.2 \times \mathrm{H}\,\% - 0.8 \times (\mathrm{O}\% + \mathrm{N}\%) \tag{1}$$

3.6. Biomass Combustion Technology

According to Wolf & Dong [29], three types of technologies for biomass combustion can be distinguished: fixed bed, fluidized bed (which can be either circulating or bubbling bed), and pulverized fuel. In our case study, considering the characteristics of the biofuel and the boiler output, fixed bed technology will be chosen for the direct combustion of the biomass once it has been dried.

Fixed-bed combustion includes furnaces with grates and feeders (stokers). The biomass is placed on a grate and moves slowly through the boiler. The required air is supplied through holes arranged along the grate. The combustible gases issued by the biomass are burned after the addition of a secondary air, usually in a combustion zone separated from the fuel bed. This technology is suitable for any type of biomass but limited to installations of up to 100 MW.

An important aspect of the grate furnace is that the combustion stages must be obtained by separating the primary and secondary combustion chambers, in order to avoid the mixing of secondary air currents and to separate the gasification and oxidation zones. The better the quality of the mixture between the combustion gases and the secondary combustion air, the less oxygen is needed to achieve complete combustion, thus achieving greater efficiency.

According to the direction of flow of the fuel and flue gases, there are three operating systems for grate-fired boilers:

1. Counter-stream flow (the flames are in the opposite position to the fuel).
2. Stream flow (the flames are in the same direction as the fuel).
3. Cross flow (the removal of the combustion gases in the middle of the furnace).

In our case study the stream flow system will be chosen. The flow in stream is applied for dry fuels or in systems where primary heated air is used. This system increases the residence time of the gases released by the fuel bed, allowing for a reduction in NO_x emissions, due to the improved contact of the combustion gases with the bed of carbonized material at the back of the grills [30].

3.7. Economic Analysis

The economic analysis has been performed at initial conditions of 267 kW as rated power, loquat properties compared to oil fuel. All these processes have been carried out with RETScreen software, which is a clean energy management software system for energy efficiency, renewable energy, and cogeneration project feasibility analysis, as well as ongoing energy performance analysis, developed by the Natural Resource of Canada office [31].

4. Results and Discussion

The physico-chemical parameters of loquat seed were analysed and compared with those of other sources of residual biomass in order to evaluate its usefulness as a solid biofuel. Afterwards, an energy, economic and environmental analysis of the installation was carried out.

4.1. Loquat Seed Values

To evaluate the quality parameters of loquat seed, 2000 g of samples from the loquat industry were analysed. Table 4 shows the mean value, the standard deviation, the maximum value, and the minimum value which determine the parametric distribution.

Table 4. Energy and chemical parameters obtained from loquat seed analysis (parameters calculated on a dry basis except for moisture).

Magnitude	Unit	Mean Value	Standard Deviation (SD)	Maximum Value	Minimum Value
Moisture	%	37.53	—	37.53	37.53
Ash content	%	2.37	0.060	2.43	2.31
HHV	MJ/kg	17.205	0.018	17.223	17.187
LHV	MJ/kg	16.007	0.090	16.097	15.917
Total carbon	%	44.03	0.006	44.036	44.024
Total hydrogen	%	5.47	0.011	5.481	5.459
Total nitrogen	%	0.63	0.037	0.667	0.593
Total sulfur	%	0.03	0.002	0.032	0.028
Total oxygen	%	46.57	2.651	49.221	43.919
Total chlorine	%	0.07	0.003	0.073	0.067

As can be seen from Table 4, one of the main disadvantages of loquat seed is its high moisture content, above 30%. This decreases the efficiency of combustion, since water needs to evaporate before heat is available, resulting in a lower heating value. Furthermore, from a technical point of view, the presence of a high moisture content produces corrosion in the equipment and generates the emission of tars that accumulate in the outlet pipes and can cause them to block. As a result, pre-drying processes must be implemented in order to obtain a moisture content below 10%.

Ash is the inorganic fraction of the biomass that remains once the fuel has been burned, mainly in the form of SiO_2 and CaO. The formation of ashes is linked to problems such as the formation of agglomerates on the walls of the grids and the formation of slag deposits which accelerate the corrosion of the installation and increase its maintenance costs. With respect to the ash content of the loquat seed, it is in the range of 2.31%–2.43%. If this value is compared with that of other standardized fuels, such as almond shell pellets (3.35%) or oak pellets (3.32%), it can be seen that in spite of its high value, it is below the ash content of other conventional fuels.

Table 5 compares the physicochemical parameters of several commercial biofuels and industrial wastes with those of loquat seed in order to evaluate the use of this by-product in the generation of thermal energy.

Table 5. Comparison between Loquat seed and other biofuels.

Parameters	Unit	Avocado Stone [32]	Olive Stone [33–35]	Pine Pellets [35,36]	Almond Shell [35,37,38]	Loquat Seed
Moisture	%	35.20	18.45	7.29	7.63	37.53
HHV	MJ/kg	19.145	17.884	20.030	18.200	17.205
LHV	MJ/kg	17.889	16.504	18.470	17.920	16.007
Ash content	%	2.86	0.77	0.33	0.55	2.37
Total carbon	%	48.01	46.55	47.70	49.27	44.03
Total hydrogen	%	5.755	6.33	6.12	6.06	5.47
Total nitrogen	%	0.447	1.810	1.274	0.120	0.63
Total sulphur	%	0.104	0.110	0.004	0.050	0.03
Total oxygen	%	42.80	45.20	52.30	44.49	46.57
Total chlorine	%	0.024	0.060	0.000	0.01	0.07
$\frac{HHV_{biomass}}{HHV_{loquat\ seed}}$	%	111.27	103.95	116.42	105.78	100.0

As shown in Table 5, the ash content of loquat seed is higher than that of other industrial waste but lower than the ash content of the mango stone. This suggests that the formation of ash deposits will be greater and therefore will need additional maintenance.

As for the higher calorific value of loquat seed (17.205 MJ/kg), it is observed that it is lower than that of other standardized solid biofuels such as almond shells or olive stones. However, if we compare it with other industrial wastes such as wheat straw (17.344 MJ/kg) or pistachio shell (17.348 MJ/kg) [31], it is shown to have similar values, showing the energy viability of this agro-industrial waste.

Another quality parameter to be considered in a biofuel is its chlorine and sulphur content. Sulphur, in addition to having a corrosive effect on the installation, is associated with the emission of greenhouse gases in the form of SO_x. Loquat seed has a 0.03% sulphur content that is lower than that of other commercially available biofuels such as almond shells (0.050%) or olive stones (0.110%), which means that SO_x emissions would be minimised if this biofuel were used.

Regarding the chlorine content, this has a significant effect on the corrosiveness of the plant due to the formation of chlorides, which have a dissociative catalytic effect on the steel pipes. Loquat seed has a low chlorine content (0.07%), like that of other standardized biofuels such as olive stone (0.06%) and lower than that of almond shells pellets (0.2%). In view of these results, corrosion problems would be minimized by using this solid biofuel.

However, in addition to its high calorific value, the main advantage of using loquat seed as biofuel is its carbon-neutral character. In fact, when the plant grows it fixes carbon from the atmosphere which is then released when it is burned, making the life cycle carbon neutral.

This residual biomass can be obtained from loquat processing industries in the surroundings, helping on the one hand to a better environmental management of these wastes and on the other hand to a reduction of greenhouse gas emissions.

4.2. Environmental Benefits

Biomass as an energy source has several advantages over other alternatives in the fight against climate change and local pollution. Biomass is a source of non-polluting energy. Plants emit CO_2 but also absorb CO_2 during their growth, so their total balance is zero (Figure 10).

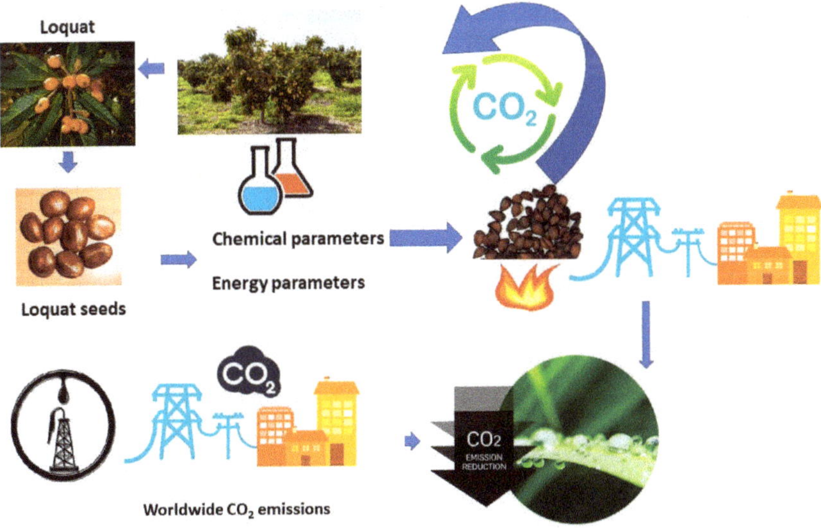

Figure 10. CO_2 cycle using loquat seed as biofuel.

Once the energy characteristics of the loquat seed are known, it is possible to calculate the CO_2 savings in the installations of the UAL indoor swimming pool, as well as the worldwide CO_2 savings in loquat producing countries.

Firstly, the potential energy obtained from the use of loquat seed as a biofuel is calculated using Equation (2). This potential energy is calculated considering the worldwide production of loquat for each country.

$$U_p = RH \times P_{loquat\ seed} \times HHV \times f_s \times F_c \tag{2}$$

where:

U_p denotes energy obtained from the loquat seed as biofuel (MWh);

RH is relative humidity (10%);

$P_{loquat\ seed}$: loquat seed production (kg);

HHV: higher heating value (17.205 MJ/kg);

f_s is the percentage of seed in a whole loquat (15%);

F_c factor conversion for units (0.000277778 Wh/J).

Figure 11 shows worldwide bioenergy potential using loquat seed as biofuel (MWh).

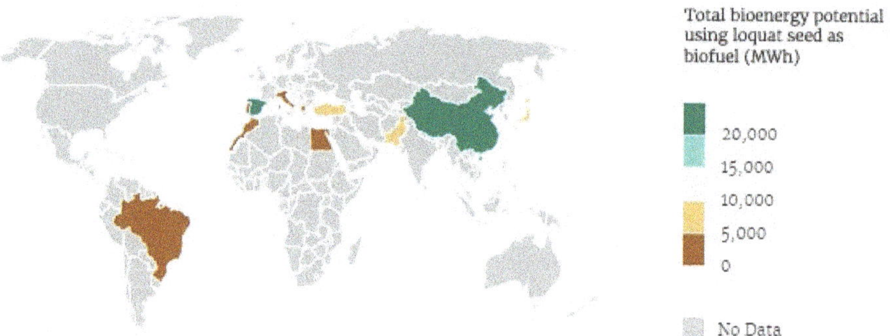

Figure 11. Worldwide bioenergy potential using loquat seed as biofuel (MWh).

The next step in the study will be to calculate the CO_2 reductions that would occur if loquat seeds were used as a biofuel instead of fuel oil.

As mentioned above, Biomass is a source of non-polluting energy, with zero CO_2 emissions. The CO_2 emission of fuel oil would be calculated as:

$$MCO_{2\ fuel\ oil} = C_{fuel\ oil} \times F_{fuel\ oil} \tag{3}$$

where:

$MCO_{2\ fuel\ oil}$: mass of carbon dioxide emitted (kg/year).

$C_{fuel\ oil}$: consumption of fuel oil per year (kWh/year).

$F_{fuel\ oil}$: carbon dioxide emission factor of fuel oil (kg/kWh).

Table 6 shows the CO_2 emission factor for biomass and fuel and the total CO_2 emission reduced annually using loquat seed as biofuel.

Table 6. CO_2 emission factor of Fuel oil and Loquat seed [39].

Boiler	CO_2 Emissions (kg/kWh)
Fuel oil	0.311
Loquat seed	0
Total CO_2 emission reduced annually (kg)	228,031.70

The change of the boiler to biomass in UAL indoor swimming pool means a reduction of 147,973.8 kg CO_2 in emissions into the atmosphere.

Figure 12 shows the worldwide CO_2 saving using loquat seed as biofuel (Tn).

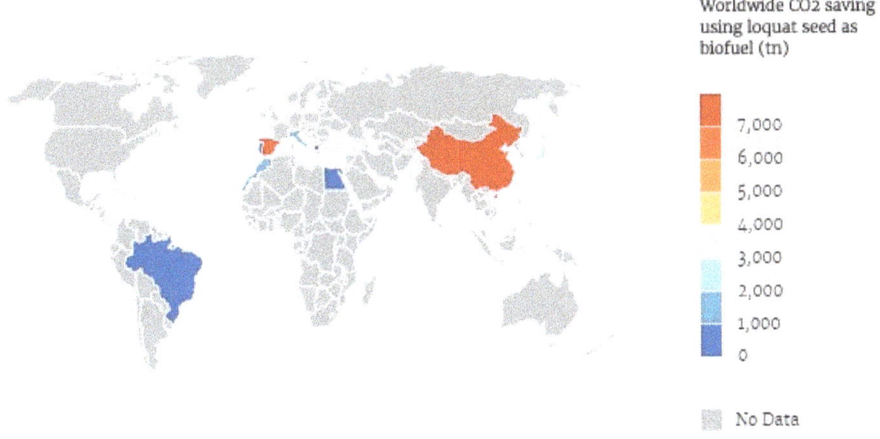

Figure 12. Worldwide CO_2 saving using loquat seed as biofuel (Tn).

The five main countries that would reduce their annual CO_2 emissions by using loquat seeds as biofuel are: China (51,184.92 Tn), Spain (10,617.54 Tn), Turkey (3454.98 Tn), Pakistan (3275.83 Tn), and Japan (2621.95 Tn). Further, the annual worldwide reduction of CO_2 emissions if loquat seed is used as biofuel would be 76,363.80 tn.

In a society committed to sustainable development, the use of biomass for heat and electricity production is an important source of renewable energy. Its increasing use as a substitute for fossil fuels can significantly reduce CO_2 emissions. However, in recent years discussion has focused on the sustainability of biomass, and the environmental implications of its use must be taken into account.

Particularly important is the particle size of the emissions that are produced, which have been identified as a relevant factor of the deterioration of air quality. Not so long ago, the existing legislation for emission control in combustion plants only took into account total particles with a diameter of less than 10 micrometers (PM_{10}). However, particles with a diameter of less than 5 micrometers (PM_5) and especially those with a diameter of less than 2.5 micrometers ($PM_{2.5}$) have the most harmful effect on health. Fine particles ($PM_{2.5}$) emitted during biomass combustion can be divided into three groups, based on their chemical composition and morphology: particles spherical organic carbon particles, sooty aggregate particles and inorganic ash particles. Because of their small size, these particles are able to reach the pulmonary alveoli and pass into the bloodstream. This is associated with an increased risk of respiratory and cardiovascular disease, especially when the environmental concentration of these particles exceeds 35.4 $\mu g/m^3$ [40–43].

This has led to a boost in research activity, both in the characterisation of emissions and in the development of control equipment, and in enacting legislation at national and European level.

For our case study, the R.D. 1073/2002 transposing the Directive 1999/30/EC on air quality, establishes in relation to PM_{10}, that the limit value of 50 $\mu g/m^3$ must not be exceeded in 24 h for more than 35 days, on the date of entry into force of 1 January 2005.

In the year 2015, new European legislation has been published (Directive (EU) 2015/2193 of the European Parliament) that involves air emissions from the combustion of solids, such as biomass, in equipment and installations with a rated thermal input of more than 1 MW and less than 5 MW, and that makes a big impact on particle emissions.

Conversely, the European Directive 2009/125/EC establishes a framework for setting ecodesign requirements for energy-related products. The national transposition of this regulation is the Royal Decree 187/2011 of 18 February.

The result of this directive is Regulation 2015/1189 of 28 April on solid fuel and wood biomass boilers of nominal output not exceeding 500 kW, which is mandatory from 1 January 2020. The environmental aspects considered important in this regulation are energy consumption and emissions generated by particulate matter (PM), organic gaseous compounds (OGC), carbon monoxide (CO), and nitrogen oxides (NO_x) in the use phase of this equipment. This regulation stipulates that seasonal particle emissions from heating may not exceed 40 mg/m^3 for automatically fed boilers and 60 mg/m^3 for manually fed ones.

In order to reduce the amount of particles emitted into the atmosphere due to the incomplete combustion of the biomass, modifications have been made to combustion equipment in recent years. In this way, for example, special emphasis has been placed on the modulation of the equipment to adapt to thermal demand, lambda probes have been used to ensure control of the most appropriate fuel-air ratio according to operating conditions and secondary and tertiary air has been introduced in different parts of the equipment.

However, the permitted emission limits are becoming increasingly restrictive, making it necessary to use of equipment that can be coupled to the stoves and boilers of biomass in order to reduce the emission of particles. In this way, research is being carried out into different technologies that can be adapted to the residential sector, such as the introduction of additives with the biomass, the use of catalytic filters or the use of electrostatic precipitators.

In low power installations (up to 1 MW) the most effective and profitable solution is the use of an electrostatic filter [44]. Its operation is based on electrically charging the particles in order to direct them out of the gas towards plates with an opposite charge, to which they adhere. Therefore, in our case study an electrostatic filter will be installed consisting of a metal rod that rotates inside the metal chimney tube, carrying a voltage of 24,000 volts, so that it ionizes the solid particles, which are attracted by the walls of the chimney tube, where they accumulate until they fall by gravity into the equipment. This technology has proven to achieve efficiencies of over 90% in $PM_{2.5}$ abatement [45].

4.3. Economical Benefits

The economic feasibility of the study of changing the fuel boiler for loquat seed as a biofuel is based on the following:

- Annual hours of operation.
- Annual consumption of fuel oil and biomass.
- LHV of fuel oil and biomass to be used.
- Current prices of fuel oil and biomass to be used.

Starting with a 267 kW boiler that will work approximately 6 h a day with an average of 297 days, the energy required is 475,800 kWh.

The high price of fossil fuels, which is also heavily taxed in many countries, is boosting the market for biomass boilers for heating generation. In order to calculate the economic benefit to be gained from the new biomass installation compared to the original fuel oil installation, the necessary fuel expenditure in both scenarios has been calculated to cover the annual energy demand. As shown in Table 7, the annual fuel oil consumption of the existing facility during 2018 was 52,239 litres, and considering a fuel price of 0.94 euros/litre, a total annual cost of 49,104.67 euros is obtained.

Table 7. Economic analysis of fuel oil installation and the Loquat seed boiler.

Parameter	Unit	Fuel oil Boiler	Biomass Boiler
Fuel		Fuel oil	Loquat seed
LHV	kWh/L	10.12	
LHV	kWh/kg		4.45
Price	€/L	0.94	
Price	€/kg		0.10
Boiler efficiency	%	90	80
Nominal power	kW	267	267
Operating hours	H	1782	1782
Thermal energy demand	kWh/year	475,800	475,800
Fuel consumption	L	52,239	
Biomass consumption	Kg		133,651.7
Annual cost	€	**49,104.67**	**13,365.17**
Annual saving	€		35,739.5
Annual saving	%		72.78
Total investment	€		**69,540.35**

The annual thermal demand of the installation (kWh/year) can be calculated as the product of the amount of fuel consumed by the lower heating value (LHV) of the same, but considering the efficiency of the boiler, this is:

$$\text{Annual thermal demand} = \text{Fuel quantity} \times \text{LHV} \times \text{Boiler efficiency} \qquad (4)$$

Taking into account the necessary thermal demand of 475,800 kwh per year, with the new biomass boiler 133,651.7 kg of loquat seeds will be consumed, and taking as a reference price of this residual biomass a value of 0.1 €/kg already treated and transported, there would be an annual saving of 35,739.5 € which means a saving of 72.78% with respect to the previous fuel oil installation.

A sensitivity analysis has been carried out with a threshold of seven years, a range of 25% and the analysis is performed on equity payback, see Table 8.

Table 8. Sensitivity analysis based in equity payback.

		Fuel Cost- Proposed Case				
Initial Costs		10,023.88	11,694.52	13,365.17	15,035.82	16,706.46
€		−25.0%	−12.5%	0.0%	12.5%	25.0%
52,155	−25.0%	*0.3*	*0.3*	*0.3*	*0.5*	*0.5*
60,848	−12.5%	*> project*	*0.3*	*0.4*	*0.5*	*0.6*
69,540	0.0%	*> project*	*> project*	**0.4**	*0.6*	*0.6*
78,233	12.5%	*> project*	*> project*	*> project*	*0.6*	*0.6*
86,925	25.0%	*> project*	*> project*	*> project*	*> project*	*0.7*

Table 8 shows that fuel cost will increases in 0.4 in 7 years if costs do not increase, however, in the worst case the costs will increases in 0.7 times if fuel cost increase in 25%.

A risk analysis is performed on net present value (NPV) and 500 combinations. In Table 9 are presented the parameters and in Figure 13 is presented a tornado chart to identify the impact of NPV on these parameters.

Table 9. Parameters.

Parameter	Unit	Value	Range	Minimum	Maximum
Initial costs	€	69,540.35	25%	52,155	86,925
Fuel cost- proposed case	€	13,365.17	25%	10,024	16,706
Fuel cost- based case	€	49,104.67	25%	36,829	61,380.84
Debt ratio	%	70%	25%	52.50%	87.50%
Debt interest rate	%	7%	25%	5.25%	8.75%
Debt term	yr	15	25%	11	19

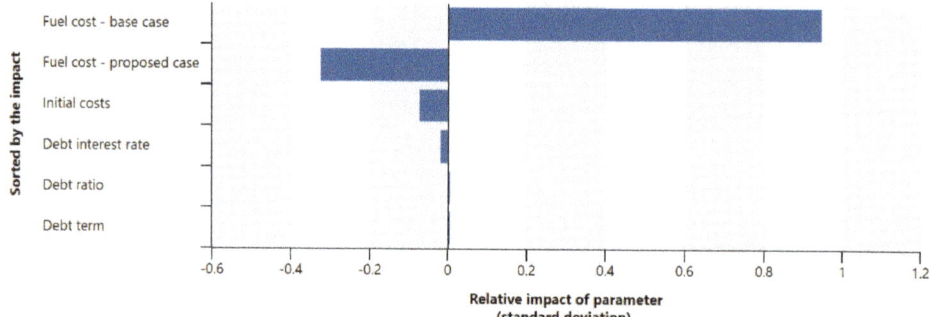

Figure 13. Impact – Net Present Value (NPV).

Figure 13 shows the economic impact of NPV on fuel cost of proposed case and base case, as it can be seen, the highest sensitivity variable is fuel cost-base case (fuel oil); debt interest rate, debt ratio and debt term have very small effects and can be ignored their uncertainty.

Financial viability presents the results provide to the decision-maker with various financial indicators for the proposed case, see Table 10.

Table 10. Financial viability.

Parameter	Unit	Value
Pre-tax-IRR-assets	%	20.6
Simple payback	yr	1.3
Net Present Value	€	162,013
Annual life cycle savings	€/yr	17,748
Benefit-Cost (B-C) ratio		25.1

The internal rate return (IRR) is 20.6%, which is higher than European central bank interest rate (0.25%), payback is 1.3 years, and B-C shows that investment has financial viability.

4.4. Biomass Storage

The Thermal Installations Regulation (2007) [46] in Spain describes a few essential requirements for solid biofuel storage systems (IT.1.3.4.1.4 Solid biofuel storage). In general, the storage site must be exclusively for this use, and this does not change the physical, chemical and mechanical characteristics of the biomass. It must also comply with a series of requirements to prevent the risk of self-combustion. The types of storage can be divided into prefabricated or built storage, either new construction or existing room, either above or below ground. In new buildings, the minimum storage capacity for biofuel will be sufficient to cover two weeks' consumption. The silo can be filled semi-automatically, with direct loading or with a pneumatic system, depending on the type of solid biofuel and size of the silo.

In the case of this study, the existing deposit room will be adapted as a silo, making some slopes of galvanized sheet and having an extraction system of about 2 m. This silo will have a floor area of 3.4 m^2 and a height of 2 m as defined in Figure 14. Therefore, a volume of 6.8 m^3 is available, which complies with the 15-day minimum fuel supply (6.18 m^3 of Loquat seeds required).

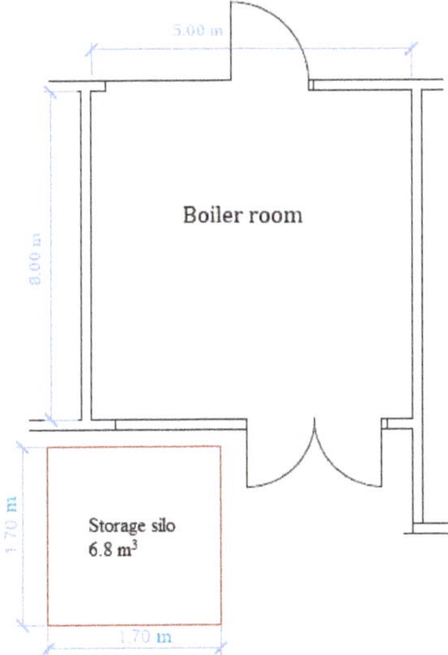

Figure 14. Location of storage silo.

4.5. Biomass Drying

An excess of humidity in the biomass, leads to a lot of problems. It is essential for biomass fuel to remain dry, or it will not burn efficiently.

The drying of the loquat seeds in the solar dryer started on 30 October 2019 and was completed on the 17th of the following month. The drying of loquat seeds with an initial moisture content of 35.20% was started and ended at 10% final moisture content after 12 days (Figure 15).

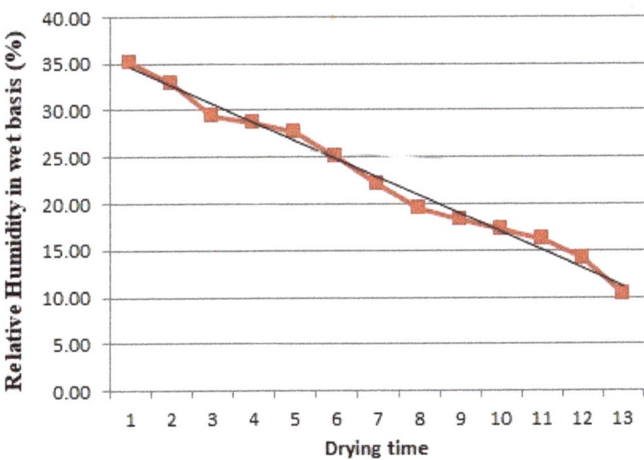

Figure 15. Evolution of the relative humidity in wet basis of the loquat seed in greenhouse dryer.

Figures 16 and 17 show the average temperature and relative humidity levels reached at different times of the day inside the greenhouse dryer.

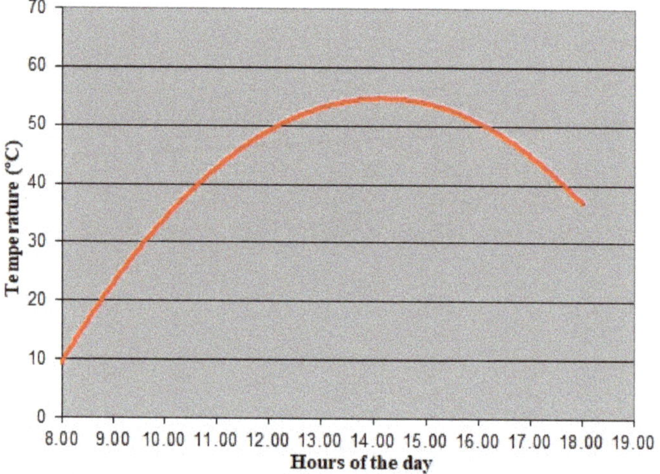

Figure 16. Average temperature variation during the day in the greenhouse dryer.

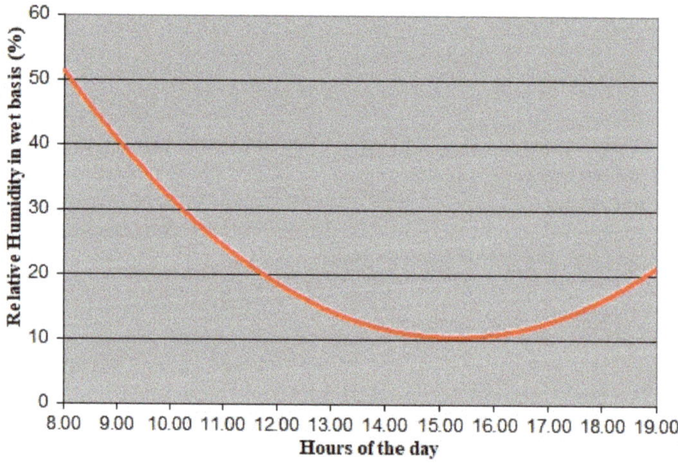

Figure 17. Average variation in relative humidity on a wet basis during the day in the greenhouse dryer.

5. Conclusions

As for the higher calorific value of loquat seed (17.205 MJ/kg), it is observed that it is lower than that of other standardized solid biofuels such as almond shells or olive stones. However, if we compare it with other industrial wastes such as wheat straw or pistachio shell, it is shown to have similar values, which demonstrates the energy potential of this residual biomass.

The use of loquat seed as an energy source has several advantages over other alternatives in the fight against climate change and local pollution that should be highlighted:

- It generates lower emissions than conventional fuel boilers, reduced sulphur and particle emissions and reduced emissions of pollutants such as CO, HC and NO_X. The change of fuel oil boiler to biomass boiler in UAL indoor swimming pool means a reduction of 147,973.8 kg CO_2 in emissions into the atmosphere and an annual saving of 35,739.5 € which means a saving of 72.78% with respect to the previous fuel oil installation.

- Reduced maintenance and hazards from toxic and combustible gas leaks.
- Reduced agricultural and forestry by-products going to landfills.
- Biomass is an inexhaustible source of energy, provided it is used sustainably.
- It helps to prevent fires and reduce the risks of forest fires and pests.
- The implantation of energy crops on abandoned land prevents erosion and soil degradation, as well as the possible use of agricultural by-products, avoiding their burning on the ground.
- It helps to clean up the mountains and to use the by-products of the agro-industries.

The five main countries that would reduce their annual CO_2 emissions by using loquat seeds as biofuel are: China (51,184.92 tn), Spain (10,617.54 tn), Turkey (3454.98 tn), Pakistan (3275.83 tn), and Japan (2621.95 tn). In addition, the annual worldwide reduction of CO_2 emissions if loquat seed is used as biofuel would be 76,363.80 tn.

The sensitivity analysis showed that the variable most sensitive is the fuel cost of base case, and the variables with minimum impact in this analysis are debt interest rate, debt ratio and debt term that have very small effects and can be ignored their uncertainty.

Financial analysis determined that the project has viability, the IRR (20.6%) is higher than European central bank interest rate (0.25%), and the B-C ratio shows that benefits are 25.1 times higher than costs.

Author Contributions: Conceptualization, M.-A.P.-M., F.M.-A., Q.H.-E. & A.-J.P.-M.; methodology, M.-A.P.-M., F.M.-A., Q.H.-E. & A.-J.P.-M.; formal analysis, M.-A.P.-M., F.M.-A., Q.H.-E. & A.-J.P.-M.; investigation, M.-A.P.-M., F.M.-A., Q.H.-E. & A.-J.P.-M.; resources, M.-A.P.-M., F.M.-A., Q.H.-E. & A.-J.P.-M.; writing—original draft preparation, M.-A.P.-M., F.M.-A., Q.H.-E. & A.-J.P.-M. All authors have read and agreed to the published version of the manuscript.

Funding: This research received no external funding.

Acknowledgments: This research has been supported by the Ministry of Science, Innovation and Universities at the University of Almeria under the programme "Proyectos de I+D de Generacion de Conocimiento" of the national programme for the generation of scientific and technological knowledge and strengthening of the R+D+I system with grant number PGC2018-098813- B-C33.

Conflicts of Interest: The authors declare no conflict of interest.

References

1. Smith, C.J.; Forster, P.M.; Allen, M.; Fuglestvedt, J.; Millar, R.J.; Rogelj, J.; Zickfeld, K. Current fossil fuel infrastructure does not yet commit us to 1.5 C warming. *Nat. Commun.* **2019**, *10*, 101. [CrossRef] [PubMed]
2. CO_2-earth 2019. Available online: https://es.co2.earth/daily-co2 (accessed on 1 April 2019).
3. Christoff, P. The promissory note: COP 21 and the Paris Climate Agreement. *Environ. Polit.* **2016**, *25*, 765–787. [CrossRef]
4. Manfren, M.; Caputo, P.; Costa, G. Paradigm shift in urban energy systems through distributed generation: Methods and models. *Appl. Energy* **2011**, *88*, 1032–1048. [CrossRef]
5. Bogner, J.; Pipatti, R.; Hashimoto, S.; Diaz, C.; Mareckova, K.; Diaz, L.; Kjeldsen, P.; Monni, S.; Faaj, A.; Gao, Q.; et al. Mitigation of global greenhouse gas emissions from waste: Conclusions and strategies from the Intergovernmental Panel on Climate Change (IPCC) Fourth Assessment Report. Working Group III (Mitigation). *Waste Manage. Res.* **2008**, *26*, 11–32. [CrossRef] [PubMed]
6. Adger, W.N.; Arnell, N.W.; Tompkins, E.L. Successful adaptation to climate change across scales. *Glob. Environ. Chang.* **2005**, *15*, 77–86. [CrossRef]
7. Perea-Moreno, A.J.; Perea-Moreno, M.Á.; Hernandez-Escobedo, Q.; Manzano-Agugliaro, F. Towards forest sustainability in Mediterranean countries using biomass as fuel for heating. *J. Clean. Prod.* **2017**, *156*, 624–634. [CrossRef]
8. Perea-Moreno, M.A.; Samerón-Manzano, E.; Perea-Moreno, A.J. Biomass as Renewable Energy: Worldwide Research Trends. *Sustainability* **2019**, *11*, 863. [CrossRef]

9.   Bridgwater, A.V. The technical and economic feasibility of biomass gasification for power generation. *Fuel* **1995**, *74*, 631–653. [CrossRef]

10.   Verma, V.K.; Bram, S.; De Ruyck, J. Small scale biomass heating systems: Standards, quality labelling and market driving factors–an EU outlook. *Biomass Bioenerg.* **2009**, *33*, 1393–1402. [CrossRef]

11.   González, J.F.; González-García, C.M.; Ramiro, A.; González, J.; Sabio, E.; Gañán, J.; Rodríguez, M.A. Combustion optimisation of biomass residue pellets for domestic heating with a mural boiler. *Biomass Bioenerg.* **2004**, *27*, 145–154. [CrossRef]

12.   Sánchez, F.; San Miguel, G. Improved fuel properties of whole table olive stones via pyrolytic processing. *Biomass Bioenerg* **2016**, *92*, 1–11. [CrossRef]

13.   Perea-Moreno, A.J.; Perea-Moreno, M.Á.; Dorado, M.P.; Manzano-Agugliaro, F. Mango stone properties as biofuel and its potential for reducing CO_2 emissions. *J. Clean. Prod.* **2018**, *190*, 53–62. [CrossRef]

14.   FAOSTAT. Agriculture Data. Available online: http://www.fao.org/faostat/en/#home (accessed on 5 April 2019).

15.   FreshPlaza 2019. Available online: https://www.freshplaza.es/article/95420/Espa%C3%83%C2%B1a-Investigaci%C3%83%C2%B3n-para-desarrollar-el-n%C3%83%C2%ADspero/ (accessed on 7 April 2019).

16.   Caballero, P.; Fernández, M.A. Loquat, production and market. In *First international symposium on loquat*; Options Méditerranéennes: Série A. Séminaires Méditerranéens; n. 58; CIHEAM: Zaragoza, Spain, 2003; pp. 11–20.

17.   Kawahito, Y.; Kondo, M.; Machmudah, S.; Sibano, K.; Sasaki, M.; Goto, M. Supercritical CO_2 extraction of biological active compounds from loquat seed. *Sep. Purif. Technol.* **2008**, *61*, 130–135. [CrossRef]

18.   Luo, Z.; Yang, S.; Xie, N.; Xie, W.; Liu, J.; Souley Agbodjan, Y.; Liu, Z. Multi-objective capacity optimization of a distributed energy system considering economy, environment and energy. *Energy Conv. Manag.* **2019**, *200*, 112081. [CrossRef]

19.   Delgado Marín, J.P.; Vera García, F.; García Cascales, J.R. Use of a predictive control to improve the energy efficiency in indoor swimming pools using solar thermal energy. *Sol. Energy* **2019**, *179*, 380–390. [CrossRef]

20.   Calise, F.; Figaj, R.D.; Vanoli, L. Energy and Economic Analysis of Energy Savings Measures in a Swimming Pool Centre by Means of Dynamic Simulations. *Energies* **2018**, *11*, 2182. [CrossRef]

21.   RITE, Complementary Technical Instruction 2007. Available online: https://www.boe.es/eli/es/rd/2007/07/20/1027 (accessed on 13 May 2019).

22.   Royal Decree 742/2013. Available online: https://www.boe.es/eli/es/rd/2013/09/27/742 (accessed on 10 May 2019).

23.   Perea-Moreno, M.-A.; Manzano-Agugliaro, F.; Hernandez-Escobedo, Q.; Perea-Moreno, A.-J. Peanut Shell for Energy: Properties and Its Potential to Respect the Environment. *Sustainability* **2018**, *10*, 3254. [CrossRef]

24.   Riedl, R.; Dahl, J.; Obernberger, I.; Narodoslawsky, M. Corrosion in fire tube boilers of biomass combustion plants. In Proceedings of the China International Corrosion Control Conference, Beijing, China, 26–28 October 1999.

25.   Montgomery, M.; Larsen, O.H. Field test corrosion experiments in Denmark with biomass fuels. Part 2: Co-firing of straw and coal. *Mater. Corros.* **2002**, *53*, 185–194. [CrossRef]

26.   Khan, A.; De Jong, W.; Jansens, P.; Spliethoff, H. Biomass combustion in fluidized bed boilers: Potential problems and remedies. *Fuel Process. Technol.* **2009**, *90*, 21–50. [CrossRef]

27.   Senelwa, K.; Sims, R.E. Bioenergy, Fuel characteristics of short rotation forest biomass. *Biomass Bioenerg.* **1999**, *17*, 127–140. [CrossRef]

28.   Nasser, R.A.-S.; Aref, I.M. Fuelwood characteristics of six acacia species growing wild in the southwest of Saudi Arabia as affected by geographical location. *BioResources* **2014**, *9*, 1212–1224. [CrossRef]

29.   Wolf, J.P. Biomass combustion for power generation: An introduction. In *Biomass Combustion Science, Technology and Engineering*; Woodhead Publishing: Cambridge, UK, 2013; pp. 3–8.

30.   Van Loo, S.; Koppejan, J. *The Handbook of Biomass Combustion and Co-Firing*; Earthscan: London, UK, 2008.

31.   RETScreen, Natural Resource of Canada. 2019. Available online: https://www.nrcan.gc.ca/maps-tools-publications/tools/data-analysis-software-modelling/retscreen/7465 (accessed on 2 October 2019).

32.   Perea-Moreno, A.-J.; Aguilera-Ureña, M.-J.; Manzano-Agugliaro, F. Fuel properties of avocado stone. *Fuel* **2016**, *186*, 358–364. [CrossRef]

33. Mata-Sánchez, J.; Pérez-Jiménez, J.A.; Díaz-Villanueva, M.J.; Serrano, A.; Núñez-Sánchez, N.; López-Giménez, F.J. Statistical evaluation of quality parameters of olive stone to predict its heating value. *Fuel* **2013**, *113*, 750–756. [CrossRef]

34. García, R.; Pizarro, C.; Lavín, A.G.; Bueno, J.L. Spanish biofuels heating value estimation. Part I: Ultimate analysis data. *Fuel* **2014**, *117*, 1130–1138. [CrossRef]

35. García, R.; Pizarro, C.; Lavín, A.G.; Bueno, J.L. Biomass sources for thermal conversion. Techno-economical overview. *Fuel* **2017**, *195*, 182–189. [CrossRef]

36. Arranz, J.I.; Miranda, M.T.; Montero, I.; Sepúlveda, F.J.; Rojas, C.V. Characterization and combustion behaviour of commercial and experimental wood pellets in south west Europe. *Fuel* **2015**, *142*, 199–207. [CrossRef]

37. Gómez, N.; Rosas, J.G.; Cara, J.; Martínez, O.; Alburquerque, J.A.; Sánchez, M.E. Slow pyrolysis of relevant biomasses in the mediterranean basin. Part 1. Effect of temperature on process performance on a pilot scale. *J. Clean. Prod.* **2016**, *120*, 181–190. [CrossRef]

38. González, J.F.; González-García, C.M.; Ramiro, A.; Gañán, J.; González, J.; Sabio, E.; Román, S.; Turegano, J. Use of almond residues for domestic heating: Study of the combustion parameters in a mural boiler. *Fuel Process. Technol.* **2005**, *86*, 1351–1368. [CrossRef]

39. Factores de Emisión de CO_2 y Coeficientes de Paso a Energía Primaria de Diferentes Fuentes de Energía Final Consumidas en el Sector de Edificios en España. Available online: https://energia.gob.es/desarrollo/EficienciaEnergetica/RITE/Reconocidos/Reconocidos/Otros%20documentos/Factores_emision_CO2.pdf (accessed on 7 September 2019).

40. Du, Q.; Cui, Z.; Dong, H.; Gao, J.; Li, D.; Yu, J.; Liu, Y. Field measurements on the generation and emission characteristics of PM2.5 generated by industrial layer burning boilers. *J. Energy Inst.* **2019**, *92*, 1251–1261. [CrossRef]

41. Liu, H.; Yang, F.; Du, Y.; Ruan, R.; Tan, H.; Xiao, J.; Zhang, S. Field measurements on particle size distributions and emission characteristics of PM10 in a cement plant of china. *Atmos. Pollut. Res.* **2019**, *10*, 1464–1472. [CrossRef]

42. Yao, S.; Cheng, S.; Li, J.; Zhang, H.; Jia, J.; Sun, X. Effect of wet flue gas desulfurization (WFGD) on fine particle (PM2.5) emission from coal-fired boilers. *J. Environ. Sci.* **2019**, *77*, 32–42. [CrossRef] [PubMed]

43. Su, C.; Madani, H.; Palm, B. Building heating solutions in china: A spatial techno-economic and environmental analysis. *Energy Conv. Manag.* **2019**, *179*, 201–218. [CrossRef]

44. Manzano-Agugliaro, F.; Carrillo-Valle, J. Conversion of an existing electrostatic precipitator casing to Pulse Jet Fabric filter in fossil power plants. *Dyna* **2016**, *83*, 189–197. [CrossRef]

45. Lind, T.; Hokkinen, J.; Jokiniemi, J.K.; Saarikoski, S.; Hillamo, R. Electrostatic precipitator collection efficiency and trace element emissions from co-combustion of biomass and recovered fuel in fluidized-bed combustion. *Environ. Sci. Technol.* **2003**, *37*, 2842–2846. [CrossRef]

46. Real Decreto 1027/2007, de 20 de julio, Por el Que se Aprueba el Reglamento de Instalaciones Térmicas en Edificios. Available online: https://www.boe.es/buscar/doc.php?id=BOE-A-2007-15820 (accessed on 14 November 2019).

Article

Sustainable Solar Energy in Mexican Universities. Case Study: The National School of Higher Studies Juriquilla (UNAM)

Quetzalcoatl Hernandez-Escobedo [1], Alida Ramirez-Jimenez [2], Jesús Manuel Dorador-Gonzalez [1], Miguel-Angel Perea-Moreno [3] and Alberto-Jesus Perea-Moreno [3,*]

[1] Escuela Nacional de Estudios Superiores Juriquilla, Universidad Nacional Autonoma de México, Queretaro 76230, Mexico; qhernandez@unam.mx (Q.H.-E.); dorador@unam.mx (J.M.D.-G.)

[2] International PhD School, Universidad de Almería, 04120 Almería, Spain; ramirezjalida@gmail.com

[3] Departamento de Física Aplicada, Radiología y Medicina Física, Universidad de Córdoba, ceiA3, Campus de Rabanales, 14071 Córdoba, Spain; k82pemom@uco.es

* Correspondence: aperea@uco.es

Received: 22 March 2020; Accepted: 10 April 2020; Published: 13 April 2020

Abstract: Universities around the world should be at the forefront of energy-saving and efficiency processes, seeking to be at the same level or preferably higher than the rest of society, and seeking the goal of 20% renewable energy by 2020. Sustainability practices have been carried out by several universities. In Mexico, the National Autonomous University of Mexico (UNAM) is a leader in this subject; in fact, the newest National School of Higher Studies - Juriquilla (ENES-J) that belongs to UNAM, located in the city of Queretaro (Mexico), is involved in its sustainability plan, with one of its main objectives being to save electric energy. UNAM has some campuses outside of Mexico City, and one of them is the National School of Higher Studies Juriquilla (ENES-J) in the state of Queretaro, where there is the Orthotics and Prosthetics Laboratory (OPL), in which has been installed a Computer Numerical Control (CNC) machine type Haas Automation model UMC-750, which has 5-axis and is an effective means to reduce the number of setups and increase accuracy for multi-sided and complex parts. This machine will be used to design, build, and assess human prosthesis. This study aimed to contribute to sustainability policies at the ENES-J from UNAM, implementing a solar photovoltaic system (PVS) to deliver electricity to the grid and contribute to reducing the electricity load at the Orthotics and Prosthetics Laboratory (OPL), as well to propose new research lines to support the sustainability policies in universities, and also proposing a financial analysis. To achieve this, in an area of 96.7 m^2, 50 solar panels type mono-Si Advance Power API-M330 with an efficiency of 17.83% and a capacity factor of 20.4% will be installed and will provide 17.25 kW of power and 345 kWh of energy. The financial analysis shows the initial costs of 46,575 USD/kW, operation and maintenance (O&M) costs (savings) of 569 USD/kW-year, a monthly electricity export rate of 0.10 USD/kWh, electricity exported to the grid of 21.5 MWh, and an electricity export revenue of 2,145 USD. To assess the environmental balance with this PVS at ENES-J, an analysis of greenhouse gases (GHG) is carried out by using the RETScreen software. In this analysis, a GHG emission factor of 0.45 tCO$_2$/MWh was found, as well as a savings of 12,089 USD per year.

Keywords: greenhouse gases; UNAM; energy saving; Mexico; photovoltaic system (PVS)

---

1. Introduction

Climate change is the greatest challenge facing humanity. All the agents involved, from leaders to citizens, must become aware of the problem and join efforts in the fight against global warming. All countries must agree on mitigation measures against climate change that involve ambitious targets for reducing greenhouse gases. To reach the Paris agreement and limit the temperature increase to 2 °C by the end of the century, or to 1.5 °C if possible, a major shift is needed in current energy policy towards a low-carbon economy, where on the one hand the share of renewables in the current energy mix is enhanced and on the other hand the energy efficiency of energy systems is increased [1–3]. Universities, within the framework of their social and environmental responsibility, cannot stay away from these objectives [4,5]. Universities, research centers, and institutions are the starting point of research in renewable energy knowledge and practice. These institutions have carried out several projects on renewable energy sources (RES), like the proposal on compressed air as reserve energy, made at the University of Auckland in New Zealand [6]. Zhou et al. [7] and the Ocean University of China (OUC) developed a methodology based on the study of the dynamic characteristics of an actual Offshore Wind Turbines (OWT) in different operational states based on sea tests to determine the negative impact on their structures. In Tokat Gaziosmanpasa University, Emeksiz and Cetin [8] analyzed the effects of tower shadow disturbance and wind shear variations. In this study, they determined that the x distance is the most correlated parameter on the tower shadow disturbance problem. Karasmanaki and Tsantopoulos [9] researched the attitude of RES students in the Department of Forestry and Management of the Environmental and Natural Resources at the Democritus University of Thrace in Greece since they are possible experts in RES, and the results show that students support renewables and have awareness about current polluting energy systems. Tran and Smith [10] developed an analysis on renewable energy systems integration and uncertainty to meet the three major types of energy consumption: Electricity (solar photovoltaic and wind), heating (combined heat and power) and cooling (electricity) at facilities on the campus of the University of Utah. They found that the uncertainty of energy loads and power generation from renewable energy heavily affects the operating cost of the district energy system.

Arnaout et al. [11] explain that the Heriot-Watt University Malaysia (HWUM) has a unique roof design that could be utilized as part of the Building-integrated photovoltaics (BIPV) system to generate electricity, thereby reducing the carbon footprint of the facility. This system is an innovative green solution that makes possible energy generation on the building facade with modification of the building material or architectural structure.

Sierra et al. [12], during characterization of an 840Wp BIPV installed at the Faculty of Sciences at the National University of Colombia, found that the energy generation with coal has a greater negative environmental impact (84%) compared with the photovoltaic system (PVS) (6%), and the use of PVS represents an emission factor of 35 gCO_2eq/kWh.

A techno-economic and environmental analysis of a PVS conducted by Sulukan [13] at the Turkish Naval Academy of the National Defense University was done. This study was carried out using the RETScreen software considering performance, efficiency, inverter efficiency, and temperature on PVS. The results of emission were a 93% reduction in greenhouse gases and a saving of 721.1 tons of crude oil.

Bilcik et al. [14] dealt with the impact of temperature on module surfaces The experiment was conducted at the University of Life Sciences in Prague and found that the performances of photovoltaic modules depend on climatic conditions.

In Morocco, Ameur et al. [15] evaluated different solar photovoltaic technologies (amorphous silicon (a-Si), Polycrystalline silicon (pc-Si) and Monocrystalline silicon (mc-Si)) connected to a low voltage three phases electrical grid of Al Akhawayn University. The results show that the polycrystalline panel is the most cost-effective technology.

Silveira et al. [16] presented an economic study about electric power generation using PSV in a Brazilian university with the aim of reducing the electric consumption. The results show that with the PSV, the tariff can be reduced by 39.9%.

Mukherji et al. [17] present a techno-economic and ecological analysis of a 50 kWp rooftop solar photovoltaic plant installed at ICFAI University, Jaipur where the plant produced

106.9 kWh/kWp/month, and the reductions in greenhouse gases obtained were 102tCO$_2$, 128 kg SO$_2$, 268 kg NOx and 7033 kg ash.

Oh and Park [18] analyzed the optimal orientation of solar panels, and the analysis was performed regarding demand and supply electricity at the Seoul National University. Their results show that orientation is very important and depends on the building's architecture. The output generation presented a low electricity demand in the evening, but monthly demand shows a pattern opposite to solar power generation.

At the Marmara University in Istanbul, a study done by Akpolat et al. [19] demonstrates that the installation of grid-connected rooftop solar photovoltaic systems of 84.75 kW can produce several benefits and an annual electrical savings of 90,298 kWh.

Another study was done by Al-Najideen et al. [20] at the faculty of Engineering-Mu'tah in University of Jordan in order to reduce the electricity demand with a 56.7 kW grid-connected PVS. The results show that this PVS will produce 97.02 MWh per year, with an investment of 117,000 USD and a payback period of 5.5 years.

A hybrid system composed by a photovoltaic (PV) panel, wind turbine, and storage batteries installed in Yildiz Technical University is analyzed by Arikan et al. [21] and determines the most accurate system sizing using the maximum power point tracking controller, a hybrid controller, and an inverter.

An Artificial Neural Network (ANN) methodology for studying and modeling the soiling effect on solar PV glass has been done by Laarabi et al. [22]. They exposed outdoor solar PV glazing at Mohammed V University in Rabat (Morocco), and found that the most influential parameter for the PV soiling rate was relative humidity, followed by wind direction.

An ensemble approach to predict solar PV power production has been proposed by Al-Dahidi et al. [23]. This ensemble approach has the capability of handling the intermittent nature of solar energy. They demonstrated it using a grid-connected solar PV system of 231 kW of capacity installed on the rooftop of the Faculty of Engineering at the Applied Science Private University of Jordan and it was determined that this methodology could allow for balancing power supplies. Since the installation of rooftop PV arrays is increasing, many standards have been designed, like the study done by Bender et al. [24] at Central Washington University (CWU) where they present one calibration done at the rooftop of CWU.

According to Wen [25], the interaction between the University-Firm-Government linkage has been discussed several times. In the special case of the solar photovoltaic industry in Taiwan, he found that this industry is essential for the development of the University-Firm-Government linkage.

As it can be seen in many universities studies about RES which have been done, with these efforts, institutions contribute to developing these systems, because in the end, the main objective of renewable energy is to create development where there is none and increase it where there is. RES faces some problems. One of them is this premise: where there is no price for carbon emissions, there is no reason to reduce carbon [26]. This premise is not related to a very important decision-making variable called sustainability, which it can be defined as the use of today's resources without compromising the ability of future generations to be at least as well off as we are [27]. Several authors coincide that sustainability is a concept with many interpretations, including economic, environmental and social ones, and complete sustainability requires all of them. This is also referred to as the triple-bottom-line [28–30].

In 1990, in France, a statement made by university presidents, chancellors, and rectors commited to environmental sustainability in higher education. This statement was called the Talloires Declaration (TD) and has ten actions to incorporate sustainability into institutions [31–34].

There is a ranking that grouped 619 worldwide universities and awards the best sustainable policies incorporated in these universities. This is the *UI GreenMetric World University Ranking*, which gives basic information about the university's policy towards a green environment. Its aim is to trigger the participating university to provide more space for developing sustainable energy [35]. From the results of this ranking, 11 Mexican universities are in different positions; the most important Mexican

University is the National Autonomous University of Mexico (UNAM) and is in 101st place from 619 universities worldwide.

UNAM developed a macro project for the transformation of the University into a model of efficient and intelligent energy use, by which it is expected to obtain electricity savings ranging from 20 and 30 percent by using solar, biomass, and hydrogen energy, as well as the promotion of green culture.

With the title of "The University City and Energy", the macro project, which can serve as an example to other communities in the country, includes 21 projects contained in 6 research lines, highlighting the creation of public lighting with solar energy, saving electric energy with the use of photovoltaic technology, and using the prototype of a multifunctional ecological vehicle and a virtual classroom for learning and teaching on the subject, among other things.

In this work, the software RETScreen is used, which will help to determine both the technical and economic feasibility of the study. RETScreen is a Clean Energy Management Software system for energy efficiency, renewable energy, and cogeneration project feasibility analysis, as well as ongoing energy performance analysis. It is developed by Canada's government through the Natural Resources Canada office.

UNAM has some campuses outside of Mexico City. One of them is the National School of Higher Studies Juriquilla (ENES-J) in the state of Queretaro, where there is the Orthotics and Prosthetics Laboratory (OPL), in which has been installed a Computer Numerical Control (CNC) machine type Haas Automation model UMC-750, which has 5-axis and is an effective means to reduce setups and increase accuracy for multi-sided and complex parts. This machine will be used to design, build, and assess human prosthesis.

The Kyoto Protocol has established three mechanisms: The Clean Development Mechanism (CDM), Joint Implementation (JI), and Emissions Trading (ET), which allow parties to pursue opportunities to cut emissions or enhance carbon sinks abroad. The cost of curbing emissions varies considerably from region to region, and therefore it makes economic sense to cut emissions where it is cheapest to do so, given that the impact on the atmosphere is the same [36]. It is important to mention that a project can be evaluated as CDM if the project is planned in a developed country.

The main objective of this work is to contribute to sustainability policies at the ENES-J from UNAM implementing a solar photovoltaic system to deliver electricity to the grid and contribute to reducing the electricity load at the Orthotics and Prosthetics Laboratory (OPL), as well to propose new research lines to support the sustainability policies in universities and proposing a financial analysis.

In the following section, the materials and methods are described. In Section 3, the results and discussion are presented. Finally, Section 4 presents the main conclusions and future research lines.

2. Materials and Methods

The average solar irradiation recorder in the area studied is 6.1 kWh/m^2/day with an average global horizontal irradiance (GHI) of 800 W/m^2. The PVS has been designed to contribute with the CNC machine type Haas Automation model UMC-750 electric load and could be applied to any site when data are known.

Figure 1 shows both, the Computer Numerical Control (CNC) machine type Haas Automation model UMC-750 arriving at the Orthotics and Prosthetics Laboratory (OPL) and an image obtained from the machine's website.

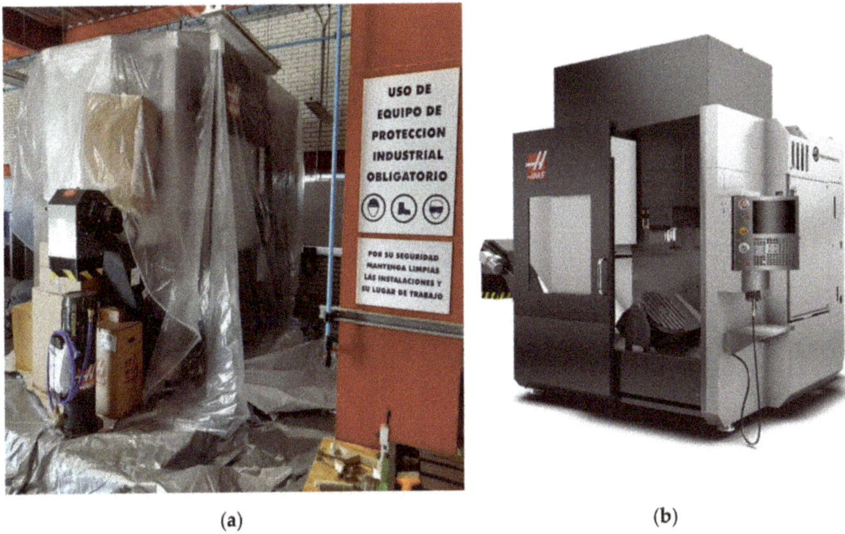

(**a**)   (**b**)

Figure 1. Computer Numerical Control (CNC): (**a**) CNC installed at the Orthotics and Prosthetics Laboratory (OPL); (**b**) CNC Haas Automation model UMC-750.

All the electric energy at the National School Higher Studies in Queretaro is delivered by the electric grid; in this case, the PVS show both economic and technical benefits.

The electric load demanded by the CNC machine is presented in Table 1.

Table 1. The CNC Haas Automation model UMC-750 electric load.

Description	Electric Load (kW)	Useful Hours per Day (h/Day)	Useful Days per Week (d/Week)
Haas Automation UMC-750	22.4	4	5

With the electric load presented in Table 1, the electric rate will be the highest one, named Domestic High Consumption Tariff (DHCT), which means that it is the one with the highest prices; the load is in AC.

The ENES-J belongs to UNAM and is located in the city of Queretaro in central Mexico. This city is a semidesert zone, with an average temperature of 16.4 °C during the year. The maximum temperature is 28 °C in April and May, and the minimum is 6 °C in January [37]. In Figure 2 is shown the geographic position of Queretaro Mexico.

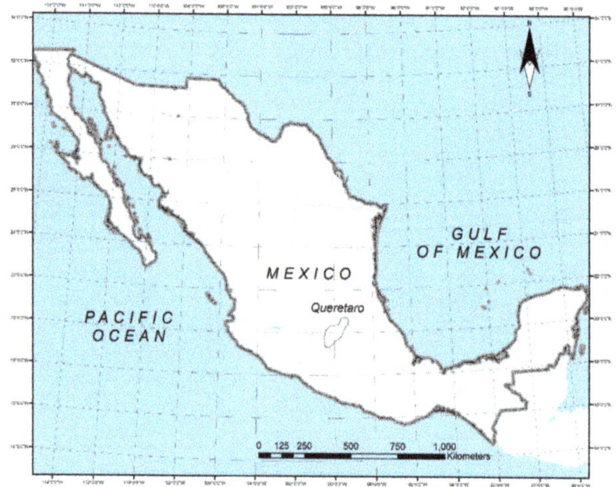

Figure 2. Queretaro's geographic location.

Data are extracted from the Synoptic Meteorological Station (SMS) that belongs to the National Weather Service (NWS) of Mexico. In the state of Queretaro, there is one SMS, and this meteorological station delivers data every 10 minutes. The process of recording data is as follows: the data acquirer is programmed to record data every 2 seconds and after 10 minutes delivers an average. The variables registered are wind speed, wind direction, air temperature, atmospheric pressure, rain, relative humidity, and solar radiation. This last one is our focus of study. Table 2 shows the main characteristics of these meteorological stations.

Table 2. AMS and SMS characteristics.

ID	Latitude (N)	Longitude (W)	Average Solar Global Radiance (W/m^2)	Average Air Temperature (°C)
Queretaro	20.5633°	−100.3694°	800	16.4

2.1. Modelling the Energy

The software RETScreen has been applied in several studies on renewable energy to determine their feasibility [38–42]. RETScreen rapidly identifies, assesses and optimizes the technical and financial viability of potential clean energy projects. This decision intelligence software platform also allows managers to easily measure and verify the actual performance of their facilities and helps find additional energy savings/production opportunities [43].

Figure 3 shows how the software looks and the variables taken into the assessment.

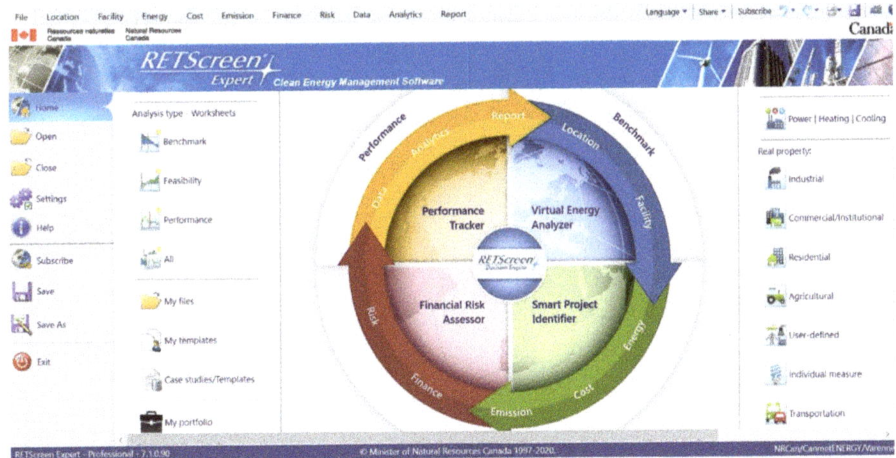

Figure 3. The workflow used in the RETScreen software.

The workflow shown in Figure 3 represents the model used by the software. The data set is the first variable needed, followed by location and facility, the energy project, costs, and emission analysis, then both financial and risk analyses are done.

Figure 4 shows the data used and technology selected in the software.

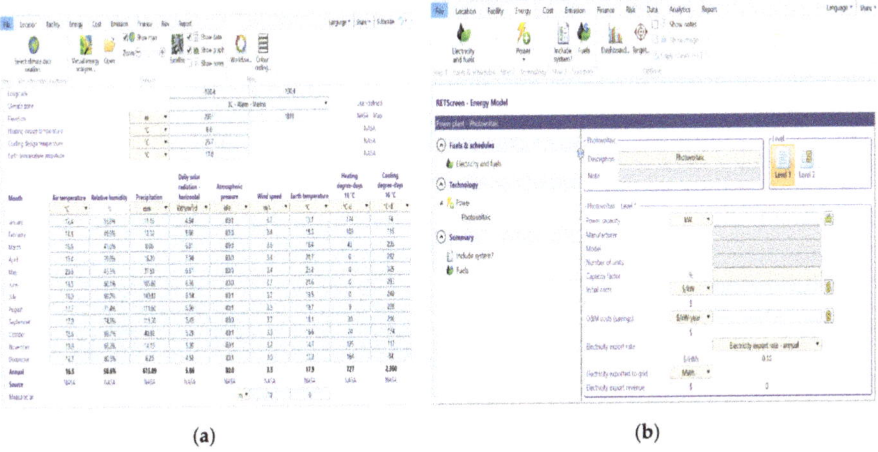

(a)   (b)

Figure 4. Initial characteristics: (**a**) Meteorological variables used; (**b**) technology used in the software.

Data used in the study are included in the location module (see Figure 4a). The technology used is added in the energy module, as seen in Figure 4b. As we can see in this last figure, the manufacturer, model, number of solar panels, initial costs, and other variables must be included.

2.2. Solar Assessment

The solar global irradiance assessment considers the variable which is most important to calculate a PVS [44]. Through the photoelectric principle, solar irradiation is converted into electrical energy and can be delivered to consumers. An inverter alternative current (AC) can be used, or a direct current (DC).

The PVS equipment includes a PV panel, a load regulator to regulate the voltage generated by the PV panel (this load regulator can perform a charge cycle on deep cycle batteries), and the protection systems.

PVS Requirements

One we defined the PVS, these recommendations must be followed:

1. Charges calculation: in this section, the charges must be defined (AC or DC), with useful time and quantity. These data can determine the installed capacity, kiloWatt, (kW) and consumption, kiloWatt-hour, (kWh).
2. Solar sizing: determining sun peak hours (SPH) or irradiation (kWh/m^2/day) with data obtained from AMSs.
3. Power, voltage, current and panel area.
4. Determining the number of PVs using Equations (1) and (2).

$$E = \frac{E_T}{R} \tag{1}$$

$$NP = \frac{E}{0.9 \, Wp \, SPH} \tag{2}$$

where E is the real energy consumption in kWh, E_T is the theoretical energetic consumption in kW, R is a proportional constant that includes losses related in the batteries use, inverters and electrical wiring, commonly its value is 0.8, NP is the number of PV, SPH are sun peak hours, and W_p is the PVS peak power.

5. Inverter selection: the inverter will be in the function of the charge and operation.

To determine the correct position of the PVS, some variables need to be considered, such as the solar declination angle (δ) that can be defined as the angle formed by the plane that contains the axis of terrestrial rotation and the plane perpendicular to the elliptic [45]. The solar declination angle is positive in the North and varies between $-23.45° \le \delta \le 23.45°$. Its highest value is on June 21, and its lowest on December 22 [46]. The expression of the solar declination angle in Equation (3) includes Julian days, n.

$$\delta = 23.45° \, \sin\left(360(n + 284)/365\right) \tag{3}$$

The angle formed with respect to the equator is called the latitude angle (ϕ) and is considered positive in the North and negative in the South, its value being between $-90° \le \phi \le 90°$.

The hour angle (ω) is formed by solar rays and the meridional plane at the site. The measure is from the meridional plane, in which the position of the Sun at 12:00 h has a $\omega = 0°$. In the East is positive and in the West is negative, the position of the Sun at 6:00 h, $\omega = 90°$, at 18:00 h, $\omega = -90°$. The hour angle is given by Equation (4).

$$\omega = \frac{360 \, (12 - t)}{24} \tag{4}$$

where t is time in hours in decimal.

The angle between the horizontal plane and PVS is known as the optimum angle (β), and this angle needs to be oriented at the South in the North hemisphere. This angle has values from $0° \le \beta \le 180°$ and can be expressed by Equation (5).

$$\beta = |\phi - \delta| \tag{5}$$

The PVS power output components such as module operating temperature and its nominal efficiency (η_{ref}) depend on the environmental conditions, as well as the optimum output (η_{PV}). The Mexican solar abridgment [47] establishes that the PV module η_{PV} is a function of η_{ref}, the temperature of the cell T_C, the module power temperature coefficient β_{ref} (which is considered between -0.3% to

−06% per °C) [48], and the standard temperature T_{stc}; this last one is provided by the manufacturer. With these conditions, the PVS efficiency can be calculated by Equation (6).

$$\eta_{PV} = \eta_{ref}\left[1 - \beta_{ref}(T_C - T_{stc})\right] \tag{6}$$

To calculate T_C, in Equation (7), is necessary to have the air temperature of the site T_a and the solar radiation I_{rad}, as well as the nominal operating cell temperature (NOCT) provided by the manufacturer.

$$T_C = T_a + \frac{NOCT - 20°C}{800\ W/m^2}I_{rad} \tag{7}$$

The PV power output module is expressed by Equation (8).

$$P_{out} = P_{max,stc}\left(\frac{I_{rad}}{G_{stc}}\right)\left[1 - \beta_{ref}(T_C - T_{stc})\right] \tag{8}$$

where G_{stc} is the irradiance at standard conditions whose value is 1 kW/m^2, and $P_{max,sct}$ is the cell maximum power at standard test conditions.

2.3. Inverter

Inverters are electric and electronic equipment developed to transform DC into AC. The inverters interconnected to the electric grid need to consider the grid electric characteristics such as voltage and frequency.

According to the Mexican standard NOM-001-SEDE-1999, the inverter nominal capacity could be between 75% to 80% of the PVS nominal capacity, because of temperature losses, electric wiring, shadowed and mismatch of the system. The inverter input voltage in PVS interconnected to the grid must be higher than 100 VDC. Additionally, it is recommended to use a maximum interconnection voltage in AC of 13% above the grid's nominal voltage.

The inverter selected is the model Advanced Solar Photonics: PV240-277V and 3 inverters are needed for the PVS. The characteristics of the inverter are presented in Table 3.

Table 3. Inverter characteristics.

Advanced Solar Photonics: PV240-277V		
Maximum AC power	3980	Wac
Maximum DC power	4076.32	Wdc
Power consumption during operation	30.8666	Wdc
Nominal AC voltage	277	Vac
Maximum DC voltage	450	Vdc
Maximum DC current	11.3231	Adc
Nominal DC voltage	360	Vdc

The inverter efficiency curve is presented in Figure 5.

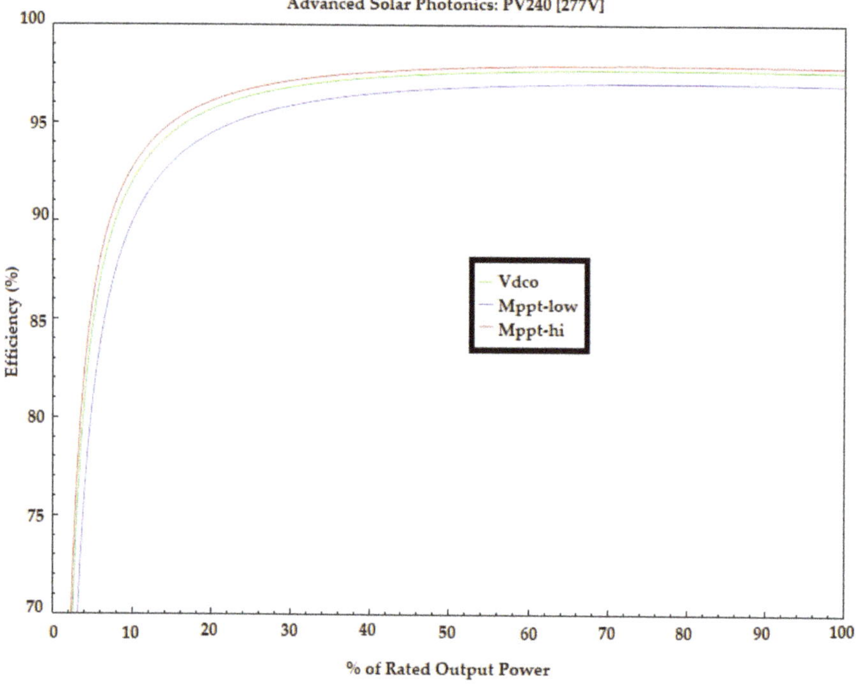

Figure 5. Inverter efficiency.

Figure 5 shows that the selected inverter has 98% efficiency. In this figure, three curves can be seen: the nominal DC voltage (Vdco), or design input voltage; the minimum MPPT DC voltage (MPPT-low), the manufacturer-specified minimum DC operating voltage; and the maximum MPPT DC voltage (MPPT-hi), the manufacturer-specified maximum DC operating voltage.

2.4. Financial Model

The financial model used calculates financial metrics of the power project based on a project's cash flows over an analysis period studied. The financial model uses the system's electrical output calculated by the performance model to calculate the series of annual cash flows.

According to Short et al. [49], the financial metrics are defined based on the definitions and methods as follows:

The present value (PV) analysis is a measure of today´s value of revenues or costs to be incurred in the future. PV is considered an important financial variable because it shows the cost assumed in moment zero. It is defined by Equation (9).

$$PV = \frac{1}{(1+d)^n} \tag{9}$$

where d is the annual discount rate, and n is the number of periods studied.

Internal rate return (IRR) is commonly used for many accept or reject decisions because it allows for a comparison with a minimum acceptable rate of return that presents an opportunity cost of capital. It is calculated through iterations until PV cash flow is equal to zero.

A simple payback period (SPB) is the number of years necessary to recover the project cost of investment under consideration and can be worked out by using Equation (10).

$$\sum_n \Delta I_n \leq \sum_n \Delta S_n \tag{10}$$

where ΔI is the non-discounted incremental investment cost, and ΔS is the non-discounted sum value of the cash flows net annual costs.

Benefit/cost ratio (B/C) shows whether, and to what extent, the benefits of a project exceed the costs. The B/C ratio is expressed by Equation (11).

$$B/C = \left(\frac{PV \text{ (all Benefits)}}{PV \text{ (all Costs)}} \right) \tag{11}$$

where PV (all Benefits) is the present value of all positive cash flow equivalents, and PV (all Costs) is the present value of all negative cash flow equivalents.

2.5. Sensitivity Analysis

According to Helton [50] and Saltelli et al., [51] a sensitivity analysis (SA) is a typical measure used to quantify the impact of parameter uncertainty on overall simulation/prediction uncertainty. Evaluating the two indices requires calculating the mean and variance in the parameter space, and this is always done by using Monte Carlo (MC) methods. The quasirandom sampling method is the most computationally-efficient one among the existing MC methods. Following Saltelli et al. [52], the mean and variance are evaluated with Equations (12) and (13),

$$V_{\theta_{\sim i}}\left(E_{\theta_{\sim i}}(\Delta|\theta_i)\right) = \frac{1}{n} \sum_{j=1}^{n} f(B_j)\left(f(A_{B,j}^i) - f(A_j)\right) \tag{12}$$

and

$$E_{\theta_{\sim i}}\left(V_{\theta_i}(\Delta|\theta_{\sim i})\right) = \frac{1}{2n} \sum_{j=1}^{n} \left(f(A_j) - f(A_{B,j}^i)\right)^2 \tag{13}$$

where $\Delta = f(.)$ denotes a model execution for its parameters. The calculation requires two independent parameter sample matrices, **A** and **B**, with the same dimension of $n \times d$, where n is the number of samples and d is the number of parameters. Matrix A_B^i is the same as matrix **A**, except that its ith column is from the ith column of matrix **B**. Suscript j denotes the jth row of the corresponding matrix [53].

2.6. Capital Asset Pricing Model (CAPM)

The Capital Asset Pricing Model (CAPM) describes the relationship between systematic risk and expected return for assets. CAPM has been developed by Sharpe [54], Lintner [55] and Ferreira et al. [56]. This model estimates the cost of equity, which allows for comparison among businesses with an economic rationale for calculations [57]. The CAMP equation is widely used for calculating the expected returns of an asset and can be expressed by Equation (14).

$$ER_i = R_f + \beta_i\left(ER_m - R_f\right) \tag{14}$$

where ER_i is the expected return of an asset, R_f is the risk-free rate, β_i is the beta of the investment (according to NASDAQ:FSLR the beta of the solar stock is 1.84), ER_m is the expected return of the market (Mexican ER_m is 8.00%), and $(ER_m - R_f)$ is the market risk premium. The Mexican risk-free rate is taken from the Bank of Mexico and is equal to 7.00%, β_i is a measure of a stock's risk given by measuring the fluctuation of its price changes relative to the overall market; the market risk premium represents the additional return over and above the risk-free rate.

2.7. Levelized Cost of Electricity (LCOE)

The levelized cost of electricity (LCOE) has been used to determine the USD per megawatt-hour ($/MWh) cost of PVS over the life of the capacity [57] and may compare different scenarios. According to Perkins [58], the expression used to calculate LCOE is given by Equation (15).

$$
\text{LCOE} = \frac{\sum_{t=1}^{n} \frac{I_t + M_t + F_t}{(1+r)^t}}{\sum_{t=1}^{n} \frac{E_t}{(1+r)^t}}
\tag{15}
$$

where I_t is the invested capital, M_t is the operating and maintenance costs, F_t is the solar photovoltaic feedstock, E_t is the energy delivered to the grid (MWh/yr), n is the lifetime of the project, and the discount rate is given by r. The Mexican Central Bank shows the value of the discount rate of 7.25% [59].

2.8. Emission Analysis

Greenhouse gases (GHG) include water vapor, ozone (O_3), carbon dioxide (CO_2), nitrous oxide (N_2O), methane (CH_4), nitrous and several classes of halocarbons. GHG allow solar radiation to enter the Earth's atmosphere but prevent the infrared radiation emitted by the Earth's surface from escaping. Instead, this outgoing radiation is absorbed by the GHG and then partially re-emitted as thermal radiation back to Earth, warming the surface [60].

According to the "National Inventory of Emissions of Greenhouse Gases and Compounds (INEGYCEI)" that presents the National Institute of Ecology and Climate Change (INECC) in accordance with Article 74 of the General Law of Climate Change, Mexico emitted 683 million tons of carbon dioxide equivalent ($MtCO_2e$) of GHG in 2018.

The Inventory is an instrument that allows for knowing the emissions of Mexico that originate from human activities throughout the national territory. It is a fundamental exercise in designing emission reduction policies, understanding the main sources and the role that ecosystems play in capturing part of these emissions.

Mexico is conducting an inventory, in accordance with scientific and technical criteria established by the Intergovernmental Panel on Climate Change (IPCC), which is a signatory of the United Nations Framework Convention on Climate Change (UNFCCC).

The most relevant gas emitted by Mexico is carbon dioxide with 71% of emissions, followed by methane with 21%. According to total emissions, 64% corresponded to the consumption of fossil fuels; 10% originated from livestock production systems; 8% came from industrial processes; 7% were issued for waste management; 6% for fugitive emissions from oil, gas and mining extraction, and 5% were generated by agricultural activities. In the inventory 148 $MtCO_2e$ absorbed by the vegetation were also counted, mainly in forests and jungles. The net balance between emissions and removals for 2018 was 535 $MtCO_2e$. It was estimated that in 2018, 112,240 tons of this short-lived climatic forcer was generated, which has negative effects on public health.

3. Results and Discussion

3.1. Photovoltaic Output Generator

The PV generator is the group of modules connected in parallel before the interconnected boxes. The output generator depends on air temperature; the natural degradation of semiconductors via the photoelectric process; the orientation and solar tilt; dust; and shadows.

The output PV is calculated under the Mexican standard NMX-J-643/1-ANCE-2011 related to photovoltaic power. The PV modules are composed of semiconductors but have some differences because they present some variations in electric parameters which are dependent on the air temperature. In order to define how the temperature impacts these PV electrical parameters, it is necessary to know the thermal coefficients: the thermal coefficient of maximum power (γ, gamma); the thermal coefficient of open-circuit voltage (β, beta), and the thermal coefficient of short circuit current (α, alpha). Table 4 shows the typical values of thermal coefficients in different PV technologies.

Table 4. The thermal coefficients.

Coefficient	Technology		
	Silicon Solar Cells	**Thin-Film Solar Cells**	**III-V Solar Cells**
γ	−0.4%/°C	−0.3%/°C	−0.4%/°C
β	−0.3%/°C	−0.2%/°C	−0.3%/°C
α	0.05%/°C	0.01%/°C	0.02%/°C

The thermal output is calculated by Equation (16).

$$\text{Thermal}_{\text{output}} = 1 + (\gamma \cdot \Delta_T) \tag{16}$$

With this information, it is possible to forecast the variation of temperature on the module during its use. Table 5 presents the average monthly temperature in Queretaro.

Table 5. The average monthly temperature in Queretaro.

	Jan	Feb	Mar	Apr	May	Jun	Jul	Aug	Sep	Oct	Nov	Dec
T(°C)	12.4	14.1	16.7	19.4	20.6	19.5	18.0	17.7	17.0	15.6	13.9	12.7

In this study, solar panels of the type mono-Si Advance Power API-M330 are evaluated. In Figure 6 is shown the current (Amps) and voltage (Volt) relationship of the module. The PVS general efficiency is calculated considering the inverter and conductors efficiency, resulting in 72.5%.

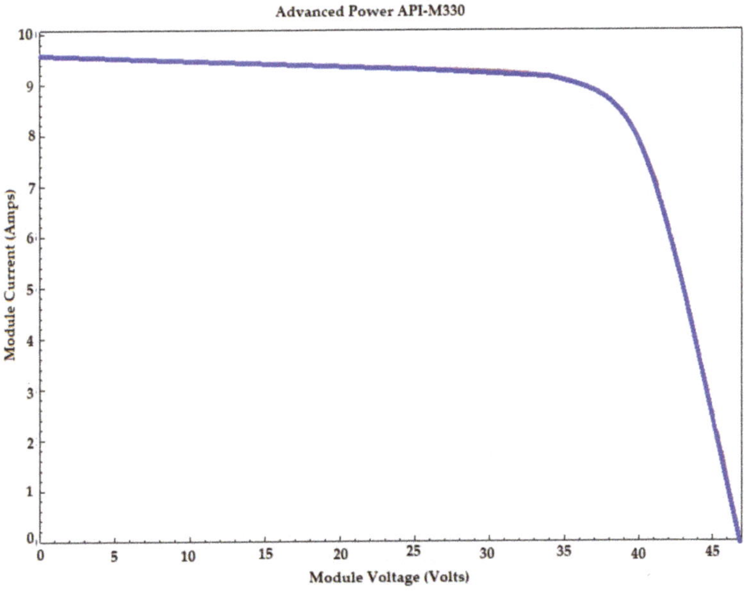

Figure 6. The photovoltaic (PV) panel amperes and voltage relationship.

The module characteristics at the reference conditions are presented in Table 6.

Table 6. Module characteristics.

Advance Power API-M330		
Nominal efficiency	17	%
Maximum power (Pmp)	329.875	Wdc
Max power voltage	37.7	Vdc
Max power current	8.8	Adc
Open circuit voltage	46.8	Vdc
Short circuit current	9.6	Adc

3.2. Solar Resource

The irradiance recorded each month in Queretaro is shown in Figure 7.

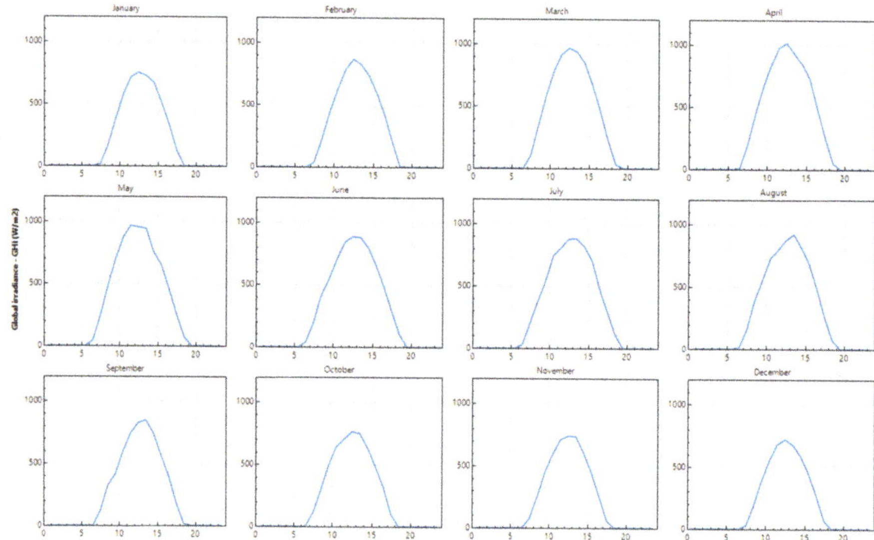

Figure 7. Monthly solar radiation in Queretaro.

Once knowing the consumption, the solar resource must be analyzed. In Table 7 is shown the irradiance in Queretaro.

Table 7. Average monthly irradiance in Queretaro.

Jan	Feb	Mar	Apr	May	Jun	Jul	Aug	Sep	Oct	Nov	Dec	Annual
						kWh/m^2/day						
4.84	5.86	6.81	7.04	6.81	6.36	6.14	6.06	5.49	5.29	5.09	4.58	5.86

Considering the efficiencies and solar resource, it is possible to calculate the PVS peak power (PVS_{pp}) by using Equation (17).

$$PVS_{pp} = \frac{\text{Daily energy consumption}}{\text{solar resource} \times \text{PVS efficiency}} \tag{17}$$

Then, the PVS_{pp} is 21.08 kW.

The comparison between the electricity delivered to the grid and the electricity load each month is shown in Figure 8.

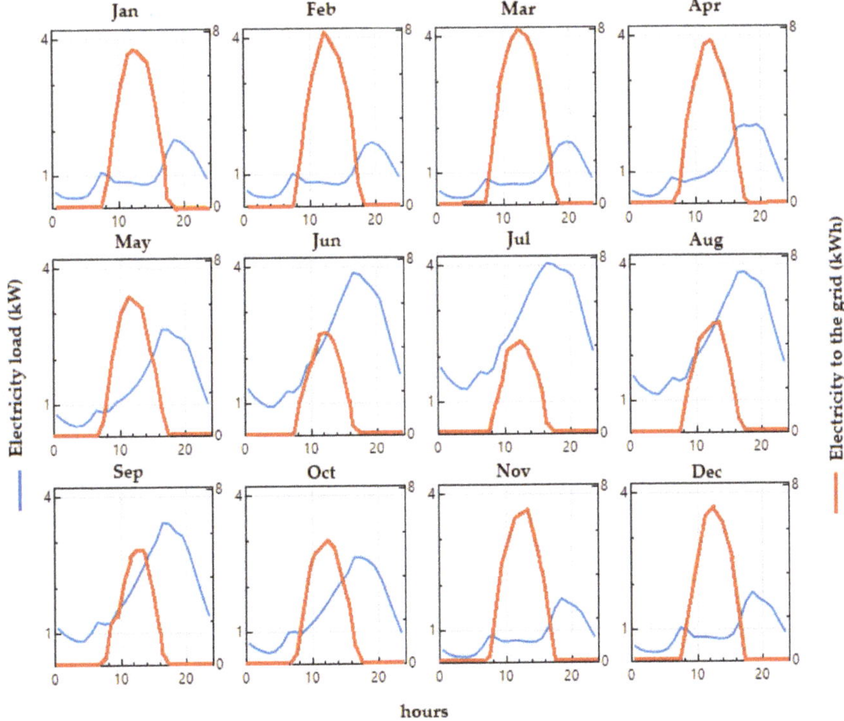

Figure 8. Electricity to the grid and electricity load.

Figure 8 presents both electricity delivered to the electric grid and the electricity load (CNC machine) per month. It can be appreciated that in June, July, and August, the electricity load is higher than the electricity delivered by PVS. This is due to higher temperatures in these months in Queretaro. This variation coincides with Bilcik et al. [14], who found that photovoltaic modules depend on climatic conditions, and as can be seen, in Queretaro these months are the warmest of the year. The electricity delivered to the grid behaves as solar radiation; even if the CNC machine works at night, the period will be short. In Queretaro, the relative humidity is 44%; in this case, this variable does not influence the PVS performance, as exposed by Arikan et al. [21].

The monthly energy production by month is presented in Figure 9. March, April, and May are the most energetic months.

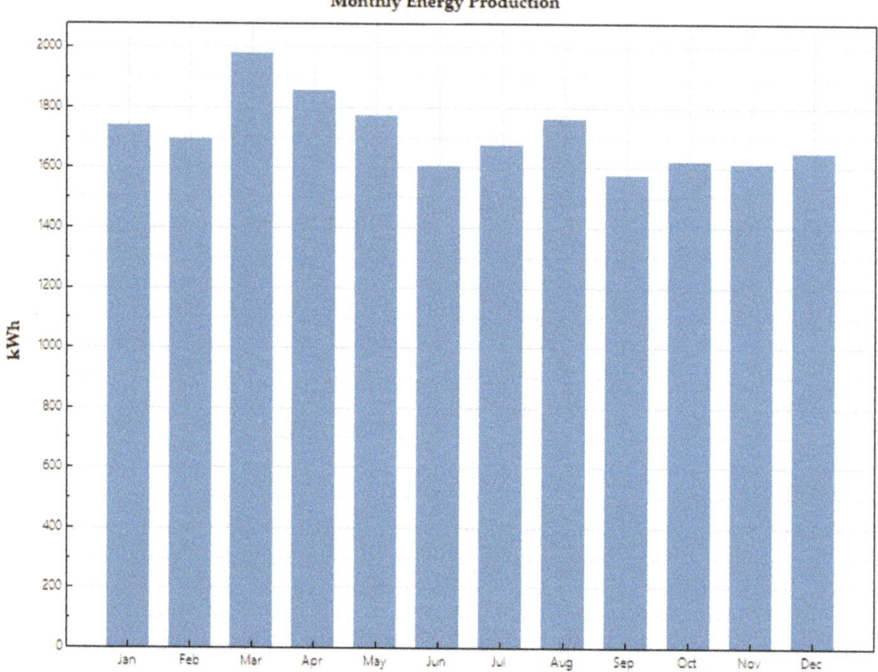

Figure 9. Monthly energy production in Queretaro.

3.3. Financial Assessment

The financial analysis has been done considering the following summary of PVS data: a capacity factor of 20.4%, initial costs 46,575 \$/kW, operation and maintenance (O&M) costs (savings) of 569 \$/kW-year, an electricity export rate-monthly of 0.10 \$/kWh, electricity exported to grid 21.5 MWh, and electricity export revenue of \$2,145, as in the work done by Al-Najideen et al. [20], they study a PVS of 56.7 kW grid connected analyzed under the initial costs and the payback period.

RETScreen has been used in different assessments. Its implementation allows it to determine the viability of PVS. Its financial model allows it to calculate from some input parameters (e.g., discount rate and debt ratio) the output items of financial viability such as IRR, SPB, and NPV. Several authors have assessed solar photovoltaic projects using the RETScreen software obtaining important results, as done by Yendaluru et al., [38] who analyzed the techno-economic feasibility of an integrating grid-tied solar PV plant in a wind farm, or Islam et al., [39] who evaluated an LED system.

In general, given the discount rate, a positive net present value indicates an economically-feasible project, while a negative net present value indicates an economically-infeasible project. It is important to evaluate the NPV along with other metrics, including capacity factor or IRR, and this is expressed as Equation (18). All these parameters allow the project decision-maker to consider various financial parameters.

$$\text{NPV} = \sum_{n=0}^{N} \frac{C_n}{(1 + d_{\text{real}})^n} \tag{18}$$

where C_n is the after-tax cash flow in year n for the residential and commercial models, N is the analysis period in years, and d_{real} is the real discount rate, because this rate excludes inflation effects.

The financial parameters obtained are presented in Table 8.

Table 8. Financial parameters.

Financial Viability		
IRR	%	22.6
Repayment of capital	year	1.8
Net present value	$	7446
Benefit-Cost ratio		6.8

As can be seen in Table 8, the parameters indicate that in 1.8 years, the cash flow will be positive. Another very interesting indicator is the cost-benefit ratio, which means that benefits are 6.8 times higher than costs.

Figure 10 shows the cumulative cash flow, which represents the net pre-tax flows accumulated from year 0. It represents the estimated sum of cash that will be paid or received each year during the entire life of the project.

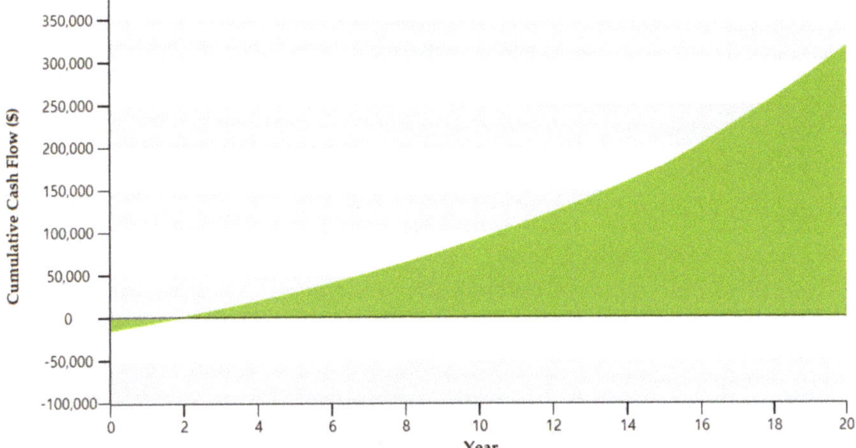

Figure 10. Cumulative cash flow.

The results of the sensitivity analysis are presented in this section. Tables 9 and 10 show what happens if the electricity price and machine hours parameters, as well as electricity exported to the grid and solar irradiance, vary, respectively.

In Tables 9 and 10, a sensitivity analysis is presented between electricity price ($)-machine hours (h) and electricity exported to grid ($)-solar irradiance (kWh/m²/day).

Table 9. The sensitivity analysis between machine hours and electricity price.

		Electricity Price ($)				
		$0.0116	**$0.0136**	**$0.0155**	**$0.0174**	**$0.0194**
	3	$0.0349	$ 0.0407	$0.0465	$0.0523	$0.0581
	4	$0.0465	$0.0543	$0.0620	$0.0698	$0.0775
Machine	5	$0.0581	$0.0678	**$0.0775**	$0.0872	$0.0969
hours	6	$0.0698	$0.0814	$0.0930	$0.1046	$0.1163
	7	$0.0814	$0.0949	$0.1085	$0.1221	$0.1356

Shaded amounts indicate the best scenario if the price increases and bolded are the optimal prices.

Table 10. The sensitivity analysis between solar irradiance and electricity price.

		Electricity Exported to Grid ($)				
		$0.0750	**$0.0875**	**$0.1000**	**$0.1125**	**$0.1250**
	4.6	$0.343	$0.400	$0.458	$0.515	$0.572
Solar	5.3	$0.400	$0.467	$0.534	$0.600	$0.667
irradiation	6.1	$0.458	$0.534	**$0.610**	$0.686	$0.763
(kWh/m^2/day)	6.9	$0.515	$0.600	$0.686	$0.772	$0.858
	7.6	$0.572	$0.667	$0.763	$0.858	$0.953

Shaded amounts indicate the best scenario if the price increases and bolded are the optimal prices.

A what-if analysis is presented in Tables 9 and 10. In Table 9 it can be observed that the electricity price increases if the hours of the machine increase as well, so that with PVS, money can be saved even if the machine works up to 7 hours per week. In Table 10, the analysis is done between solar irradiance and electricity exported to the grid. As can be seen, if the solar irradiation increases, the price of exported energy does, so that, if a CNC machine works more than 5 h per day and there are more than 6.1 kWh/m^2/day, the PVS will contribute to saving money.

The CAMP result is 8.84%, which is the expected return of the asset. It is bigger than the Mexican discount rate, which is 7.25%.

LCOE analysis shows that the cost of utility electricity is 15.5 cents/kWh, and with this price the energy produced by PVS will reduce future costs, compared with the cost of an electricity tariff that is 5 USD per kWh/month [61].

3.4. Emissions

Greenhouse gases (GHG) include water vapor, carbon dioxide (CO_2), methane (CH_4), nitrous oxide (N_2O), ozone (O_3) and several classes of halocarbons (that is, chemicals that contain carbon together with fluorine, chlorine, and bromine). Greenhouse gases allow solar radiation to enter the Earth's atmosphere but prevent the infrared radiation emitted by the Earth's surface from escaping. Instead, this outgoing radiation is absorbed by the greenhouse gases and then partially re-emitted as thermal radiation back to Earth, warming the surface.

The most relevant greenhouse gases to the energy project analysis are CO_2, CH_4, and N_2O. These GHG can also have a significant impact on global warming.

Table 11 shows the proposed case system with the GHG summary in comparison with a combined cycle power plant.

Table 11. The PVS GHG summary.

Fuel Type	Fuel Mix %	CO$_2$ Emission Factor kg/GJ	CH$_4$ Emission Factor kg/GJ	N$_2$O Emission Factor kg/GJ	Fuel Consumption MWh	GHG Emission Factor tCO$_2$/MWh
Solar PV	100	0.0	0.0	0.0	0	0.45
Combined cycle power plant	100	278.9	0.0108	0.0072	21.5	1.012

As can be seen, the system proposed has no GHG emission and has a GHG emission factor of 0.45 tCO$_2$/MWh.

The results obtained in this study establish a relationship with the works done by Lintner [55], where he determined that a PVS installed in a university could deliver enough energy to contribute to reducing the energy demand; or with Mukherji et al. [17], where in their results they concluded that could reduce greenhouse emissions.

4. Conclusions and Future Research Lines

University sustainability covers both the set of activities aimed at the appropriate use of resources in such a way as to guarantee the permanence and development of the University as an institution and the effect that the university's activity can have on the sustainability of society.

Universities continue to have a major responsibility in contributing to a more sustainable world; their actions in favor of sustainability and integrity should be a model for all sectors of society.

Many previous studies have shown the importance of the use of renewable energies within universities to achieve energy, economic, and environmental sustainability.

With the implementation of PVSs in Mexican universities, UNAM contributes both to its own sustainability plan and Mexico's sustainability.

In our case study, a PVS will be installed at ENES-J, which will be interconnected to the electrical grid and will support the electric demand of the Computer Numerical Control (CNC) type Haas Automation model UMC-750 at the Orthotics and Prosthetics Laboratory (OPL). The CNC will work 5 days a week for 4 h a day, with a peak load of 22.4 kW, and an energy required of 448 kWh per week.

UNAM's sustainability plan includes energy savings. To help achieve this, in the facilities of ENES-J in Queretaro in an area of 96.7 m^2, 50 solar panels of type mono-Si Advance Power API-M330 with an efficiency of 17.83% and a capacity factor of 20.4% will be installed, and they will provide 17.25 kW of power and 345 kWh of energy. The financial assessment shows initial costs of 46,575 $/kW, O&M costs (savings) of 569 $/kW-year, electricity export rate-monthly of 0.10 $/kWh, electricity exported to the grid of 21.5 MWh, and electricity export revenue of $2,145. Using this PVS, the ENES-J will save $12,089 per year, equivalent to 24,566 liters of fuel or 23.6 of barrels of crude oil not consumed. A sensitivity analysis, or what-if analysis, was done between machine hours and electricity price and solar irradiance and electricity exported to the grid. In the first analysis it can be proven that if the CNC machine works more hours per day, the use of PVS can save money; in the second analysis, the electricity exported to grid increases if solar irradiation does, so that if the CNC machine works more than 5 h per day and there are more than 6.1 kWh/m^2/day, the PVS will contribute to saving money.

It was found that the levelized cost of electricity (LCOE) is 15.5 cents/kWh, which allows for reducing the costs during the lifetime of the project.

The gross annual reduction in GHG emissions will be 10.2 tCO$_2$, with a GHG emission factor of 0.45 tCO$_2$/MWh. Considering these results, the implementation of the PVS is feasible, contributing to the ENES-J sustainability plan.

Comparing this technology with a combined cycle power plant, the contribution to reducing GHG represents a reduction of CO$_2$ emission factor of 278.9 kg/GJ, or CH$_4$ emission factor of 0.0108 kg/GJ and N$_2$O emission factor of 0.0072 kg/GJ.

As future research, it would be interesting to use other renewable energy sources such as biomass, wind, and thermal energy, not only for individual use, but also to improve energy sustainability at Universities.

Author Contributions: Conceptualization, Q.H.-E., A.R.-J., J.M.D.-G., M.-A.P.-M. and A.-J.P.-M.; methodology, Q.H.-E., A.R.-J., J.M.D.-G., M.-A.P.-M. and A.-J.P.-M.; formal analysis, Q.H.-E., A.R.-J., J.M.D.-G., M.-A.P.-M. and A.-J.P.-M.; investigation, Q.H.-E., A.R.-J., J.M.D.-G., M.-A.P.-M. and A.-J.P.-M.; resources, Q.H.-E., A.R.-J., J.M.D.-G., M.-A.P.-M. and A.-J.P.-M.; writing—original draft preparation, Q.H.-E., A.R.-J., J.M.D.-G., M.-A.P.-M. and A.-J.P.-M. All authors have read and agreed to the published version of the manuscript.

Funding: This research received no external funding.

Acknowledgments: The authors acknowledge RETScreen software for providing a license.

Conflicts of Interest: The authors declare no conflict of interest.

References

1. Perea-Moreno, A.J.; Perea-Moreno, M.Á.; Dorado, M.P.; Manzano-Agugliaro, F. Mango stone properties as biofuel and its potential for reducing CO$_2$ emissions. *J. Clean. Prod.* **2018**, *190*, 53–62. [CrossRef]

2. Perea-Moreno, M.-A.; Hernandez-Escobedo, Q.; Perea-Moreno, A.-J. Renewable energy in urban areas: Worldwide research trends. *Energies* **2018**, *11*, 577. [CrossRef]

3. Perea-Moreno, M.-A.; Manzano-Agugliaro, F.; Perea-Moreno, A.-J. Sustainable energy based on sunflower seed husk boiler for residential buildings. *Sustainability* **2018**, *10*, 3407. [CrossRef]

4. AbouHamad, M.; Abu-Hamd, M. Framework for construction system selection based on life cycle cost and sustainability assessment. *J. Clean. Prod.* **2019**, *241*, 118397. [CrossRef]

5. Perea-Moreno, M.-A.; Manzano-Agugliaro, F.; Hernandez-Escobedo, Q.; Perea-Moreno, A.-J. Sustainable thermal energy generation at universities by using loquat seeds as biofuel. *Sustainability* **2020**, *12*, 2093. [CrossRef]

6. Saputro, E.A.; Farid, M.M. Performance of a small-scale compressed air storage (CAS). *Int. J. Energy Res.* **2019**, *43*, 6233–6242. [CrossRef]

7. Zhou, L.; Li, Y.; Liu, F.; Jiang, Z.; Yu, Q.; Liu, L. Investigation of dynamic characteristics of a monopile wind turbine based on sea test. *Ocean Eng.* **2019**, *189*, 106308. [CrossRef]

8. Emeksiz, C.; Cetin, T. In case study: Investigation of tower shadow disturbance and wind shear variations effects on energy production, wind speed and power characteristics. *Sustain. Energy Technol. Assess.* **2019**, *35*, 148–159. [CrossRef]

9. Karasmanaki, E.; Tsantopoulos, G. Exploring future scientists' awareness about and attitudes towards renewable energy sources. *Energy Policy* **2019**, *131*, 111–119. [CrossRef]

10. Tran, T.T.D.; Smith, A.D. Stochastic optimization for integration of renewable energy technologies in district energy systems for cost-effective use. *Energies* **2019**, *12*, 533. [CrossRef]

11. Arnout, M.A.; Go, Y.I.; Saqaff, A. Pilot study on building-integrated PV: Technical assessment and economic analysis. *Int. J. Energy Res.* **2020**. [CrossRef]

12. Sierra, D.; Aristizabal, A.J.; Hernandez, J.A.; Ospina, D. Life cycle analysis of a building integrated photovoltaic system operating in Bogota, Colombia. *Energy Rep.* **2020**, *6*, 10–19. [CrossRef]

13. Sulukan, E. Techno-economic and environmental analysis of a photovoltaic system in Istanbul. *Pamukkale Univ. J. Eng. Sci.-Pamukkale Univ. Muhendis. Bilimleri Derg.* **2020**, *1*, 127–132. [CrossRef]

14. Bilck, M.; Bozikova, M.; Malinek, M. The influence of selected external factors on temperature of photovoltaic modules. *Acta Technol. Agric.* **2019**, *22*, 122–127. [CrossRef]

15. Ameur, A.; Sekkat, A.; Loudiyi, K.; Aggour, M. Performance evaluation of different photovoltaic technologies in the region of Ifrane, Morocco. *Energy Sustain. Dev.* **2019**, *52*, 96–103. [CrossRef]

16. Silveira, A.G.; Santos, D.F.L.; Montoro, S.B. Economic potential for generation of electric power by photovoltaic system in a public university. *Navus-Rev. De Gest. e Tecnol.* **2019**, *9*, 49–65. [CrossRef]

17. Mukherji, R.; Mathur, V.; Bhati, A.; Mukherji, M. Assessment of 50 kWp rooftop solar photovoltaic plant at The ICFAI University, Jaipur: A case study. *Environ. Prog. Sustain. Energy* **2019**, e13353. [CrossRef]

18. Oh, M.; Park, H.D. Optimization of solar panel orientation considering temporal volatility and scenario-based photovoltaic potential: A case study in Seoul National University. *Energies* **2019**, *12*, 3262. [CrossRef]

19. Akpolat, A.N.; Dursun, E.; Kuzucuoğlu, A.E.; Yang, Y.; Blaabjerg, F.; Baba, A.F. Performance analysis of a grid-connected rooftop solar photovoltaic system. *Electronics* **2019**, *8*, 905. [CrossRef]

20. Al-Najideen, M.I.; Alrwashdeh, S.S. Design of a solar photovoltaic system to cover the electricity demand for the faculty of Engineering- Mu'tah University in Jordan. *Resour.-Effic. Technol.* **2017**, *3*, 440–445. [CrossRef]

21. Arikan, O.; Isen, E.; Kekezoglu, B. Performance analysis of stand-alone hybrid (wind-photovoltaic) energy system. *Pamukkale Univ. J. Eng. Sci.-Pamukkale Univ. Muhendis. Bilim. Derg.* **2019**, *25*, 571–576. [CrossRef]

22. Laarabi, B.; May Tzuc, O.; Dahlioui, D.; Bassam, A.; Flota-Banuelos, M.; Barhdadi, A. Artificial neural network modeling and sensitivity analysis for soiling effects on photovoltaic panels in Morocco. *Superlattices Microstruct* **2019**, *127*, 139–150. [CrossRef]

23. Al-Dahidi, S.; Ayadi, O.; Alrbai, M.; Adeeb, J. Ensemble approach of optimized artificial neural networks for solar photovoltaic power prediction. *IEEE Access* **2019**, *7*, 81741–81758. [CrossRef]

24. Bender, W.; Waytuck, D.; Wang, S.; Reed, D.A. In situ measurement of wind pressure loadings on pedestal style rooftop photovoltaic panels. *Eng. Struct.* **2018**, *163*, 281–293. [CrossRef]

25. Wen, D.W. University-firm-government interactions in a knowledge-importing economy: Implications based on the creation of the solar photovoltaic industry in Taiwan. *Technol. Anal. Strateg. Manag.* **2019**, *31*, 1184–1198. [CrossRef]

26. Schwarz, P.M. *Energy Economics*; Routledge: Abingdon, UK, 2017; ISBN 978-1-315-11406-4.

27. Piacentini, R.D.; Della Ceca, L.S. The use of environmental sustainability criteria in industrial processes. *Dry. Technol.* **2017**, *35*, 1–3. [CrossRef]

28. Szekely, F.; Dossa, Z.; Hollender, J. *Beyond the Triple Bottom Line: Eight Steps towards a Sustainable Business Model*; The MIT Press: Cambridge, MA, USA, 2017; ISBN 978-0-262-33907-0.

29. Zafrilla, J.-E.; Arce, G.; Cadarso, M.-Á.; Córcoles, C.; Gómez, N.; López, L.-A.; Monsalve, F.; Tobarra, M.-Á. Triple bottom line analysis of the Spanish solar photovoltaic sector: A footprint assessment. *Renew. Sustain. Energy Rev.* **2019**, *114*, 109311. [CrossRef]

30. Ahmed, W.; Sarkar, B. Management of next-generation energy using a triple bottom line approach under a supply chain framework. *Resour. Conserv. Recycl.* **2019**, *150*, 104431. [CrossRef]

31. Hammond, J.; Tarabay, M. Higher Education for Sustainability in the Developing World: A Case Study of Rafik Hariri University in Lebanon. *Eur. J. Sustain. Dev.* **2019**, *8*, 379–403. [CrossRef]

32. Zutshi, A.; Creed, A.; Connelly, B.L. Education for sustainable development: Emerging themes from adopters of a declaration. *Sustainability* **2019**, *11*, 156. [CrossRef]

33. Zutshi, A.; Creed, A. Declaring Talloires: Profile of sustainability communications in Australian signatory universities. *J. Clean. Prod.* **2018**, *187*, 687–698. [CrossRef]

34. Brito, M.R.; Rodriguez, C.; Aparicio, J.L. Sustainability in teaching: An evaluation of university teachers and students. *Sustainability* **2018**, *10*, 439. [CrossRef]

35. Criteria & Indicators | UI GreenMetric. Available online: http://greenmetric.ui.ac.id/criteria-indicator/ (accessed on 27 November 2019).

36. What is the Kyoto Protocol? Available online: https://unfccc.int/kyoto_protocol (accessed on 16 January 2020).

37. Instituto Nacional de Estadística y Geografía (INEGI). Available online: https://www.inegi.org.mx/ (accessed on 27 November 2019).

38. Yendaluru, R.S.; Karthikeyan, G.; Jaishankar, A.; Babu, S. Techno-economic feasibility analysis of integrating grid-tied solar PV plant in a wind farm at Harapanahalli, India. *Environ. Prog. Sustain. Energy* **2019**, e13374. [CrossRef]

39. Islam, G.; Darbayeva, E.; Rymbayev, Z.; Dikhanbayeva, D.; Rojas-Solorzano, L. Switching-off conventional lighting system and turning-on LED lamps in Kazakhstan: A techno-economic assessment. *Sustain. Cities Soc.* **2019**, *51*, 101790. [CrossRef]

40. Liu, C.; Xu, W.; Li, A.; Sun, D.; Huo, H. Energy balance evaluation and optimization of photovoltaic systems for zero energy residential buildings in different climate zones of China. *J. Clean. Prod.* **2019**, *235*, 1202–1215. [CrossRef]

41. Jahangiri, M.; Nematollahi, O.; Haghani, A.; Raiesi, H.A.; Shamsabadi, A.A. An optimization of energy cost of clean hybrid solar-wind power plants in Iran. *Int. J. Green Energy* **2019**, *16*, 1422–1435. [CrossRef]

42. Samu, R.; Fahrioglu, M.; Ozansoy, C. The potential and economic viability of wind farms in Zimbabwe. *Int. J. Green Energy* **2019**, *16*, 1539–1546. [CrossRef]

43. Canada, N.R. RETScreen. Available online: https://www.nrcan.gc.ca/energy/retscreen/7465 (accessed on 2 December 2019).

44. Castro, L.M.; Rodríguez-Rodríguez, J.R.; Martin-del-Campo, C. Modelling of PV systems as distributed energy resources for steady-state power flow studies. *Int. J. Electr. Power Energy Syst.* **2020**, *115*, 105505. [CrossRef]

45. Kallioglu, M.A.; Durmus, A.; Karakaya, H.; Yilmaz, A. Empirical calculation of the optimal tilt angle for solar collectors in northern hemisphere. *Energy Sources Part A-Recovery Util. Environ. Eff.* **2020**, *42*, 1335–1358. [CrossRef]

46. Perea-Moreno, A.-J.; Hernandez-Escobedo, Q.; Garrido, J.; Donaldo Verdugo-Diaz, J. Stand-alone photovoltaic system assessment in warmer urban areas in Mexico. *Energies* **2018**, *11*, 284. [CrossRef]

47. Tejada, A.; Gómez-Azpeitia, G. *Prontuario Solar de México*; Universidad de Colima: Colima, Mexico, 2015.

48. Spataru, S.; Hacke, P.; Sera, D.; Packard, C.; Kerekes, T.; Teodorescu, R. Temperature-dependency analysis and correction methods of in situ power-loss estimation for crystalline silicon modules undergoing potential-induced degradation stress testing. *Prog. Photovolt. Res. Appl.* **2015**, *23*, 1536–1549. [CrossRef]

49. Short, W.; Packey, D.; Holt, T. *A Manual for the Economic Evaluation of Energy Efficiency and Renewable Energy Technologies (No. NREL/TP-462-5173)*; National Renewable Energy Laboratory: Golden, CO, USA, 1995.

50. Helton, J.C. Uncertainty and sensitivity analysis techniques for use in performance assessment for radioactive waste disposal. *Reliab. Eng. Syst. Saf.* **1993**, *42*, 327–367. [CrossRef]

51. Saltelli, A.; Tarantola, S.; Campolongo, F.; Ratto, M. *Sensitivity Analysis in Practice: A Guide to Assessing Scientific Models*; John Wiley & Sons: Hoboken, NJ, USA, 2004.

52. Saltelli, A.; Annoni, P.; Azzini, I.; Campolongo, F.; Ratto, M.; Tarantola, S. Variance based sensitivity analysis of model output. Desing and estimator for the total sensitivity index. *Comput. Phys. Commun.* **2010**, *181*, 259–270. [CrossRef]

53. Ye, M.; Hill, M.C. Global Sensitivity analysis for uncertain parameters, models, and scenarios. In *Sensitivity Analysis in Earth Observation Modelling*; Elsevier: Amsterdam, The Netherlands, 2017; Chapter 10; pp. 177–210.

54. Sharpe, W. Capital asset prices: A theory of market equilibrium under conditions of risk. *J. Financ.* **1964**, *19*, 425–442.

55. Lintner, J. Security prices, risk, and maximal gains from diversification. *J. Financ.* **1965**, *20*, 587–615.

56. Ferreira, J.R.; Securato, J.R.; Reed, D.; Lopes, F. Comparing results of the implied cost of capital and capital asset pricing models for infrastructure firms in Brazil. *Util. Policy* **2019**, *56*, 149–158. [CrossRef]

57. Bosch, J.; Staffell, I.; Hawkes, A.D. Global levelized cost of electricity from offshore wind. *Energy* **2019**, *189*, 116357. [CrossRef]

58. Perkins, G. Techno-economic comparison of the levelized cost of electricity generation from solar PV and combustion of bio-crude using fast pyrolysis of biomass. *Energy Convers. Manag.* **2018**, *17*, 1573–1588. [CrossRef]

59. Mexico Central Bank Discount Rate. Available online: https://www.indexmundi.com/mexico/central_bank_discount_rate.html (accessed on 12 February 2020).

60. Secretaria de Energia, Mexico. Available online: https://www.gob.mx/inecc/acciones-y-programas/inventario-nacional-de-emisiones-de-gases-y-compuestos-de-efecto-invernadero (accessed on 16 January 2020).

61. Comision Federal de Electricidad, CFE, Mexico. Available online: https://app.cfe.mx/Aplicaciones/CCFE/Tarifas/TarifasCRECasa/Tarifas/TarifaDAC.aspx (accessed on 14 March 2020).

Article

Zapote Seed (*Pouteria mammosa L.*) Valorization for Thermal Energy Generation in Tropical Climates

**Miguel-Angel Perea-Moreno [1],*, Quetzalcoatl Hernandez-Escobedo [2],
Fernando Rueda-Martinez [3] and Alberto-Jesus Perea-Moreno [4]**

[1] Graduate School of Engineering and Technology, Universidad Internacional de La Rioja (UNIR),
    Av. de la Paz, 137, Logroño, 26006 La Rioja, Spain
[2] Escuela Nacional de Estudios Superiores Juriquilla, UNAM, 76230 Queretaro, Mexico;
    qhernandez@unam.mx
[3] Facultad de Ingeniería, Universidad Veracruzana Campus Coatzacoalcos, 96535 Coatzacoalcos, Mexico;
    frueda@uv.mx
[4] Departamento de Física Aplicada, Radiología y Medicina Física, ceiA3, Campus de Rabanales,
    Universidad de Córdoba, 14071 Córdoba, Spain; aperea@uco.es
* Correspondence: miguelangel.perea@unir.net; Tel.: +34-941-210211

Received: 15 April 2020; Accepted: 20 May 2020; Published: 23 May 2020

Abstract: According to the Law for the Use of Renewable Energies and the Financing of Energy Transition, Mexico's goal for 2024 is to generate 35% of its energy from non-fossil sources. Each year, up to 2630 tons of residual biomass from the zapote industry are dismissed without sustainable use. The main purposes of this study were to determine the elemental chemical analysis of the zapote seed and its energy parameters to further evaluate its suitability as a solid biofuel in boilers for the generation of thermal energy in a tropical climate. Additionally, energy, economic, and environmental assessments of the installation were carried out. The results obtained show that zapote seed has a higher heating value (18.342 MJ/kg), which makes it appealing for power generation. The Yucatan Peninsula is the main zapote-producing region, with an annual production of 11,084 tons. If the stone of this fruit were used as biofuel, 7860.87 MWh could be generated and a CO_2 saving of 1996.66 tons could be obtained. Additionally, replacing a 200 kW liquefied petroleum gas (LPG) boiler with a biomass boiler using zapote seed as a biofuel would result in a reduction of 60,960.00 kg/year of CO_2 emissions. Furthermore, an annual saving of $7819.79 would be obtained, which means a saving of 53.19% relative to the old LPG installation. These results pave the way toward the utilization of zapote seed as a solid biofuel and contribute to achieving Mexico's energy goal for 2024 while promoting sustainability in universities.

Keywords: sustainability; renewable energy; zapote seed; universities; biomass boiler

---

1. Introduction

The increase in the worldwide population, which is expected to exceed 11 billion by the end of the century, together with economic and social development, has led to an increase in energy demand [1,2]. Despite progressive growth and the falling costs of renewable energy sources, coal, oil, and gas remain the mainstay of the steadily growing energy consumption on a global scale [3]. The energy transition to a low-carbon economy is also threatened by geopolitical and commercial tensions, and by declining investment in clean energy, according to a study published by Capgemini in collaboration with De Pardieu Brocas Maffei and Vaasa ETT. As a result, the concentrations of the main greenhouse gases (GHG) that trap heat in the atmosphere once again reached record levels in 2018: carbon dioxide (CO_2) increased by 147%, methane (CH_4) by 259%, and nitrous oxide (N_2O) by 123%.

These increases make climate change more acute as temperatures rise and extreme weather events multiply [4].

Climate change represents by far one of the major threats facing our society [5]. On 12 December 2015, at the Conference of Parties 21 (COP21) in Paris, all signatory countries came to a historic agreement to promote and accelerate the necessary measures in the fight against climate change, which included accelerating the energy transition to a sustainable low-carbon economy. The purpose of the Paris Agreement was to establish a coordinated global response to the threat of climate change and to conclude a global agreement on the reduction of greenhouse gases (GHGs), which intends to limit the increase in global temperature this century to below 2 °C compared to pre-industrial levels and to maintain ongoing efforts to further reduce the rise in temperature to 1.5 °C. Furthermore, the agreement seeks to enhance the capacity of countries to address the effects of climate change and to ensure that economic streams produce low greenhouse gas (GHG) emissions [6,7].

According to the International Energy Agency, more than a third of the world's energy is consumed in buildings, 70% of which is used to meet heating and cooling requirements [8]. Buildings are one of the main sources of greenhouse gas emissions, and for this reason, more effort is needed to understand the energy consumption patterns in buildings and to establish strategies for energy saving at the district level [9,10]. To reduce these emissions (up to 20% since 1990) and promote the use of renewable energies, among other measures, the European Directive 2010/31/EU introduced the definition of a nearly zero energy building (NZEB) and appeals for deployment in public institutions by 31 December 2018, whereas it extends to the identical date in 2020 for privately owned buildings [11–13]. The principal feature of low-energy buildings is that their energy demand must be equal to their energy generated, that is, they must be able to produce the same energy or more than they will consume over a whole year. Furthermore, the energy produced must be in situ or in the closest environment and using renewable energies [14].

Mexico has clear goals regarding the use and exploitation of renewable energies, and Mexican universities, in the context of their social and environmental liability, cannot remain on the sidelines of this commitment. According to the Law for the Use of Renewable Energies and the Financing of Energy Transition, Mexico's goal for 2024 is to generate 35% of its energy from non-fossil sources. To achieve this goal, Mexico would have to increase the share of renewables by 400 percent in less than 10 years. In Mexico, there are a large number of renewable resources with enormous potential for energy generation. We can highlight the high levels of solar radiation, a multitude of dams and reservoirs for the generation of hydroelectric energy, potential for the development of geothermal energy, areas with intense wind activity, and the generation of large quantities of agricultural and industrial waste, as well as solid urban waste [15,16].

Renewable energies are almost unlimited resources that nature provides us with and that we must take advantage of due to their sustainable nature. Another important advantage is that these resources are indigenous and increase a country's energy resilience by decreasing its dependence on energy imports. They also increase the diversification of the energy supply and boost the growth of new technologies and job creation [17]. Under the term of renewable energy, different heterogeneous categories of technologies can be grouped. Some technologies can generate electricity and heat, as well as mechanical energy, while others can produce solid, liquid, or gaseous fuels that meet multiple energy needs. On the other hand, while some technologies can be implemented within rural and urban environments at the same place of consumption (centralized), others are generated at a distance and therefore require large transport networks (decentralized) [18,19].

Among the available renewable energies, the use of biomass for heating has experienced great growth in the last decade. Biomass is becoming more and more important as an alternative to traditional fossil fuels, such as coal or oil derivatives, since it is a more accessible and a better-distributed geographical resource, its theoretical balance of CO_2 emissions is considered to be zero, and it is renewable, among other advantages [20]. Biomass can be defined as the organic substances that have their origin in the carbon compounds formed in photosynthesis and that can be used as an energy source. Under this definition, many materials can be considered biomass. Solid biofuel sources are usually divided into those of primary nature and those of secondary nature. In the former, which includes energy crops and forest biomass extracted for energy purposes, their management and use are mainly oriented toward energy production. In the sources of secondary origin, in what is generally called residual biomass, we find agricultural by-products (straw, cane, and fruit prunings), forest by-products (remains from forestry interventions, such as branches and trees with no commercial value), and industrial by-products (sawdust, fruit shells and stones, etc.). Both primary and secondary biomass can be used either via direct combustion or thermochemical processes, such as pyrolysis and gasification [21]. Additionally, both primary and secondary sources can be used for the generation of thermal energy in boilers and district heating networks, as well as for the generation of electrical energy in biomass plants.

Heating and domestic hot water generation are the most widespread thermal applications of biomass, though it is also used for cooling and power generation. Biomass can be used to generate the thermal energy necessary to power an air-conditioning system (heating and cooling), either at a centralized level (single-family house or block of flats) or at a decentralized district level (several buildings) [22,23]. Implementing a distribution network to supply not only housing estates and other residential dwellings but also public buildings, industries, etc., is known as "district heating", and involves transporting the thermal energy over a distance through ducts that are not structurally integrated into a single building [24].

A recent Reuters Business Insight report estimates that the contribution of bioenergy to the world's primary energy supply could reach 50% by 2050. Bioenergy projects are largely conditioned by the location and availability of biomass resources, although in recent years, the market for standardized biofuels has developed considerably. Within the applications of biomass for heating, its use in boilers to replace fossil fuels has experienced a great boom in the last decade, mainly due to the significant energy and economic savings obtained [25].

Among the most commonly used fuels for biomass boilers are pellets, mostly from the wood and forestry industries that generate waste in the form of sawdust and chips. However, due to the increase in the price of this biofuel in recent years, it is necessary to investigate new forms of bioenergy that allow for reducing costs on the one hand and alleviating the pressure on the wood industry on the other. The European standard EN 14961-1 promotes the use of new forms of biofuels, and studies have already shown the potential of certain fruit shells and stones for generating heat [26–29].

Today, Mexico is a country that depends heavily on fossil fuels for the production of its primary energy. If we pay attention to its energy production between the years 2007 and 2017, this amounted to 8935 PJ, of which 88.2% was referred to fossil fuels, 7.2% was renewable energy (1.31% obtained from biomass), 3.4% was from coal, and 1.2% was nuclear energy [30].

If we focus on primary energy production in Mexico from biomass, the main sources are firewood, occupying 5.2%, and sugarcane bagasse with 0.9%. A large part of the energy production from sugar cane bagasse (50%) is used by the sugar industry for the production of its energy [31].

According to the Mexican Secretary of Energy (SENER), 45% of the waste obtained from crops is used as fertilizer or for cooking and heating houses in rural areas [31].

Honorato-Salazar et al. [32] studied the available agricultural residues in Mexico from twenty crops. They estimated that there is between 17.5 and 58.1 megatons of dry matter produced per year (Mt DM/yr). The corresponding bioenergy potential ranges from 313.4 PJ/yr to 1039.4 PJ/yr.

The zapote mamey (*Pouteria sapota*, also called *Lucuma mammosa*) is a tree species of the Sapotaceae family. The zapote mamey tree is an ornamental, perennial tree that can reach a height of up to 40 m and a diameter at breast height of more than 1 m. The trunk is straight and may have buttresses. The external bark is cracked and falls off into rectangular pieces, brownish-gray to brownish-black in color, with a thickness of 10 to 20 mm. The leaves are arranged in a spiral and agglomerate at the tips of the branches. The fruits are berries that are up to 20 cm long, ovoid, reddish-brown in color, and rough in texture. The mesocarp is sweet, fleshy, orange to reddish, with small amounts of latex when unripe. It usually contains a seed that is up to 10 cm long, ellipsoid, and black to dark brown. The percentage in weight of the seed ranges from 8% to 16% of the total weight of the fruit [33].

Its exact origin is difficult to determine since it was already being cultivated throughout tropical America before the arrival of Europeans. Its natural range probably extends from southern Mexico to Nicaragua, Belize, and northern Honduras. In Mexico, the original areas of distribution are likely to have been in Veracruz, Tabasco, and northern Chiapas; however, it is now found in all southern Mexican states. It is cultivated from Florida (USA) to Brazil and Cuba. It has also been introduced to the Philippines, Indonesia, Malaysia, and Vietnam. It is a plant that is widely distributed in its place of origin, from southern Mexico to Guatemala and Belize, where it is found both wild and cultivated [34].

In Mexico, production has been concentrated since 2006, mainly in the states of Yucatan and Guerrero, but the states of Chiapas, Michoacan, Veracruz, and Campeche are consolidating their participation in production every year. The trend in the production of Mamey in Mexico has had a greater momentum from 2006 to 2008; after that period of growth, the upward trend has been constant but slow [33]. Figure 1 shows the zapote production in the main states of Mexico.

Figure 1. Zapote production in the main states of Mexico.

The fruit is eaten raw or made into smoothies, ice cream, and fruit bars. It can be used to produce jam and jelly. Some beauty products use pressed oilseed to produce what is known as sapayul oil. However, the food and cosmetics zapote industries produce large amounts of waste, which is thrown away without making environmentally responsible use of it.

The main purposes of this study were to determine the elemental chemical analysis of the zapote seed and its energy parameters to further evaluate its suitability as a solid biofuel in boilers for the generation of thermal energy in a tropical climate. Additionally, energy, economic, and environmental assessments of the installation were carried out.

2. Case Study

The heavy chemistry laboratory of the Faculty of Engineering of the University of Veracruz in Mexico was taken as a case study (Figure 2). The Veracruzana University (UV) was founded in 1944 and became autonomous in 1996. It is located in the state of Veracruz on the Gulf of Mexico. It has five regional headquarters: Xalapa, Veracruz, Orizaba-Córdoba, Poza Rica-Tuxpan, and Coatzacoalcos-Minatitlán. In terms of the number of students it serves, it is among the five largest public state universities of higher education in Mexico.

Figure 2. Location of the Faculty of Engineering of the University of Veracruz.

The dimensions of the chemistry laboratory (Figure 3) are $35 \times 15 \times 3$ m, and among the equipment, the laboratory has two evaporators and a reactor for the dehydration of food. These equipment are fed by one 200 kW liquefied petroleum gas (LPG) boiler to produce steam. It consumes 200 L of LPG to produce 400 kg/h of saturated steam. Figure 4 shows the evaporator–condenser assembly.

Figure 3. University of Veracruz heavy chemistry lab.

Figure 4. Evaporator–condenser assembly.

Figure 5 shows the LPG boiler used to produce the steam needed.

Figure 5. Liquefied petroleum gas (LPG) boiler used for steam production.

Initially, an energy audit was conducted, which served as the basis for the technical study. Subsequently, all the required data was gathered to assess the feasibility of substituting the current LPG boiler with a biomass boiler that would use zapote seeds as fuel.

2.1. Climatological Data

The city of Coatzacoalcos in the state of Veracruz (Mexico) has a tropical, warm, and humid climate characterized by a short dry season and heavy rainfall. The average annual temperature is 25 °C and the average rainfall is 2471 mm per year. Table 1 shows the most relevant climatic data for the city of Coatzacoalcos in the state of Veracruz.

Table 1. Climatic data of Veracruz state.

Parameters	Values
Longitude	94°28′32.55″ W
Latitude	18°8′39.77″ N
Altitude	10 m
Average maximum annual temperature	29.2 °C
Average minimum annual temperature	22.9 °C
Average annual temperature	25.3 °C
Average temperature of the dry season	24.3 °C
Average temperature of the wet season	26.8 °C
Relative humidity	78%

2.2. Overview of the Existing Heating System

The 200 kW LPG boiler operates for approximately 4 h daily and an annual average of 300 days. Therefore, the thermal power demanded is 240,000 kWh.

Figure 6 shows the operating diagram of the LPG boiler.

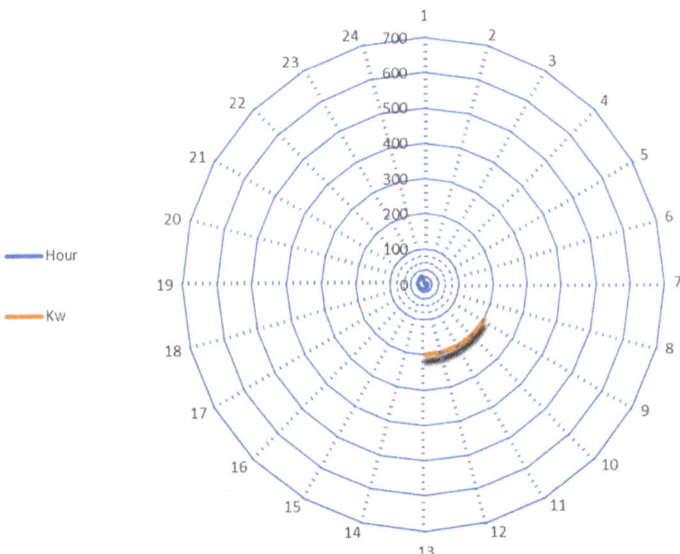

Figure 6. Running chart of the LPG boiler.

3. Materials and Methods

Regarding the use of zapote seeds as biofuel, 2000 g of this waste was collected for the determination of its chemical composition and the evaluation of its energy parameters (Figure 7).

Figure 7. Zapote mamey (*Pouteria mammosa L*).

The UNE-EN 14961-1 standard "Solid biofuels–Specifications and fuel classes—Part 1: General requirements," was used to assign the biomass quality parameters and the standard UNE-EN 15148 was used for the determination of the volatile matter. These standards have been developed by the Spanish Association for Standardisation and Certification (AENOR).

Table 2 shows the standards and measuring equipment used.

Table 2. Standards and the measuring equipment used in this study.

Parameter	Unit	Standards	Equipment	Standard Deviation (SD)
Higher heating value	MJ/kg	EN 14918	Calorimeter Parr 6300 (Parr Instrument Company, Illinois, United States)	0.02
Total sulphur	%	EN 15289	Analyzer LECO TruSpec S 630-100-700 (LECO Instrumentos S.L., Madrid, Spain)	0.002
Total hydrogen	%	EN 15104	Analyzer LECO TruSpec CHN 620-100-400 (LECO Instrumentos S.L., Madrid, Spain)	0.03
Total chlorine	mg/kg	EN 15289	Titrator Mettler Toledo G20	6.73
Total carbon	%	EN 15104	Analyzer LECO TruSpec CHN 620-100-400 (LECO Instrumentos S.L., Madrid, Spain)	0.12
Total nitrogen	%	EN 15104	Analyzer LECO TruSpec CHN 620-100-400 (LECO Instrumentos S.L., Madrid, Spain)	0.009
Ash	%	EN 14775	Muffle Furnace NABERTHERM LVT 15/11 (Nabertherm GmbH, Lilienthal, Germany)	0.02
Moisture	%	EN 14774-1	Drying Oven Memmert UFE 700 (Memmert GmbH, Schwabach, Germany)	According to EN14774-1
Volatile matter and fixed carbon	%	EN 15148	Muffle Furnace NABERTHERM LVT 15/11 (Nabertherm GmbH, Lilienthal, Germany)	0.02

3.1. Humidity

The moisture content was determined by drying the fuel at a temperature of 105 ± 2 °C in a MEMMERT UFE 700 oven and carrying out three to five complete hourly air swaps. After a period of permanence that should not exceed 24 h to avoid releasing volatiles, the sample was weighed again. The difference in mass corresponded to the sample's moisture content.

The percentage of moisture in biofuel is a major factor when determining its energy efficiency. This is because for the heat to be available, the water must first be evaporated, where the greater the moisture percentage, the lesser the calorific value of the fuel. Furthermore, a high moisture content leads to the appearance of slag that tends to accumulate in the flue pipes, leading to corrosion and clogging problems. Most combustion reactions are complete with biomass moisture contents below 30%, with a maximum heat transfer at 10% moisture [35].

3.2. Elemental Composition

Except for the fraction corresponding to moisture and ash, the biomass is mainly made up of carbon (C), hydrogen (H), nitrogen (N), oxygen (O), sulfur (S), and chlorine (Cl) compounds. The elemental composition provides the relative amounts of these chemical elements.

For the calculation of the carbon, hydrogen, and nitrogen content, a given mass of the sample is burnt in an oxygen-containing atmosphere, resulting in ash and gaseous products of combustion (carbon dioxide, water vapor, elemental nitrogen, sulfur oxides, and hydrogen halides) [36]. Using instrumental techniques and the LECO Tru Spec CHN 620-100-400 analyzer, the mass fraction of carbon, hydrogen, and nitrogen in the gas stream was quantitatively determined.

The ultimate analysis of the biomass is required to evaluate the performance of biofuel and also allows for the estimation of its calorific value through the use of analytical expressions. It is also indicative of its environment-related burden since nitrogen turns to N_2 and NO_x, the last of which is one of the main greenhouse gases. The carbon content also helps to evaluate CO_2 emissions.

3.3. Ash Content

Ash is the inorganic and non-combustible fraction of biomass formed due to the minerals it contains, some of which are sodium, potassium, chlorine, and phosphorus. The ash is considered an indicator of the efficiency of the biofuel because the more residue it leaves when burned, the greater the problems of clogging and corrosion in the installation [37].

For its determination, a muffle oven (NABERTHERM LVT 15/11) was used, quantifying the residual mass after the sample was burned under controlled conditions of weight, time, and temperature.

3.4. Chlorine and Sulfur Content

The chlorine and sulfur contents are other indicators of the quality of a biofuel. After combustion, chlorine is transformed into chlorides, which cause corrosion problems in the installation. On the other hand, sulfur is transformed into sulfur oxide, which is one of the main precursors of SO_x emissions into the atmosphere [38].

The standard EN 15289 was followed for its determination. A calorimetric pump (Tirator Mettler Toledo G20) was utilized to determine the chlorine content in the combustion washing water using silver nitrate. For the sulfur, a tubular furnace (LECO TruSpec S 630-100-700 Analyzer) was used for the combustion of the sample at high temperature (1350 °C) and further gas quantification.

3.5. Higher Heating Value (HHV) and Lower Heating Value (LHV)

By breaking the bonds of organic compounds, by direct combustion of biomass, or by combustion of products obtained from it through physical or chemical transformations, carbon dioxide and water, among other substances, are obtained and energy is released in the process. The quality of a biofuel depends on the amount of heat it can release in the energy conversion process; this amount of heat relative to the unit of mass is the power or calorific value.

Depending on how the calorific value of the biomass is measured, the higher heating value (HHV) and lower heating value (LHV) can be defined. The HHV is the total amount of heat released in the combustion of 1 kg of biomass considering the energy needed to evaporate the water since it is assumed that the water vapor will be completely condensed, while the LHV is the heat released during the combustion of 1 kg of biofuel without considering the energy consumed to evaporate the water since it is considered that water vapor does not condense and therefore this energy is not transformed into useful heat power and is lost.

Whereas the HHV is determined by experimentation using a calorimetric pump according to the UNE-EN 14918 standard, the LHV can be calculated from it by using the following expression [39–41]:

$$\mathrm{LHV}\left(\frac{\mathrm{kJ}}{\mathrm{kg}}\right) = \mathrm{HHV}\left(\frac{\mathrm{kJ}}{\mathrm{kg}}\right) - 212.2 \times \mathrm{H\%} - 0.8 \times (\mathrm{O\%} + \mathrm{N\%}). \tag{1}$$

3.6. Volatile Matter (VM)

The volatile matter content is determined by the loss of mass, less that due to moisture, when the biofuel undergoes a process of heating due to high temperature without being in contact with air. It can come from the organic or inorganic fraction of the biomass. Volatile matter conditions the design of the installation. The higher the volatile content of the biomass, the better the ignition at low temperatures will be since the reactivity is improved and the combustion process is enriched.

3.7. Fixed Carbon (FC)

Fixed carbon is the non-volatile part found in the crucible when the volatile materials have been determined. The quantity of fixed carbon is not obtained from the weight of the waste but is obtained using the equation:

$$FC = 100 - (\%Ash + \%VM).$$

The VM/FC ratio is known as the combustibility index. If the VM/FC ratio increases, the degree of reactivity of the biomass also increases; therefore, the greater this quotient, the easier the ignition and the lower time needed for complete combustion.

3.8. RETScreen Analysis

The analysis was carried out using the RETScreen software, which has been used to analyze and assess several bioenergy projects. RETScreen is software developed by the Natural Resource of Canada, which helped with the analysis of zapote seed as a source of energy [42]. The biomass heating project analysis that RETScreen uses is presented in Figure 8.

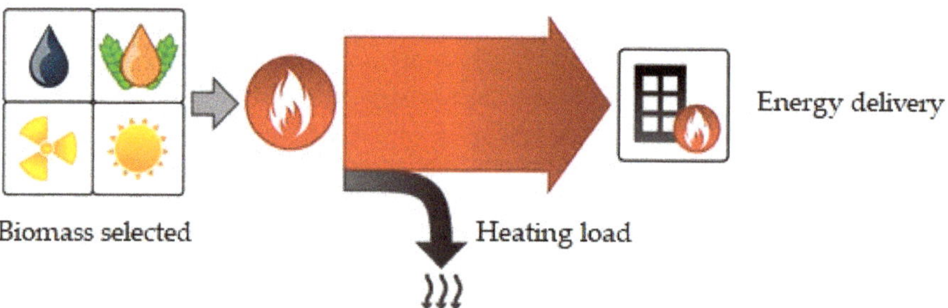

Figure 8. Bioenergy process in RETScreen.

This process starts by specifying the biomass selected, which in this case was zapote seed. The model calculates the building's (or buildings') heating requirements. The next phase considers the heating load, where in this stage, all information about zapote seed is given. Finally, the energy delivered and fuel consumption is calculated. This final stage uses the load and demand information of the biomass heating system and/or the peak load heating system.

4. Results and Discussion

In this section, the results of the zapote seed's physical, chemical, and energy parameter analyses are presented and compared with the quality parameters of other standardized solid biofuels to evaluate the use of this agro-industrial waste as a solid biofuel in thermal energy generation systems. An energy, environmental, and economic balance of the thermal system is also presented.

4.1. Zapote Seed Values

In this section, we present the quality indicators of the zapote seed, for which 2000 g of a sample of this waste from surrounding industries were collected and analyzed in the laboratory.

Table 3 shows the mean value, the standard deviation, the maximum value, and the minimum value, which determined the parametric distribution.

Table 3. Chemical and energy indicators derived from zapote seed profiling (parameters calculated on a dry basis, except for moisture).

Magnitude	Unit	Mean Value	Standard Deviation (SD)	Maximum Value	Minimum Value
Moisture	%	56.83	—	56.83	56.83
Ash content	%	2.42	0.070	2.49	2.35
HHV	MJ/kg	18.342	0.022	18.364	18.320
LHV	MJ/kg	17.021	0.012	17.033	17.009
Total carbon	%	48.12	0.007	48.127	48.113
Total hydrogen	%	6.05	0.012	6.062	6.038
Total nitrogen	%	0.83	0.052	0.882	0.778
Total sulfur	%	0.05	0.002	0.052	0.048
Total oxygen	%	45.02	2.542	47.562	42.478
Total chlorine	%	0.06	0.003	0.063	0.057
Volatile matter	%	85.73	0.060	85.79	85.67
Fixed carbon	%	11.85	0.060	11.91	11.79

HHV: Higher heating value, LHV: Lower heating value.

As shown in Table 3, the main problem with zapote seed for use as a biofuel is its high moisture content of over 50%. Due to this, instability in the thermochemical process is produced that diminishes the efficiency of the combustion since it is necessary to evaporate the water first such that the heat is available; as a result, the net calorific value of the biomass is reduced and causes problems of jamming and corrosion in the thermal installation due to the emission of tars. Furthermore, environmental pollution increases as a result of inefficient conversion processes in which carbon monoxide and volatile organic compounds are generated.

The endothermic limit of biomass is 65% humidity; above this value, not enough energy is released to satisfy the evaporation requirement and produce heat. Above a 50% moisture content, some combustion systems will require a support fuel, such as natural gas or LPG. Therefore, in most heat transfer processes, humidities below 30% are required, with an optimum performance occurring at 10% humidity.

In our case study, a drying system needed to be implemented to reduce the humidity of the seed to around 10% before it was used for direct combustion in a boiler. The drying of the biomass can be done naturally in piles or forced with industrial drying systems.

On the other hand, the use of biomass via thermochemical conversion generates a solid residue or ash that comes from the mineral matter of the original plants, as well as from that which may have been added during the collection process. As it is an inert fraction, the energy available in any fuel is reduced in proportion to its ash content.

Therefore, ash is the inorganic fraction of biomass that is mainly composed of metallic oxides, such as calcium oxide (CaO), sodium oxide (Na_2O), potassium oxide (K_2O), phosphorus oxide (P_2O_5), aluminum oxide (Al_2O_3), iron oxide (Fe_2O_3), silicon oxide (SiO_2), and magnesium oxide (MgO), with the oxides being either in the form of salts or organically associated compounds.

Calcium oxide (CaO) and silica (SiO_2) affect the abrasion of the plant components due to the dragging of the ashes toward the gas outlet. Alkaline metals (K and Na) react with phosphorus and silica, which can cause slag deposits that prevent the correct transfer of heat, even generating problems in the correct operation of the installation. Furthermore, alkaline and alkaline-earth metals influence the melting temperature of the ashes, and consequently, the greater or lesser adhesion of the deposits that are formed. Potassium and sodium also react with chlorine in the gas phase (HCl) to form corresponding chlorides that can be deposited on the pipe surfaces and cause corrosion.

In the case of the zapote seed, its ash content varies between 2.49% and 2.35%, which indicates that despite being a moderate value, it is below other standardized solid fuels, such as almond shell pellets (3.35%) or oak wood pellets (3.32%) [43].

The deviation in the results depends on multiple parameters. In addition to the number of repetitions and samples analyzed, it also depends on the part of the plant analyzed. For example, waste fractions, as in the case study, usually have a higher ash content (due to impurities that may be introduced during collection and transport). On the other hand, the reproducibility of the results depends significantly on the type of plant, the growth conditions, the use of fertilizers, and the type of soil, which is related to the level of soluble inorganic elements (e.g., sandy soils give rise to less ash compared to clay soils). It can be seen that C3 plants, such as zapote, tend to have a higher water intake compared to C4 plants, and therefore have a higher ash content.

Table 4 contrasts the physicochemical indicators of the quality of various commercial biofuels and industrial wastes versus those of zapote seed to assess the applicability of this by-product in boilers.

Table 4. Comparison between zapote seed and other biofuels.

Parameters	Unit	Mango Stone [26]	Olive Stone [29,35,44]	Pine Pellets [43,44]	Almond Shell [44–46]	Zapote Seed
Moisture	%	59.70	18.45	7.29	7.63	56.83
HHV	MJ/kg	18.049	17.884	20.030	18.200	18.342
LHV	MJ/kg	17.271	16.504	18.470	17.920	17.021
Ash content	%	2.14	0.77	0.33	0.55	2.42
Total carbon	%	48.263	46.55	47.70	49.27	48.12
Total hydrogen	%	3.481	6.33	6.12	6.06	6.05
Total nitrogen	%	1.041	1.810	1.274	0.120	0.830
Total sulphur	%	0.086	0.110	0.004	0.050	0.050
Total oxygen	%	49.022	45.20	52.30	44.49	45.02
Total chlorine	%	0.070	0.060	0.010	0.200	0.060
Volatile matter	%	83.77	78.3	83.5	82.0	85.73
VM/FC ratio	%	5.63	3.84	5.49	5.19	7.23
$\frac{HHV_{biomass}}{HHV_{zapote\ seed}}$	%	98.40	97.50	109.20	99.23	100.0

FC: Fixed carbon, VM: Volatile matter.

As reflected in the comparison, zapote seed had a higher ash content than other standardized fuels, which would suggest that the problems associated with corrosion and maintenance costs of the installation could be greater.

When analyzing the calorific value of the zapote seed, we found that it was in the range of 18.320–18.364 MJ/kg, which was higher than the energy value of other standardized fuels, such as olive stone (17.884 MJ/kg) or almond shell (18.200 MJ/kg), which shows the energy potential of this residual biomass. The last row of the table shows the variation of HHV as a percentage, varying from −2.5% to +9.2%.

Other parameters that indicate the quality of biofuel are its chlorine and sulfur contents. Sulfur has a corrosive effect on the installation and causes the emission of SO_x gases that contribute to global warming and cause acid rain. The sulfur content of the zapote seed (0.05%) was very low, similar to the sulfur content of almond shell and lower than that of olive stone (0.110%), which shows that corrosion problems and SO_x emissions would be lessened by the use of this biofuel.

The chlorine content has a decisive influence on the corrosion of the plant since the reaction with potassium and sodium produces chlorides that have a dissociative catalytic action in the system. The chlorine content of the zapote seed (0.06%) was very low, similar to that of other agro-industrial wastes, such as olive stone (0.06%) or mango stone (0.07%), but lower than that of almond shells (0.2%), which shows that corrosion problems in thermal installations could be lessened by using this biofuel.

Another important quality parameter of biofuel is its volatile matter content. As can be seen, zapote seed had a higher volatile matter content than other solid biofuels. This high volatility and low fixed carbon content suggest that it produces less corrosion and boiler fouling. On the other hand, the combustibility index or VM/FC ratio was significantly higher than that of other solid fuels. This fact was due to greater reactivity of the fuel and facilitated ignition, even at low temperatures, reducing the residence state until the complete combustion.

It is necessary to take into account that not only is the zapote seed interesting because of its energy potential but also for its carbon neutrality. This is because it is assumed that the plant is capable of retaining more carbon during its life cycle than it releases during combustion, making the carbon cycle neutral.

Zapote waste can be collected from industries processing this fruit in the surrounding area in such a way that it both contributes to the sustainable environmental treatment of this waste and a significant fall in CO_2 emissions.

4.2. Energy and Environmental Assessment

Biomass is considered to be theoretically CO_2 neutral because its combustion does not contribute to the increase of the greenhouse effect. This is because the carbon that is released during the combustion of biomass is the same that was continuously absorbed and released by the plants for their growth (Figure 9). The CO_2 emissions generated from biomass energy do not alter the balance of atmospheric carbon concentration since it comes from carbon removed from the atmosphere in the same biological cycle, and therefore do not increase the greenhouse effect. Its use contributes to reducing CO_2 emissions into the atmosphere whenever it replaces fossil fuel.

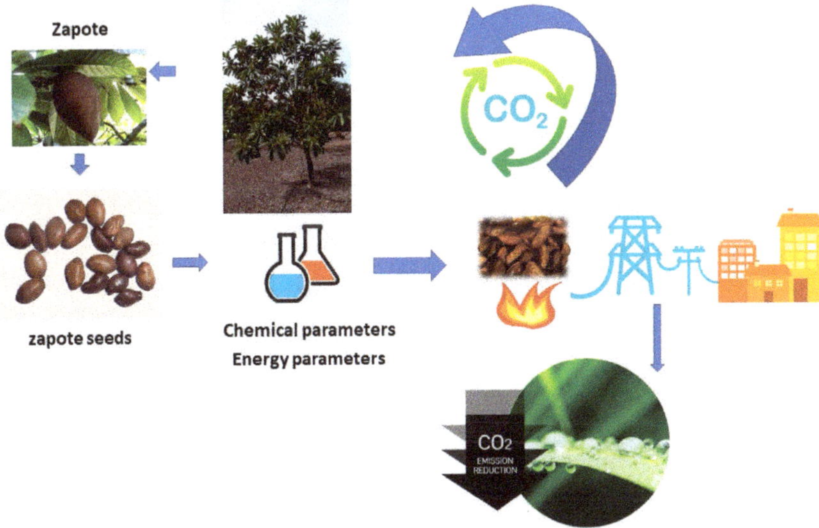

Figure 9. CO_2-saving scheme using zapote seed as biofuel.

Once the energy parameters of the zapote seed had been calculated, the next phase of this study was to calculate the savings in CO_2 emissions in the heavy chemistry laboratory of the Faculty of Engineering of the University of Veracruz, derived from the use of this seed as biofuel. Furthermore, the savings in CO_2 emissions that would be produced in the states of Mexico that produce zapote were calculated.

The first step was to calculate the energy (MWh) that could be obtained from the zapote seeds, considering the residues that would be obtained from all the zapote production in the main producing states of Mexico. This was done using Equation (2) and the results are given in Figure 10:

$$E_p = RH \times P_{zapote} \times HHV \times f_s \times F_c,\tag{2}$$

where:

E_p: energy obtained from the zapote seed (MWh);
RH: considering 10% humidity in zapote seed;
P_{zapote}: zapote production (kg);
HHV: higher heating value (18.342 MJ/kg);
f_s: % of seed in a whole zapote (15%);
F_c: unit conversion factor (0.000277778 Wh/J).

Figure 10. Bioenergy potential in the main producer states of Mexico using zapote seed as biofuel (MWh).

Once the energy that would be obtained by using zapote seed as a biofuel was calculated, the savings in CO_2 emissions that this would represent compared to the use of LPG were calculated. As discussed in a previous section, biomass is a clean, environmentally friendly source of energy since it produces zero CO_2 emissions. The mass (kg/year) of CO_2 emitted by the combustion of liquefied petroleum gas (LPG) was calculated using:

$$MCO_{2\,LPG} = C_{LPG} \times F_{LPG},\tag{3}$$

where:

$MCO_{2\,LPG}$: kg/year of CO_2 emitted by the combustion of liquefied petroleum gas (LPG).

C_{LPG}: consumption of liquefied petroleum gas (LPG) per year (kWh/year).

F_{LPG} : CO_2 emission factor of liquefied petroleum gas (LPG) (kg/kWh).

Table 5 shows the CO_2 emission factor for biomass and LPG and the total CO_2 emissions reduced annually using zapote seed as biofuel.

Table 5. CO_2 emission (kg/kWh) of LPG and zapote seed [47].

Boiler	CO_2 Emissions (kg/kWh)
Liquefied petroleum gas (LPG)	0.254
Zapote seed	0
Total savings of CO_2 emissions in the heavy chemistry laboratory (kg/year)	60,960.00

The use of a biomass boiler in the heavy chemistry laboratory of the Faculty of Engineering of the University of Veracruz in Mexico would save 60,960.00 kg/year of CO_2 emissions.

Figure 11 shows the CO_2 savings in the main producer states of Mexico using zapote seed as biofuel (t).

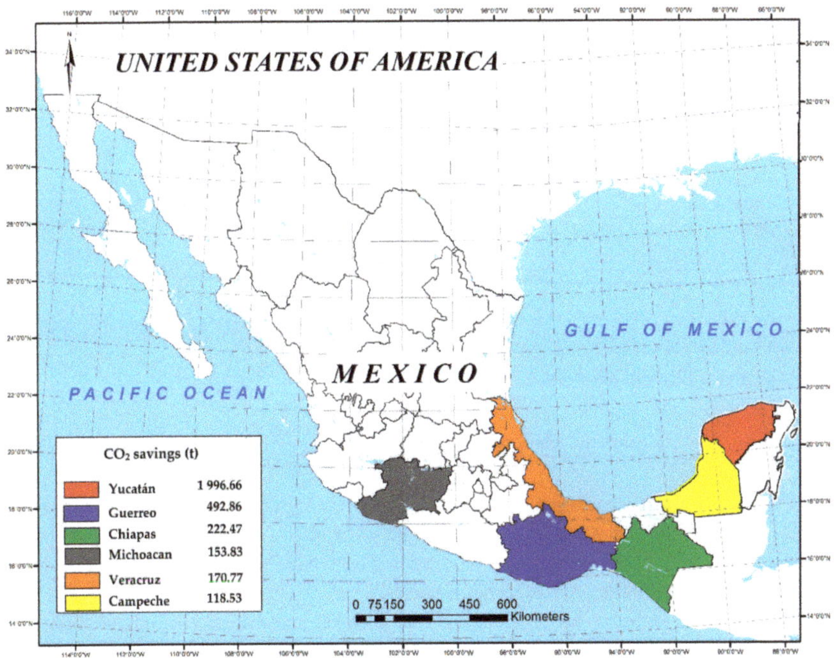

Figure 11. CO_2 Savings in the main producer states of Mexico using zapote seed as biofuel (t).

4.3. Economical Profit

The economic viability assessment of the project to replace the former gas boiler with a biomass boiler fed by zapote seed was made by taking into consideration the following data:

- The number of annual operating hours of the installation.
- Calculation of the LPG (L) and biomass (kg) required to satisfy the energy demand.
- Lower heating value (LHV) of both the zapote seed and LPG.
- Current market prices of both types of fuel (biomass and LPG).

The thermal demand of the installation was calculated by taking into account the fact that the 200 kW LPG boiler works four hours a day for an average of 300 days a year. Under this scenario, the annual consumption of the installation is 240,000 kW/h.

The market for biomass boilers for heat generation at both residential and industrial levels is booming due to the rising cost of fossil fuels, which are also severely taxed in most countries. To determine the economic savings that would be achieved with the new biomass heating plant compared to the old LPG boiler, it was first necessary to calculate the amount of fuel required in both cases to meet the annual thermal demand. From the values of the calorific value of LPG and its density, it was calculated that 37,696.73 L of LPG are needed annually to meet the energy demand. Considering the current market price of LPG ($0.39/L) [48], a total annual cost of $14,701.72 was calculated for the LPG installation. Table 6 shows the economic analysis of the LPG installation and the zapote seed boiler.

Table 6. Economic analysis of the LPG installation and the zapote seed boiler.

Parameter	Unit	LPG Boiler	Biomass Boiler
Fuel		Liquefied petroleum gas (LPG)	Zapote seed
LHV	kWh/kg	13.10	4.73
Price	$/L	0.39	
Price	$/kg		0.108
Boiler efficiency	%	90	80
Nominal power	kW	200	200
Operating hours	h	1200	1200
Thermal energy demand	kWh/year	240,000	240,000
LPG consumption	L	37,696.73	
Biomass consumption	kg		63,424.95
Annual cost	$	14,701.72	6,881.93
Annual saving	$		7,819.79
Annual saving	%		53.19
Total investment	$		67,530.48

Once the calorific value of the biomass (LHV) had been determined and knowing the annual thermal demand of the installation (kWh), the amount of biomass necessary to satisfy it could be obtained from the following expression:

$$\textbf{Biomass consumption} = \frac{\textbf{Annual thermal demand}}{\textbf{LHV} \times \textbf{Boiler efficiency}}. \tag{4}$$

From this expression, it was calculated that 63,424.95 kg of zapote seed will be consumed, and considering that the price per kilo of this residual biomass, once treated and transported to the point of consumption, is $0.108/kg, an annual saving of $7,819.79 will be made, which represents a 53.19% saving compared to the old LPG installation.

Some financial parameters were calculated, such as the internal rate of return (IRR), payback, and benefit/cost ratio (B/C).

The IRR represents the ideal interest yield provided by the project, considering all costs and benefits. The IRR can be calculated when all the cash flow is at time 0 and the costs are equal to the incomes.

Payback is the time that the investment takes to recoup its initial investment; this analysis considers the cash flows from its inception, as well as the leverage of the project.

B/C is the ratio of the net benefits to the costs of the project. Net benefits represent the present value of the annual revenue and savings minus the annual costs, while the cost is defined as the project equity.

Table 7 presents the financial results and its feasibility.

Table 7. Financial analysis.

Parameter	Unit	Biomass Boiler
IRR	%	7%
Payback	years	9.8
B/C		1.13

B/C: Benefit/cost ratio, IRR: Internal rate of return.

As seen in Table 7, the financial analysis describes the viability of the project, where the IRR is predicted to be 7%, which is a competitive market rate; the payback period is predicted to be 9.8 years to recover its investment, and the B/C is predicted to be 1.13, which means that the benefits are 13% higher than the costs.

A sensitivity or what-if analysis was also done and is presented in Table 8.

Table 8. Sensitivity analysis of zapote seed. The bold amounts indicate the range of profitability of the project.

		Quantity of Biomass (kg)				
		47,568.71	55,496.83	63,424.95	71,353.07	79,281.19
	0.081	$ 3,853.07	$ 4,495.24	$ 5,137.42	$ 5,779.60	$ 6,421.78
	0.095	$ 4,495.24	$ 5,244.45	$ 5,993.66	$ 6,742.86	$ 7,492.07
Zapote Seed Price ($)	0.108	$ 5,137.42	$ 5,993.66	**$ 6,849.89**	$ 7,706.13	$ 8,562.37
	0.122	$ 5,779.60	$ 6,742.86	$ 7,706.13	$ 8,669.40	$ 9,632.66
	0.135	$ 6,421.78	$ 7,492.07	$ 8,562.37	$ 9,632.66	$ 10,702.96

In Table 8, it can be seen that $6,849.89 is the base of the project, that is, the feasibility of the project requires 63,424.95 kg of zapote seed at 0.108 USD per kilogram; if the price and quantity increase in different scenarios, then the project can be rejected. The bold amounts indicate the range of profitability of the project.

4.4. Energy Policy

The implications of the Mexican Law of Promotion and Development of Bioenergetics (LPDB) that can be taken into account in this paper are the promotion and development of bioenergetics to contribute to the energy diversification and sustainable development as conditions that guarantee support for the Mexican countryside. This law has three main objectives [49]:

1. Promote the production of inputs for bioenergetics from agricultural, forestry, algae, biotechnological, and enzymatic processes in the Mexican countryside without jeopardizing the country's food security and sovereignty.
2. Develop the production, commercialization, and efficient use of bioenergetics to contribute to the reactivation of the rural sector, the generation of employment, and a better quality of life for the population, particularly those with high and very high marginality.
3. To seek the reduction of polluting emissions and greenhouse gases into the atmosphere in accordance with international instruments contained in the Treaties to which Mexico is a party.

According to the Secretary of Agriculture and Rural Development, Ministry of Agriculture and Rural Development (SAGARPA), Mexico is the leader in the production of zapote mamey, with only five states in Mexico producing it [30]. The benefits shown in this study can be proposed to SAGARPA and use the LPDB to increase the market opportunity for small producers and increase the production overall. This necessitates an adequate credit policy for this sector, which includes low interest rates, especially during the first few years, as well as flexibility in the terms and forms of payment.

Considering the results of this study, beginning with a sustainable proposal to universities along with the five producer states of zapote mamey regarding their boiler equipment, this experience can be shared with other institutions with similar conditions.

5. Conclusions

Biomass projects are an interesting alternative to fossil fuels where resources are widely available. Due to its carbon-neutral condition and low price, biomass boilers are having an increasing presence in international markets. According to the Law for the Use of Renewable Energies and the Financing of Energy Transition, Mexico's goal for 2024 is to generate 35% of its energy from non-fossil sources.

Zapote mamey is one of the most widely grown fruits in Mexico, with an annual production of 17,515 t. Zapote industry waste could be used to generate power and reduce CO_2 emissions; while the CO_2 emissions of LPG are 0.254 kg/kWh, the zapote seed emissions will be theoretically 0. Replacing a 200 kW LPG boiler with a biomass boiler fed with dry zapote seeds in the heavy chemistry laboratory of the Faculty of Engineering of the University of Veracruz in Mexico would mean a reduction in CO_2 emissions of 60,960.00 kg/year.

The use of zapote seed as a biofuel has important advantages in the fight against climate change:

- It reduces the emissions of CO_2, NO_x, and HCl and emissions of sulfur and particles.
- Reduces boiler maintenance and the danger of toxic fuel gas leaks.
- Jobs are generated in the agro-food industry to produce food, waste is used to generate thermal energy, and the planting of these crops aids the fight against erosion and soil degradation.

Comparing the annual costs of these two sources of energy, the amount saved was predicted to be $US7,819.79, which means an annual saving of 53.19%. The total investment of the project was estimated to be $US67,530.48. According to financial results, the project was predicted to be attractive, with its financial variables showing a predicted IRR of 7%, a payback period of 9.8 years, and a B/C ratio of 1.13.

A sensitivity or what-if analysis was carried out because the price and quantity of zapote seed are sensitive to changes. In this case, the project can be accepted if the maximum quantity required increases to 79,281.19 kg but the price decreases 25% (to 0.081 USD), or the maximum price increases 25% to 0.135 USD and the quantity decreases 25% (to 47,568.71 kg).

Regarding policy implications, this study shows that there is a market opportunity for small producers and overall increases in production, which necessitates an adequate credit policy for this sector, including low interest rates, especially during the first few years, as well as flexibility in the terms and forms of payment.

These results pave the way toward the utilization of zapote seed as a solid biofuel, contribute to achieving Mexico's energy goal for 2024, and promote sustainability in universities.

One of the main limitations of this study was the humidity of the zapote seeds. In future work, the possibility of designing dryers for these seeds, built in an environmentally sustainable way, will be studied.

Author Contributions: Conceptualization, M.-A.P.-M., Q.H.-E., F.R.-M., and A.-J.P.-M.; methodology, M.-A.P.-M., Q.H.-E., F.R.-M., and A.-J.P.-M.; formal analysis, M.-A.P.-M., Q.H.-E., F.R.-M., and A.-J.P.-M.; investigation, M.-A.P.-M., Q.H.-E., F.R.-M., and A.-J.P.-M.; resources, M.-A.P.-M., Q.H.-E., F.R.-M., and A.-J.P.-M.; writing—original draft preparation, M.-A.P.-M., Q.H.-E., F.R.-M., and A.-J.P.-M. All authors have read and agreed to the published version of the manuscript.

Funding: This research received no external funding.

Acknowledgments: The authors thank MS. Florentino Sanchez Padilla for his support in obtaining the university laboratory data.

Conflicts of Interest: The authors declare no conflict of interest.

References

1. Perea-Moreno, M.-A.; Manzano-Agugliaro, F.; Perea-Moreno, A.-J. Sustainable energy based on sunflower seed husk boiler for residential buildings. *Sustainability* **2018**, *10*, 3407. [CrossRef]
2. Sharma, S.; Basu, S.; Shetti, N.P.; Aminabhavi, T.M. Waste-to-energy nexus for circular economy and environmental protection: Recent trends in hydrogen energy. *Sci. Total Environ.* **2020**, *713*, 136633. [CrossRef] [PubMed]
3. Smith, C.J.; Forster, P.M.; Allen, M.; Fuglestvedt, J.; Millar, R.J.; Rogelj, J.; Zickfeld, K. Current fossil fuel infrastructure does not yet commit us to 1.5 C warming. *Nat. Commun.* **2019**, *10*, 101. [CrossRef] [PubMed]
4. Wold Energy Markets Observatory 2019. Available online: https://www.capgemini.com/news/world-energy-markets-observatory-2019/ (accessed on 25 March 2020).
5. Jung, J.; Petkanic, P.; Nan, D.; Kim, J.H. When a girl awakened the world: A user and social message analysis of greta thunberg. *Sustainability* **2020**, *12*, 2707. [CrossRef]
6. Lima, M.A.; Mendes, L.F.R.; Mothé, G.A.; Linhares, F.G.; de Castro, M.P.P.; da Silva, M.G.; Sthel, M.S. Renewable energy in reducing greenhouse gas emissions: Reaching the goals of the Paris agreement in Brazil. *Environ. Dev.* **2020**, *33*, 33. [CrossRef]
7. Johannsdottir, L.; McInerney, C. Calls for Carbon Markets at COP21: A conference report. *J. Clean. Prod.* **2016**, *124*, 405–407. [CrossRef]
8. International Energy Agency (IEA). World Energy Outlook 2019. Available online: https://www.iea.org/reports/world-energy-outlook-2019 (accessed on 26 March 2020).
9. De la Cruz-Lovera, C.; Perea-Moreno, A.-J.; De la Cruz-Fernández, J.-L.; Alvarez-Bermejo, J.A.; Manzano-Agugliaro, F. Worldwide research on energy efficiency and sustainability in public buildings. *Sustainability* **2017**, *9*, 1294. [CrossRef]
10. Chen, S.; Zhang, G.; Xia, X.; Setunge, S.; Shi, L. A review of internal and external influencing factors on energy efficiency design of buildings. *Energy Build.* **2020**, *216*, 109944. [CrossRef]
11. Hamburg, A.; Kuusk, K.; Mikola, A.; Kalamees, T. Realisation of energy performance targets of an old apartment building renovated to nZEB. *Energy* **2020**, *194*, 116874. [CrossRef]
12. Kim, D.; Cho, H.; Koh, J.; Im, P. Net-zero energy building design and life-cycle cost analysis with air-source variable refrigerant flow and distributed photovoltaic systems. *Renew. Sustain. Energy Rev.* **2020**, *118*, 109508. [CrossRef]
13. Ferrara, M.; Monetti, V.; Fabrizio, E. Cost-optimal analysis for nearly zero energy buildings design and optimization: A critical review. *Energies* **2018**, *11*, 1478. [CrossRef]
14. Harkouss, F.; Fardoun, F.; Biwole, P.H. Optimal design of renewable energy solution sets for net zero energy buildings. *Energy* **2019**, *179*, 1155–1175. [CrossRef]
15. Pérez-Denicia, E.; Fernández-Luqueño, F.; Vilariño-Ayala, D.; Manuel Montaño-Zetina, L.; Alfonso Maldonado-López, L. Renewable energy sources for electricity generation in mexico: A review. *Renew. Sustain. Energy Rev.* **2017**, *78*, 597–613. [CrossRef]
16. Vidal-Amaro, J.J.; Østergaard, P.A.; Sheinbaum-Pardo, C. Optimal energy mix for transitioning from fossil fuels to renewable energy sources - the case of the mexican electricity system. *Appl. Energy* **2015**, *150*, 80–96. [CrossRef]
17. Ram, M.; Aghahosseini, A.; Breyer, C. Job creation during the global energy transition towards 100% renewable power system by 2050. *Technol. Forecast. Soc. Chang.* **2020**, *151*, 119682. [CrossRef]
18. Perea-Moreno, M.-A.; Hernandez-Escobedo, Q.; Perea-Moreno, A.-J. Renewable energy in urban areas: Worldwide research trends. *Energies* **2018**, *11*, 577. [CrossRef]
19. Mousavi, N.; Kothapalli, G.; Habibi, D.; Das, C.K.; Baniasadi, A. A novel photovoltaic-pumped hydro storage microgrid applicable to rural areas. *Appl. Energy* **2020**, *262*, 114284. [CrossRef]
20. Perea-Moreno, M.A.; Samerón-Manzano, E.; Perea-Moreno, A.J. Biomass as renewable energy: Worldwide research trends. *Sustainability* **2019**, *11*, 863. [CrossRef]
21. Paredes-Sánchez, J.P.; López-Ochoa, L.M.; López-González, L.M.; Las-Heras-Casas, J.; Xiberta-Bernat, J. Evolution and perspectives of the bioenergy applications in Spain. *J. Clean. Prod.* **2019**, *213*, 553–568. [CrossRef]

22. Las-Heras-Casas, J.; López-Ochoa, L.M.; Paredes-Sánchez, J.P.; López-González, L.M. Implementation of biomass boilers for heating and domestic hot water in multi-family buildings in Spain: Energy, environmental, and economic assessment. *J. Clean. Prod.* **2018**, *176*, 590–603. [CrossRef]

23. Carlon, E.; Schwarz, M.; Prada, A.; Golicza, L.; Verma, V.K.; Baratieri, M.; Gasparella, A.; Haslinger, W.; Schmidl, C. On-site monitoring and dynamic simulation of a low energy house heated by a pellet boiler. *Energy Build.* **2016**, *116*, 296–306. [CrossRef]

24. Perea-Moreno, A.J.; Perea-Moreno, M.Á.; Hernandez-Escobedo, Q.; Manzano-Agugliaro, F. Towards forest sustainability in Mediterranean countries using biomass as fuel for heating. *J. Clean. Prod.* **2017**, *156*, 624–634. [CrossRef]

25. Saidur, R.; Abdelaziz, E.A.; Demirbas, A.; Hossain, M.S.; Mekhilef, S. A review on biomass as a fuel for boilers. *Renew. Sustain. Energy Rev.* **2011**, *15*, 2262–2289. [CrossRef]

26. Perea-Moreno, A.J.; Perea-Moreno, M.Á.; Dorado, M.P.; Manzano-Agugliaro, F. Mango stone properties as biofuel and its potential for reducing CO_2 emissions. *J. Clean. Prod.* **2018**, *190*, 53–62. [CrossRef]

27. Sánchez, F.; San Miguel, G. Improved fuel properties of whole table olive stones via pyrolytic processing. *Biomass Bioenergy* **2016**, *92*, 1–11. [CrossRef]

28. Perea-Moreno, A.-J.; Aguilera-Ureña, M.-J.; Manzano-Agugliaro, F. Fuel properties of avocado stone. *Fuel* **2016**, *186*, 358–364. [CrossRef]

29. García, R.; Pizarro, C.; Lavín, A.G.; Bueno, J.L. Spanish biofuels heating value estimation. Part I: Ultimate analysis data. *Fuel* **2014**, *117*, 1130–1138. [CrossRef]

30. Secretaría de Energía (SENER), 2018. Sistema de Información Energética. Balance Nacional de Energía: Producción Deenergía Primaria. Available online: http://sie.energia.gob.mx/bdiController.do?action=temas.Dfgdfg (accessed on 9 May 2019).

31. Secretaría de Energía (SENER), 2012. Iniciativa Para el Desarrollode las Energías Renovables en México. Energía de la Biomasa. PwC. Available online: http://rembio.org.mx/wp-content/uploads/2014/10/iniciativa-renovable-sener.pdf (accessed on 10 May 2020).

32. Honorato-Salazar, J.A.; Sadhukhan, J. Annual biomass variation of agriculture crops and forestry residues, and seasonality of crop residues for energy production in Mexico. *Food Bioproc. Technol.* **2020**, *119*, 1–19. [CrossRef]

33. Comité Sistema Producto Mamey de Guerrero, A.C. Available online: http://dev.pue.itesm.mx/sagarpa/estatales/EPT%20COMITE%20SISTEMA%20PRODUCTO%20MAMEY%20GUERRERO/PLAN%20RECTOR%20QUE%20CONTIENE%20PROGRAMA%20DE%20TRABAJO%202012/PR%20Mamey%20Gro.%202012.pdf (accessed on 23 March 2020).

34. Alexiades, M.N.; Shanley, P. *Productos Forestales, Medios de Subsistencia y Conservación: Estudios de Caso Sobre Sistemas de Manejo de Productos Forestales no Maderables*; CIFOR: Bogor, Indonesia, 2004.

35. Mata-Sánchez, J.; Pérez-Jiménez, J.A.; Díaz-Villanueva, M.J.; Serrano, A.; Núñez-Sánchez, N.; López-Giménez, F.J. Statistical evaluation of quality parameters of olive stone to predict its heating value. *Fuel* **2013**, *113*, 750–756. [CrossRef]

36. Farias, R.D.; García, C.M.; Palomino, T.C.; Andreola, F.; Lancellotti, I.; Barbieri, L. Valorization of agro-industrial wastes in lightweight aggregates for agronomic use: Preliminary study. *Environ. Eng. Manag. J.* **2017**, *16*, 1691–1700. [CrossRef]

37. Grabke, H.J.; Reese, E.; Spiegel, M. The effects of chlorides, hydrogen chloride, and sulfur dioxide in the oxidation of steels below deposits. *Corros. Sci.* **1995**, *37*, 1023–1043. [CrossRef]

38. Riedl, R.; Dahl, J.; Obernberger, I.; Narodoslawsky, M. Corrosion in fire tube boilers of biomass combustion plants. In Proceedings of the China International Corrosion Control Conference, Beijing, China, 26–28 October 1999.

39. Khan, A.; De Jong, W.; Jansens, P.; Spliethoff, H. Biomass combustion in fluidized bed boilers: Potential problems and remedies. *Fuel Process. Technol.* **2009**, *90*, 21–50. [CrossRef]

40. Senelwa, K.; Sims, R.E. Bioenergy, Fuel characteristics of short rotation forest biomass. *Biomass Bioenergy* **1999**, *17*, 127–140. [CrossRef]

41. Nasser, R.A.-S.; Aref, I.M. Fuelwood characteristics of six acacia species growing wild in the southwest of Saudi Arabia as affected by geographical location. *BioResources* **2014**, *9*, 1212–1224. [CrossRef]

42. RETScreen, Natural Resource of Canada. 2019. Available online: https://www.nrcan.gc.ca/maps-tools-publications/tools/data-analysis-software-modelling/retscreen/7465 (accessed on 2 October 2019).

43. Arranz, J.I.; Miranda, M.T.; Montero, I.; Sepúlveda, F.J.; Rojas, C.V. Characterization and combustion behaviour of commercial and experimental wood pellets in south west Europe. *Fuel* **2015**, *142*, 199–207. [CrossRef]

44. García, R.; Pizarro, C.; Lavín, A.G.; Bueno, J.L. Biomass sources for thermal conversion. Techno-economical overview. *Fuel* **2017**, *195*, 182–189. [CrossRef]

45. Gómez, N.; Rosas, J.G.; Cara, J.; Martínez, O.; Alburquerque, J.A.; Sánchez, M.E. Slow pyrolysis of relevant biomasses in the mediterranean basin. Part 1. Effect of temperature on process performance on a pilot scale. *J. Clean. Prod.* **2016**, *120*, 181–190. [CrossRef]

46. González, J.F.; González-García, C.M.; Ramiro, A.; Gañán, J.; González, J.; Sabio, E.; Román, S.; Turegano, J. Use of almond residues for domestic heating: Study of the combustion parameters in a mural boiler. *Fuel Process. Technol.* **2005**, *86*, 1351–1368. [CrossRef]

47. Factores de Emisión de CO2 y Coeficientes de Paso a Energía Primaria de Diferentes Fuentes de Energía Final Consumidas en el Sector de Edificios en España. Available online: https://energia.gob.es/desarrollo/EficienciaEnergetica/RITE/Reconocidos/Reconocidos/Otros%20documentos/Factores_emision_CO2.pdf (accessed on 8 January 2020).

48. Gobierno de México. Precios al Público de Gas LP Reportados Por Los Distribuidores. Available online: https://www.gob.mx/cre/documentos/precios-al-publico-de-gas-lp-reportados-por-los-distribuidores (accessed on 2 April 2020).

49. Mexican Law of Promotion and Development of Bioenergetics (LPDB). Available online: http://www.diputados.gob.mx/LeyesBiblio/pdf/LPDB.pdf (accessed on 11 May 2020).

 *sustainability* 

Article

The Potential Role of Stakeholders in the Energy Efficiency of Higher Education Institutions

Rubén Garrido-Yserte * and María-Teresa Gallo-Rivera *

Department of Economics and Business Management and Institute of Economy and Social Analysis, University of Alcalá, Victoria Square, 2, 28802 Alcalá de Henares, Madrid, Spain
* Correspondence: ruben.garrido@uah.es (R.G.-Y.); maria.gallo@uah.es (M.-T.G.-R.);
  Tel.: +34-918855193 (R.G.-Y.); +34-918855112 (M.-T.G.-R.)

Received: 29 September 2020; Accepted: 15 October 2020; Published: 27 October 2020

Abstract: Higher education institutions (HEIs) have a huge potential to save energy as they are significantly more energy-intensive in comparison with commercial offices and manufacturing premises. This paper provides an overview of the chief actions of sustainability and energy efficiency addressed by the University of Alcalá (Madrid, Spain). The policies implemented have shifted the University of Alcalá (UAH) to become the top-ranking university in Spain and one of the leading universities internationally on environmentally sustainable practices. The paper highlights two key elements. First, the actions adopted by the managerial teams, and second, the potential of public–private collaboration when considering different stakeholders. A descriptive study is developed through document analysis. The results show that energy consumption per user and energy consumption per area first fall and are then maintained, thereby contributing to meeting the objectives of the Spanish Government's Action Plan for Energy Saving and Efficiency (2011–2020). Because of the research approach, the results cannot be generalized. However, the paper fulfils an identified need to study the impact of HEIs and their stakeholders on sustainable development through initiatives in saving energy on their campuses and highlights the role of HEIs as test laboratories for the introduction of innovations in this field (monitoring, sensing, and reporting, among others).

Keywords: energy consumption; higher education institution; energy efficiency indicators; green campus; social responsibility; Spain

---

1. Introduction

Currently, higher education institutions (HEIs) worldwide have incorporated sustainability development strategies into their study programs, research, operations, dissemination, assessment, and reporting. Undoubtedly, the international implementation scheme of the United Nations (UN) a Decade of Education for Sustainable Development (2015–2014) has boosted the integration of the principles of sustainable development (SD) in all aspects of HEIs [1]. The institutionalization and incorporation of these strategies has been progressive and has had to overcome the initial resistance of the main stakeholders [2,3]. HEIs are significantly much more energy-intensive in comparison with commercial offices and manufacturing premises, so improving their energy efficiency becomes a fast and cost-effective method of achieving targets in the reduction of gas emissions and can foster their economic growth. The main advantages of these strategies are cost reductions, environmental protection, better public health, and economic sustainability [4,5].

The energy efficiency and carbon reduction initiatives of HEIs are conditioned by both national programs and the main trends in HEIs (increases of students, associated complexity of research and teaching activities, and the intensive use of equipment, resulting in an increase of energy demand and consumption on campus) [6].

The main influences on the policies and initiatives of HEIs concerning energy efficiency and carbon reductions come from both external and internal sources. The former includes European Union (EU) directives, such as the EU Emissions Trading Scheme (EU ETS), the Environmental Performance of Buildings Directive (EPBD) (2010/31/EU) on energy efficiency in buildings, and the 2012/27/EU energy efficiency directive. The national regulatory framework in the case of Spain includes the Action Plan For Energy Saving and Efficiency 2011–2020, the application of a Technical Building Code (RD 314/2006), the Basic Procedure for Energy Performance Certification of New Buildings (RD 47/2007), the Regulations on Heating Installations in Buildings (RD 1027/2007), and the Building Certification Regulations (RD 235/2013) which replaced the RD 47/2007 [7].

The latter sources include increased energy costs, initiatives in the arena of corporate social responsibility, attention to legal obligations, maintenance of economic competitiveness, concerns for the environment, the capacity to access financing sources, and the interest for corporate image [6]. Both kinds of factors determine the presence of an "energy efficiency gap" in the organizations, referring to a potential improvement of energy efficiency or, in other terms, the difference between a cost-minimizing level of energy efficiency and the level of energy efficiency actually implemented. This gap is also conditioned by economic factors and, especially, by organizational and institutional factors. The main organizational barriers for implementation of energy efficiency interventions in HEIs include access to financing sources, bounded rationality, hidden costs, imperfect information, risk and uncertainty, and non-aligned incentives [4].

Furthermore, the insights and perceptions of staff and students are crucial to the success of energy efficiency interventions. Generating a greater responsibility of the occupants towards energy conservation requires three main aspects to be taken care of: transmitting a clear conservation message, improving their participation in energy reduction schemes, and providing correct information about the use of energy [8].

The University of Alcalá (UAH), considered to be one of Europe's oldest universities, is aware of the environmental impact of its activities and, especially, of its energy consumption. The UAH is committed to sustainable development and is actively involved in finding solutions to environmental problems. It promotes the efficient use of energy through programs of energy conservation, boosting use of renewable energy and encouraging actions to raise awareness and involvement among the entire academic community. In the last few years, the UAH has been involved in several energy-saving measures to minimize its impact, which is of special relevance as stated in its 2003 Environmental Policy Statement.

In the last few years, a group of different external studies and audits have been carried out by experts to grasp more fully the University's energy needs and to set ambitious goals for cutting down consumption and replacing current energy sources with those that are cleaner and more sustainable. These policies have allowed the UAH to become, according to the Green Metrics World University Ranking, which assesses the sustainability policies of universities worldwide, the top ranking university in Spain and one of the leading universities internationally for environmentally sustainable practices.

However, in recent years, it seems that, despite good results, progress has stagnated. The stabilization in the number of students and the implementation of inefficient 24 × 7 services in heritage buildings have led people to demand new measures and investments in the future. A strong management commitment is needed to find the most innovative solutions in the field of energy efficiency. Innovative public procurement would seem to be an option that could be re-explored in an area where the most innovative solutions are in the private sector.

This paper provides an overview of the UAH's chief actions in sustainability matters and energy efficiency to date and introduces the main future action lines.

Two research questions were proposed to analyze the potential of implementing energy saving initiatives in HEIs, taking into account the perspective of different actors:

RQ1: What kind of actions can be implemented on university campuses in order to achieve effective energy saving along with a substantial improvement of the perceptions and the impacts of the different university community actors? What is the role of public–private partnerships in the context of financial or budgetary constraints?

RQ2: Which are the main indicators of energy saving that could be considered in an impact evaluation of HEIs on sustainable development?

Nowadays the relevance of the energy efficiency is undoubtable. The importance of the energy efficiency in HEIs is key to optimize the resources for the benefit of their academic community as well as the environment. The relevance of HEIs is threefold. First, it raises awareness about the importance of energy efficiency activities and the implementation of energy saving measures. Secondly, HEIs are places of high electricity and gas consumption due to their multiple installations, for many hours a day, many days a year, and with a large number of people. Thirdly, HEIs could be good laboratories for the introduction of industry innovations. They have an adequate dimension with activities similar to a small city; having all the elements of an urban concentration (roads, street lighting, service and transport vehicles, energy-intensive facilities, etc.) and its inhabitants (students, teachers and staff) are, a priori, more favorable to innovation and monitoring can be more effective.

This paper has a novelty that focuses on a key sector, HEIs, and contributes to the comprehension on how universities may implement effective measures of energy efficiency involving their main stakeholders, by analyzing the UAH's energy efficiency program results and its national and international recognized good practices, referent in energy saving in HEIs campuses.

Section 2 offers a review of the specialized literature in energy efficiency and carbon reduction programs, especially in the context of HEIs, in addition to the key HEI trends that influence the patterns of energy consumption and saving. The data and the methodology used as well as the case description are presented in Section 3. Section 4 analyzes the results obtained regarding UAH's interventions in respect of energy efficiency and presents an approximation of UAH's contribution to climate change. The discussion of the UAH's future challenges to improve energy efficiency is presented in Section 5. Finally, Section 6 presents the concluding remarks.

2. Background and Literature Review

2.1. Current Situation of Energy Efficiency and Carbon Reduction in Spain

The term energy efficiency is much used in public policy and refers to a set of "policies that seek to reach a balance in sustainable development, competitiveness, and secure supply, mainly by promoting energy efficiency and the use of renewable energies and other directives and documents directed at the energy sector" [7].

Over the years, the European Commission has implemented a set of directives and regulations to strengthen the EU energy policy. What have become known as 20/20/20 targets are part of the current EU strategic energy policy goals, which include increasing energy efficiency to achieve 20% savings in EU energy consumption by 2020.

Quantitative indicators are used often to assess the progress toward energy efficiency targets. The energy intensity ratio (consumption of energy related to its gross domestic product) shows the extent to which energy consumption has become more sustainable, illustrating the reduction in energy used to generate one unit of activity in the case of GDP remaining constant. Other indexes are also used to provide a more realistic evaluation of improvements in energy efficiency [9].

The different EU Member States have made different levels of progress toward these goals. Overall, the trend is positive, but their efforts are not enough to achieve the EU goal of 20% energy saving. The Spanish progress in energy efficiency issues, and particularly in the building sector, is slower than in other EU Member States and has been intensified by the economic crisis, the demotivation of its citizens, the decentralization of power in energy matters, and the huge stock of houses for sale. In the household sector, Spain achieved an energy efficiency gain of 12.6% in 2000–2010, which is below the EU-27 average of 15.3%; moreover, evidence has shown that Spanish housing markets capitalize the value of energy efficiency [7,10].

The last report of Regulatory Indicators for Sustainable Energy (RISE 2018) published [11], indicates that Spain is located in the 22nd position in the energy efficiency ranking, well behind some countries of similar socioeconomic level. This reveals that, in practice, public policies relating to the efficient consumption of energy are not yet a priority in Spain's energy policies.

2.2. The Role of HEIs for Sustainable Development and Key Trends in the HEIs' Sector

Universities can be regarded as "small cities", and their activities can have great impacts in terms of environmental pollution and degradation. Universities also have special societal responsibility with regard to knowledge, skills, values, and social awareness of sustainability [12,13].

The contributions of HEIs to the Sustainable Development Goals (SDG) stem from their five functions as educators—providing knowledge; trainers—providing professional training to manage the needs of government and other stakeholders; researchers—producing data and analysis for policymakers; facilitators—contributing to regional development and international cooperation; and enablers—improving social and intellectual development and the well-being of society. Specifically, in the area of sustainability, HEIs have an impact in preparing individuals to meet the needs of modern economies across global, national, regional, and local levels; i.e., creating a competitive, adaptable, sustainable, and knowledge-based economy—particularly a green economy—with benefits in terms of social equity and well-being, while reducing their environmental impacts [14].

The initiatives to implement Sustainable Campuses or Green Campuses have grown in the last ten years across the world. However, there are different definitions of what a sustainable university campus is. Therefore, interpretations and strategies to achieve a sustainable campus vary from one university to other.

Next are presented some definitions about what a sustainable or a green campus is.

"A green campus is one which addresses environmental challenges in all its fields of activity: administration, research and education. Higher education institutions are in an outstanding position to act as incubators, role models and multipliers for sustainable development among researchers, people in leadership positions, and in wider society" [15].

" . . . a healthy campus environment, with a prosperous economy through energy and resource conservation, waste reduction, and efficient environmental management, promotes equity and social justice in its affairs and exports these values at community, national, and global levels" [12].

Considering this framework, for some universities, a sustainable campus is defined by either an environmental plan or an environmental statement. Others consider the signing of a national or international declaration aimed to tackle the sustainability challenge, and others conduct their own approaches, including green building initiatives, ISO 14001, and Environmental Management Systems (EMSs) [12,15]. In the latter, the universities that have the practice of diffusion of sustainability reports have improved their visibility, mobilizing resources to improve their endowments and raising funds for future sustainability activities. Therefore, the sustainability reporting could become a tool with huge potential to increase the engagement of the main stakeholders [16,17].

In any case, the evidence indicates that a three-pronged strategy that include university EMSs; public participation and social responsibility with the real implication of key stakeholders, and promoting sustainability in teaching and research, could be the framework to foster a more suitable approach to achieving campus sustainability in a systematic and integrated way [12,18].

It is evident that the HEIs' sustainability practices have been increasing, and they are prominent especially in Europe, the US, and Canada in addition to Australia, Asia, South America, and Africa. Some guides and good practices in HEIs are analyzed in literature and by international organizations as The Organisation for Economic Co-operation and Development (OECD) and International Alliance of Research Universities (IARU) [13,19,20]. Furthermore, well know are the energy efficiency interventions in HEIs as Coimbra, Portugal [21,22], the United Kingdom [6,8], the United Kingdom and South Africa [23], Canada [24], and Universiti Teknologi Malaysia [25].

Although the role of HEIs in environmental, economic, and social sustainability is key, the level of engagement is different; there is a strong focus on activities related to environmental sustainability, whereas there is a weaker focus on social and economic sustainability [13]. In the environmental sustainability area, the implementation of EMSs is more frequent, and they comprise either whole campuses or part of them and those that include either all environmental issues or just some key issues, such as either water or energy. Some HEIs have connected their financial systems with financial incentives to boost sustainability behaviors; and the improvements in the teaching and research activities related to sustainability development are relevant.

In terms of economic sustainability, in the less-developed area, there are practices related to connecting the financial system with new purchasing procedures and developing financial incentives to promote sustainability behaviors. These include new forms of contracts with providers, premiums to staff who participate in the sustainability projects, contribution in innovation through building and infrastructure planning, and increasing research capacity to address the organizational challenges related to sustainability matters.

Regarding social sustainability, there are different level of social responsibility of HEIs related to staff and student well-being and the relationship with local communities; practices related to prevention and mental healthcare, volunteer activities, outreach programs, and approaches of Corporate Social Responsibility into HEIs with the participation of key stakeholders, including the local governments and interaction with small and medium size business enterprises (SMEs).

In the Spanish case, the engagement of universities in sustainability issues has received a decisive stimulus since 2002, when the Spanish University Rector's Committee (CRUE) approved the creation of a working group of Environmental Quality and Sustainable Development in a group of thirty universities (Sustainability-CRUE). In turn, in 2009, this group created a special commission focused on environmental quality, sustainable development, and risk prevention (CADEP-CRUE). Since then this commission has developed seminars to exchange experiences, encourage good practices, and develop joint projects [26,27]. Since January 2014, the Sustainability-CRUE Chair has been Professor Galván, Rector of the UAH. Furthermore, the creation of an Inter-university network of teaching and research in education for sustainability by the CADEP-CRUE represents an important point of reference about materials and teaching experiences in the field of sustainability in university teaching [28].

In the 8th Seminar of the CADEP-CRUE (2009) about [27], the Spanish universities recognized that HEIs have huge potential to saving energy "buildings are not the ones that waste energy, but people in their daily use". The availability of accurate statistical data on energy consumption of centers and buildings, would allow for developing reliable indicators before initiating any intervention, measuring the starting situation and after the intervention. It is also appropriate that each university has an energy manager to secure the commitment from institutional managers, to get better energy certifications for its buildings, and to both develop and manage their energy efficiency plans. Besides, projects must include a post-construction energy maintenance and management plan. The university contracting and bidding process must include all the guarantees necessary to make buildings sustainable. Finally, a close relationship between universities and companies is fundamental for sustainable energy and also serves as a test bed for new technologies and methodologies.

Overall, studies show that Spanish universities' practices about sustainability and energy efficiency in different areas, such as in the fields of strategic planning, practices, curricula, and reporting, show a low rate of progress, which seems like clear indication of there being several obstacles to overcome [29,30]. In fact, there are few analyses of the extent to which Spanish universities have implemented the sustainability approach in their functions. Highlights include the case of Universitat Politecnica de Catalunya, which has carried out actions to reduce energy and raw material use and drive waste out of the system [13] and the University of Santiago de Compostela's ecological footprint since 2005 [26]. The incorporation of sustainability in curriculum in the University of Valencia [31] is also worth mentioning.

3. Methodology

Documental data. To understand the Energy Efficiency Program carried out at UAH in Madrid, Spain and its main results, Section 4 presents the UAH's Environmental Quality Program and its main strategic lines, in addition to the energy saving and conservation initiatives developed since 2006.

Quantitative analysis. The analyses of the buildings' energy consumption performance indicators are presented in order to offer a characterization of the results obtained in terms of energy savings. These indicators are the energy consumption in terms of electricity (Kwh) and natural gas consumption (Kwh).

In addition, a set of Energy Efficiency Indexes (*EEI*) were estimated to explore the evolution of electricity and natural gas consumption per user (Kwh/user) and per area (Kwh/m^2) according to Equations (1) and (2). The data used referred to period between 2003 and 2017.

$$EEI = \frac{Total\ energy\ used\ (Kwh)}{Total\ users} \tag{1}$$

$$EEI = \frac{Total\ energy\ used\ (Kwh)}{Gross\ floor\ area\ (m^2)} \tag{2}$$

Additionally, the UAH's contribution to climate change, from estimation of its CO_2 emissions, is discussed. The available data include the estimations until to 2014 and currently the University is working to update the results until 2018.

Last, the evolution of the international position of UAH in Green Metrics World University [32] and Coolmyplanet Rankings [33] on environmentally sustainable practices, is analyzed.

Because the methodology approach is descriptive, the results may lack generalizability. However, the UAH's case analyzed has potential, being referent of HEIs policies of energy savings at national and international level.

Case description. In addition to teaching, advancing knowledge, and so forth, the UAH social commitment entails becoming a benchmark for the reaching of a sustainability model, not only in the Spanish university but also across all social, technological, and economic sectors. Its many functions include fostering habits of environmental respect and decreasing the environmental impact of human activities.

With this goal in mind, in 2003, the UAH issued its "Environmental Policy Statement" (UAH-EPS) with a view to making environmental issues part of its planning, execution, and assessment strategies. This environmental policy comprises several strategic lines:

- To prevent, reduce, and eliminate negative environmental impacts that may result from the university's activities.
- To rationalize consumption and promote increasing efficiency in the use of material and energy resources.
- To promote waste prevention and appraisal (recycling, recovery, and re-use).
- To inform, train, and make the university community aware by promoting active participation in environmental management and in enhancing the quality of the university environment.
- To monitor continuously the environmental repercussions of university activity and assess achievement of the established aims and goals.
- To maintain channels of dialog and cooperation with public and private bodies engaged in environmental issues.
- To tailor the environmental policy to the new requirements proposed by university associations at home and abroad, always with a view to continuous improvement.
- To promote in the territory of influence a policy of Environmental Excellence in Development by acting as a catalyst and assessor of such a policy in collaboration with other public and private bodies.

Because of this institutional commitment, the UAH has implemented an "Environmental Quality Program" (UAH-EQP) with the aim of identifying and controlling factors affecting the environment so as to affect continuous improvement in the University's environmental management. Under this program, several actions have been taken in various fields. These include energy saving and efficiency, economizing on water, waste management, sustainable mobility, and environmental enhancement.

The UAH-EQP also aims to promote the economic and human development of territory that is compatible with an optimum environmental quality. This social commitment is reflected in the different actions designed to make clear the environmental engagement of the organization, being aware of the environmental impact generated by its activity and, especially, its energy consumption, which is why it has been working for years to improve this aspect and reduce energy consumption.

The energy and thermal consumption of the university are distributed in an equitable way. The electric consumption is mainly due to illumination both inside and outside the buildings (in the External Scientific-Technological Campus in Figure 1 in addition to buildings and facilities also serves the outdoor lighting); office equipment and investigation; air conditioning, especially refrigeration (not available in classrooms); sanitary hot water (ACS); and other energy consuming equipment, such as kitchen appliances and cafeteria. Most of the thermal resources (natural gas) are used for heating systems. The UAH's has three campuses (the historical campus, the scientific-technological external campus and the Guadalajara's campus). The major energy efficiency interventions have been addressed in the UAH's Scientific–Technological External Campus. In this campus, the University can undertake different measures (road lighting, transformation centers, buildings, etc.). The other two are integrated within the city so that efficiency actions can only be undertaken within the buildings. However, the results obtained refer to all three campuses. Hence, the UAH's energy profile has to take into account the diversity in the typology of buildings (historical and modern), the heterogeneity in terms of teaching and research activities and the environment surrounding them (urban core and campus with its own urban infrastructure).

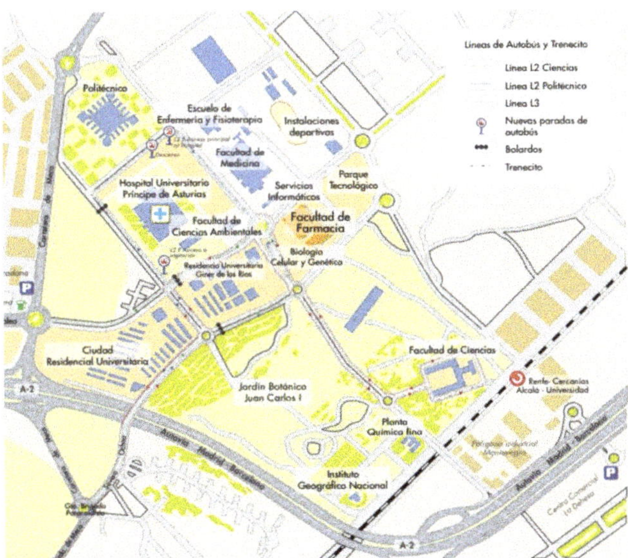

Figure 1. University of Alcalá's (UAH) Scientific–Technological External Campus. © The University of Alcalá.

4. Results

4.1. The Trend in Energy Consumption Saving of UAH

Since 2005 to now, the University has managed to stabilize electricity consumption, and it has registered a downturn from 2006, due chiefly to energy-saving and efficiency initiatives (see Figure 2).

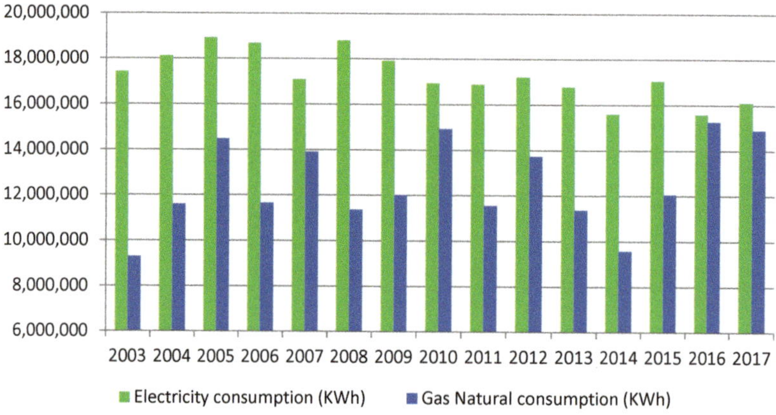

Figure 2. UAH's evolution of electricity and natural gas consumption. Source: Own elaboration.

The following is a synthesis of the actions taken regarding energy saving and efficiency:

1. Initiatives to raise awareness of energy saving among the university community, led chiefly by our EcoCampus office (See Ecocampus UAH. Office of Participation, Analysis and Enviromental Initiatives) [34].
2. The introduction of environmental criteria in public tender procedures (green contracting) and, particularly, when adjudicating contracts, encouragement and positive assessment of tenders made by companies that have an energy efficient management system certified by some official body.
3. Energy audits of all buildings have been performed and efficient rehabilitation work undertaken, and installations such as boilers have been replaced with others running on natural gas.
4. In addition, initiatives have been implemented in lighting systems (including sectoring of lighting in all buildings, exploitation of natural light, reduction of lighting in lifts, lighting management systems, and presence detectors in zones used occasionally) and in information systems (reduction of hidden energy consumption or stand by effect).
5. Radiators have been fitted with thermostatic valves, and solar shield panels have been installed, in addition to air curtains over either main or frequently transited entrances—all with a view to making a relevant cut in energy consumption and raising awareness among the university community.

Moreover, important efforts have been made to increase our renewable energy pool by means of thermal solar technology for ACS, the use of plant waste as biomass, the use of alternative energies for pumping, and so forth. The UAH also has an energy generating facility (Trigeneration) in its Engineering School, while its Chemistry Building boasts the most important geothermal installation in any public building in the Madrid Region and the largest of its kind in any European university (Figure 3).

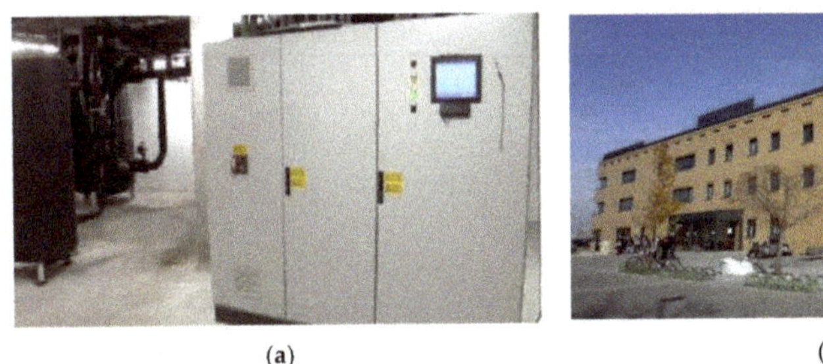 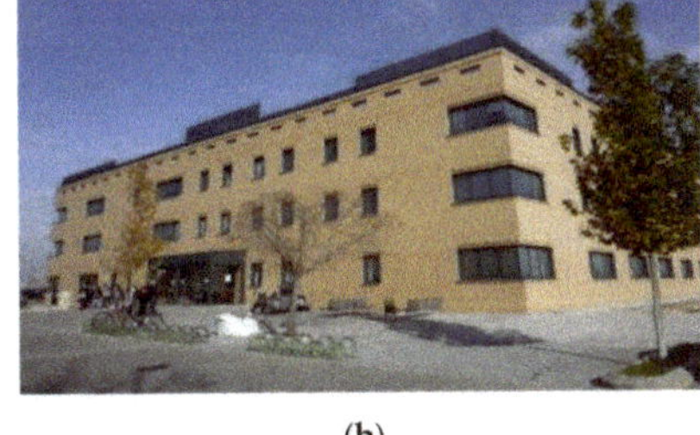

(a)   (b)

Figure 3. (**a**) Trigeneration Plant UAH's Polytechnic School; (**b**) UAH's Chemistry Faculty, a building of high efficiency due to the geothermal installation. © The University of Alcalá.

In recent years, however, efforts seem to have run out of steam, and there is a certain substitution effect between the two basic energy sources. There are several reasons for this. The first is the commissioning of energy-intensive facilities, such as the Learning and Research Resource Centre, an 11,000 m^2 heritage building, open 24 h a day, 7 days a week. Secondly, the regulatory barriers to operate the Trigeneration Plant and its technological obsolescence.

Both demand investment policies that are conditioned by budgetary restrictions or, in contrast, by new public–private partnership actions, which allow investment in efficiency by paying with the savings generated in future years.

4.2. Main Indicators of Energy Consumption Saving of UAH in Terms of Consumption per User and Consumption per Area

All these actions have allowed the University's energy consumption per user and per area first to fall and then to be maintained, thereby contributing to meeting the objectives of the Spanish Government's Action Plan for Energy Saving and Efficiency (2011–2020) but with a worrying change of trend in recent years (see Figures 4 and 5).

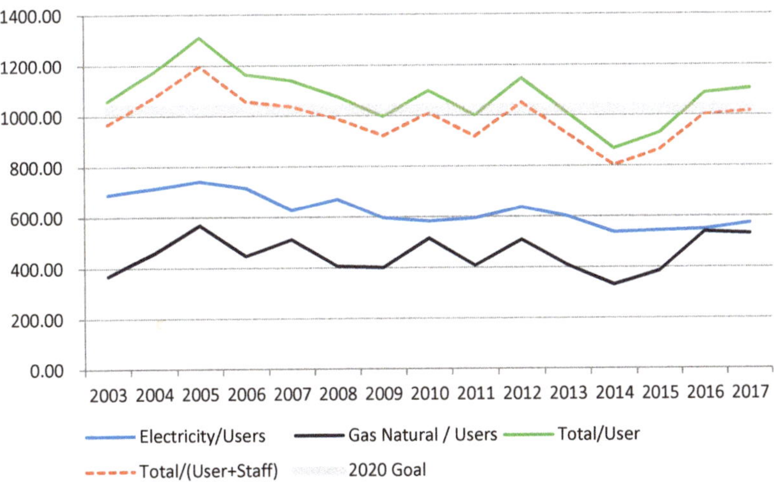

Figure 4. Energy consumption per user (2003–2017) and goals for 2020. Source: Own elaboration.

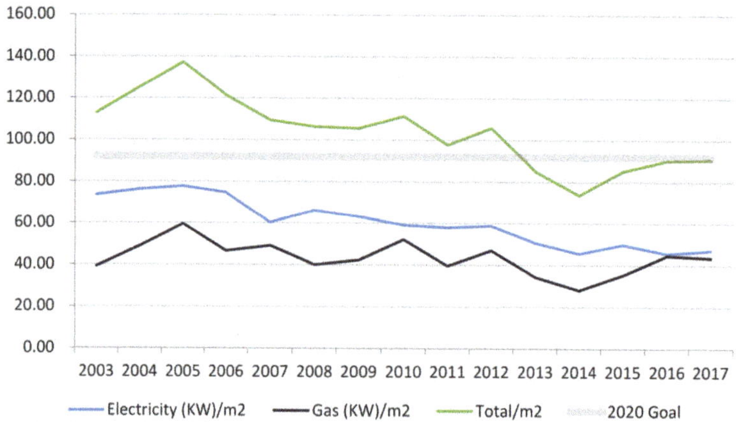

Figure 5. Energy consumption per area (2003–2017) and goals for 2020. Source: Own elaboration.

The number of students has fluctuated since 2010 showing a certain downward trend, with degrees now lasting 4 years instead of 5. The new energy-intensive services (24X7) increase their consumption more than proportionally to the square meters they occupy, which leads to an increase in consumption per square meter (Figure 6).

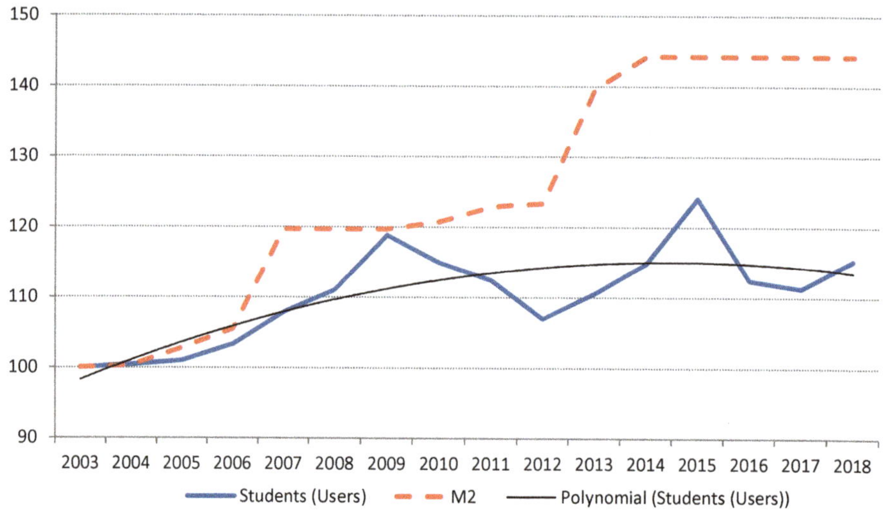

Figure 6. UAH's Students (clients) and Square Meters (2003=1000). Source: Own elaboration.

The UAH also boasts Spain's first solar-powered vehicle recharging point in its "Rey Juan Carlos" Botanical Garden. Two other recharging points have been opened recently at other sites (in the Pharmacy Faculty and in Malaga School in the City Campus). The Botanical Garden is, itself, a further example of sustainability, a 260,000 m^2 green space with more than 120,000 plant species that has become a first-class teaching and experimental resource for students and the general public alike (Figure 7).

Figure 7. (a) UAH's Botanical Garden; (b) UAH's Solar-Powered Vehicle Recharging Points. © The University of Alcalá.

4.3. *The Evolution of UAH's Footprint*

All the UAH's initiatives carried out in the last 17 years represent a significant contribution toward building a more sustainable society, both locally and globally. In terms of sustainability, the UAH has implemented actions and measures that aim at preventing, reducing, and eliminating any negative environmental impact deriving from the university activities. However, it wants to go one step further. As a public institution devoted to education, it wants to have a transforming effect on policy and to contribute, albeit modestly, to solving global problems, including, of course, global warming.

To estimate its emissions, the UAH followed the guidelines of ISO 14064-1: 2006 and of the World Resources Institute's GHG Protocol and took into account these sources of emissions [35,36]:

- Source 1. Direct GHG emissions: Associated with combustion of natural gas in boilers, with use of refrigerant gases, and with combustion of vehicle fuel.
- Source 2. Indirect GHG emissions: Associated with electricity consumption and use of direct thermal energy.
- Source 3. Other indirect GHG emissions: Associated with travelling in buses paid for by the UAH staff, but also air and railroad travelling by UAH staff, and with fuel consumption of maintenance vehicles belonging to UAH subcontractors.

In 2014, 59.14% of the UAH's GHG net emissions were direct emissions from source 1 (2193.67 TCO2 tons of source 1), 16.37% were indirect emissions due to the consumption of electrical and thermal energy (source 2) (607.89 TCO2), and the remaining 24.49% were emissions from source 3. Emissions deriving from electricity consumption are undoubtedly the most significant (55.4% of TCO2 tons), followed by gas consumption (24.3%). That is why efficient energy policies are so important, since they reduce quite considerably the direct and indirect emissions caused by university activity (see Figure 8) (During 2020, work is underway to update the data of 2018. The restrictions resulting from the pandemic have delayed these tasks and no data are available).

In 2010, due to energy efficiency policies, the UAH was already able to consider itself a low-carbon organization, meeting even then the objectives fixed for reduction by 2040. Currently, the UAH is continuing to work on updating its carbon footprint. However, its commitment to directly reducing the emissions due to electricity usage (100 percent of renewable sources) or the compensation policies for emissions, as in the case of gas (all gas consumed is compensated) has enabled UAH to reach the 2050 targets already, with reductions above 80% in 2008–2011 and additional reductions of 25% in 2012–2014.

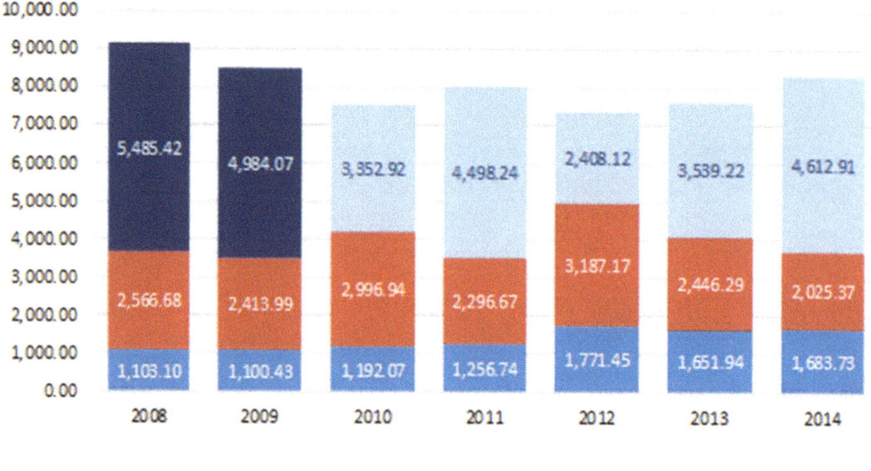

Figure 8. UAH's Emissions TCO2, 2008–2014. Source: Carbon footprint, UAH 2011 y 2014. Gas Natural-Fenosa—UAH [35,36].

These results encourage UAH to work even harder to improve efficiency and sustainability and to act as an example for the students, who are the generation that will have to build a more socially, economically, and environmentally sustainable future. International distinctions, such as being rated the third most sustainable university in the world by the non-profit, San Francisco-based organization, Coolmyplanet Universities Ranking [33], or its consistently high position—always among the top 40—in the University of Indonesia's Green Metric World University Rankings [32], stimulate UAH to make further improvements in the same direction (see Table 1).

Table 1. UAH position in Green Metrics and Coolmyplanet Rankings.

Ranking	Year	UAH Position/Total Universities	UAH Position Regarding Spanish Universities
	2019	19/780	2nd Spanish University
	2018	16/719	1st Spanish University
	2017	16/619	1st Spanish University
	2016	26/516	2nd Spanish University
Green Metrics	2015	37/407	2nd Spanish University
	2014	28/361	1st Spanish University
	2013	12/301	1st Spanish University
	2012	31/215	1st Spanish University
	2011	31/178	1st Spanish University
Coolmyplanet, 50 Most Environmentally Sustainable Universities in the world	2016	3/50	1st Spanish University

Source: Own elaboration based on Green Metric World University Rankings [32] and Coolmyplanet Universities Ranking [33].

5. Discussion

The UAH's experience over 17 years of initiatives between 2003 and 2019 suggest that the energy efficiency policies should be stepped up in accordance with four basic lines of action:

- To bring to completion the policies for energy sustainability through renewable energies.
- To go further in saving and management issues on the principle that the most sustainable energy is the energy that is not consumed.
- To integrate technology and innovation into the strategic management of sustainable energy.
- To set up a model of management and ongoing improvement to facilitate the monitoring and certification of the efficiency policies.

To this end, in 2014 a cooperation contract was signed by the public sector (the University) and the private sector to execute a global, integrated project for the interior lighting of the UAH's buildings and installations and its external campus's road. The idea was to improve energy efficiency in terms of lighting, to reduce energy consumption, and/or to create energy generating systems for either sale or self-consumption, thereby helping to cut the UAH's energy bill. This effort has resulted in a twelve-year contract worth EUR 7 million of private funding to be financed essentially through savings achieved in energy consumption. When all initiatives are 100 percent operative, there will be an annual saving of KWH 3.9, amounting to a saving of more than EUR 200,000 in direct costs to the University over the ten years that the project lasts. The project rests on four basic pillars (Figure 9) [37].

Figure 9. UAH's strategy regarding energy efficiency. Source: Own elaboration.

Pillar 1. Improvements and renovation work.

A series of initiatives has been envisaged on three fronts:

1. The replacement and/or renewal of both indoor and outdoor lighting installations on the External Campus. This has meant replacing more than 70,000 light fittings; it was completed in 2016.
2. The installation of building automation control and management devices (detectors, pushbuttons, sensors, light regulators, and so on) to enhance performance and efficiency while ensuring permanent user comfort.
3. Facility improvements and renovation work.

Pillar 2. Centralized energy management.

Initiatives are under way to:

1. Implement centralized energy management: quality control, uses, licenses, certifications, and regulations.
2. Develop a centralized remote Management Platform for all our facilities' building automation devices and lighting.
3. Implement a full guarantee maintenance system, with the aim of achieving operational perfection in all elements of new installations and of maintaining the performance at initial levels over time.

Pillar 3. Renewable energies.

The UAH has a significant tradition of commitment to renewable energies: geothermal energy, trigeneration, and solar panels are only some examples. Despite the scarcity of funds for investment and the absence of a stable regulatory framework, the UAH remains steadfast in its commitment to renewable energies on two fronts:

1. Encouraging direct use of renewable energies, with the installation of 4-kilowatt mini-wind farm installation (completed in 2016) that will generate electricity to be used in the Science Building, as well as a solar photovoltaic plant for the sports hall and a biomass plant at University Residential City (CRUSA) facilities. In addition, a free-cooling system is to be installed in the computing services' CPD. Furthermore, photovoltaic LED light fittings running entirely on solar power are being installed. UAH aims to promote the use of electric vehicles by increasing the number of top-up points and requiring our suppliers to use this type of vehicle.

2. Promoting consumption compensation for both electricity and gas. Since 2010 to now, UAH electrical energy has come from renewable sources, and it is certified as doing so by the National Commission of Markets and Competition. As for gas, since 2016, all emissions due to UAH consumption have been compensated by the supplier.

Pillar 4. Certification.

Introduction of an ISO 50001 EMS, to develop and implement the energy policy and manage all elements of UAH activities and all products or services that impinge on energy use (energy aspects).

The EMS, in conjunction with ISO 14001 (environment) and ISO 9001 (quality) management models, aims to achieve ongoing improvement in energy use and associated costs, the reduction of greenhouse gas (GHG) emissions, the proper use of natural resources, and the promotion of renewable energies.

The EMS is recommended for organizations wishing to show publicly the implementation of an EMS, their better use of either renewable or surplus energies, and the systematization of their processes in consonance with their energy policy. Thus, for the UAH, it is of vital importance that this ISO 50001 EMS be implemented, as it will be its energy and environmental blueprint for the years to come.

6. Conclusions

Energy efficiency has been one of the basic management strategies in the UAH up until 2018. This has meant that, at present, the UAH is a world leader in Energy and Climate Change, according to the 2019 edition of the Green Metrics ranking.

There are two key elements to achieving this sustained result over time: first, the actions of the UAH management team, and, second, the public–private collaboration considering different stakeholders.

The energy savings and efficiency actions include the creation of the EcoCampus office to raise awareness of energy saving among the university community, the introduction of environmental criteria in public tender procedures (green contracting), the energy audits of all buildings and the efficient rehabilitation works, the implementation of initiatives in lighting systems and in information systems, the installation of thermostatic valves in radiators, and solar shield panels, and the increase of its renewable energy pool by means of thermal solar technology for ACS, the use of plant waste as biomass, the use of alternative energies for pumping.

They have allowed the University's energy consumption first to fall and then to be maintained, thereby contributing to meeting the objectives of the Spanish Government's Action Plan for Energy Saving and Efficiency (2011–2020) but with a worrying change of trend in recent years. The UAH's has managed to stabilize electricity consumption, and it has registered a downturn by 13.8% from 2006 until 2017. The electricity consumption per user and per m2 has fallen by 20% and by 37%, respectively, in the same period. In recent years, however, the efforts developed to increase its renewable energy pool seem to have run out of steam. The functioning of Learning and Research Resource Centre, an 11,000 m^2 heritage building, open 24 hours a day, 7 days a week and the regulatory barriers to operate the Trigeneration Plant and its technological obsolescence are the main factors explain that negative trend. Its required investment policies that are conditioned by budgetary restrictions or by new public–private partnership actions, which allow investment in efficiency by paying with the savings generated in future years.

Regarding the UAH's footprint, the available data show that in 2010 the UAH was already in a position to consider itself a low-carbon organization, meeting even then the objectives fixed for reduction by 2040. Currently, the UAH is continuing to work on updating its carbon footprint for the period 2015–2018. However, its commitment to directly reducing the emissions due to electricity usage (100 percent of renewable sources) or the compensation policies for emissions, as in the case of gas (all gas consumed is compensated) has enabled UAH to reach the 2050 targets already, with reductions above 80% in 2008–2011 and additional reductions of 25% in 2012–2014.

On the other hand, the public–private partnership signed in 2014 made it possible to mobilize very substantial financial resources of the order of EUR 7 million.

It allowed the remodeling of obsolete installations (lighting, computer servers) by introducing the latest available technologies and contributing to a management scheme in accordance with international certifications ISO 14,000 and ISO 50,000. However, efforts need to be redoubled and public–private collaboration maintained. Legislative changes and changes in University government teams have not contributed in recent years.

The UAH's energy saving actions findings from a thorough document analysis and the energy efficiency indicators are in the line of other similar studies as for example those referred to University of Coimbra [21], UK higher education institutions [4], Canadian post-secondary institutes [24] or in the case of a Malaysian public university [25]. In some of these studies a set of difficulties to analyze the impacts of energy efficiency intervention on HEIs are highlighted including lack of methodology, ambiguity with respect to energy consumption indicators, problems in establishing assessment boundaries, lack of clear targets for carbon reductions within the HEIs, among others.

The case analyzed of the UAH's energy efficiency has been guide by the Spanish Government's Action Plan for Energy Saving and Efficiency (2011–2020) to offer an objective analysis. However, carrying out a survey among the responsible of the energy efficiency program in the UAH could have given more understanding about the efficacy of internal programs carried out, the limitations faced, and the need to develop future actions as sectorization strategies of energy consumption and the modernization of infrastructures to introduce sensing mechanisms. The knowledge of the users' behavior is also key to undertake saving policies aimed at reducing further the consumption of all energy sources. The use of data and centralized management will allow, in the future, for advancing in this direction. For this reason, a survey among the academic community could give a more accurate comprehension about their patterns of energy-environmental behaviors and take action to raise awareness of energy efficiency in the campus.

Historically, the UAH has shown that a Sustainability Strategy not only corresponds to its basic Social Responsibility Policies, but that it is a profitable option, stabilizing consumptions and seeking, with industry, the most sustainable options in the market.

Efficiency actions in World Heritage buildings is a challenge for the future. The will of both university management and public authorities (archaeological and construction regulations) is necessary to provide these facilities with the most efficient technologies.

For example, geothermal energy installation on the scientific campus has proven to be an excellent solution for energy saving. This can be done on the historic campus but requires a shared vision from all stakeholders: University managers (General, Construction Officer, and Contracts Officer); local services and authorities that oversee the protection of Heritage.

Although the results cannot be generalized due to the limitations of the methodology used, the findings show that internal and external strategies including searching for alliances are essential elements to improving the contribution of the University to the fulfillment of the Sustainable Development Goals, not only because of its direct actions but also because of the multiplying effect that the University has as an agent of social change.

Author Contributions: R.G.-Y. served as general manager of UAH during 2010–2018 contributed data collection, method, data analysis as well as discussion and conclusion. M.-T.G.-R. wrote introduction, background, case description as well as discussion and conclusion. All authors have read and agreed to the published version of the manuscript.

Funding: This research received no external funding.

Sustainability **2020**, *12*, 8908

Conflicts of Interest: The authors declare no conflict of interest.

References

1. Findler, F.; Schönherr, N.; Lozano, R.; Reider, D.; Martinuzzi, A. The impacts of higher education institutions on sustainable development A review and conceptualization. *Int. J. Sustain. High. Educ.* **2019**, *20*, 23–38. [CrossRef]
2. Lozano, R. Incorporation and institutionalization of SD into universities: Breaking through barriers to change. *J. Clean. Prod.* **2006**, *14*, 787–796. [CrossRef]
3. Lozano, R. Diffusion of sustainable development in universities' curricula: An empirical example from Cardiff University. *J. Clean. Prod.* **2010**, *18*, 637–644. [CrossRef]
4. Maiorano, J.; Savan, B. Barriers to energy efficiency and the uptake of green revolving funds in Canadian universities. *Int. J. Sustain. High. Educ.* **2015**, *16*, 200–216. [CrossRef]
5. United Nations Economic Commission for Europe (UNECE). *Best Policy Practices for Promoting Energy Efficiency. A Structured Framework of Best Practices in Policies to Promote Energy Efficiency for Climate Change Mitigation and Sustainable Development*; UNECE Energy Series 43; United Nations Economic Commission for Europe (UNECE): New York, NY, USA; Geneva, Switzerland, 2015.
6. Altan, H. Energy efficiency interventions in UK higher education institutions. *Energy Policy* **2010**, *38*, 7722–7731. [CrossRef]
7. Tolón-Becerra, A.; Lastra-Bravo, X.B.; Fernández-Membrive, V.J.; Flores-Parra, I. Opportunities in Spanish energy efficiency. Current situation, trends and potential in the building sector. *Energy Procedia* **2013**, *42*, 63–72. [CrossRef]
8. Whittle, C.G.; Jones, C. User perceptions of energy consumption in university buildings: A University of Sheffield case study. *J. Sustain. Educ.* **2013**, *5*, 27.
9. Project, O.-M. *Energy Efficiency Trends in the Industrial Sector in the EU. Lessons from the ODYSSEE-MURE project*; Intelligent Energy Europe, European Commission: Brussels, Belgium, 2012.
10. De Ayala, A.; Galarraga, I.; Spadaro, J.V. The price of energy efficiency in the Spanish housing market. *Energy Policy* **2016**, *94*, 16–24. [CrossRef]
11. Foster, V.; Portale, E.; Bedrosyan, D.; Besnard, J.S.G.; Parvanyan, T. *Policy Matters: Regulatory Indicators for Sustainable Energy*; The World Bank: Washington, DC, USA, 2018.
12. Alshuwaikhat, H.M.; Abubakar, I. An integrated approach to achieving campus sustainability: Assessment of the current campus environmental management practices. *J. Clean. Prod.* **2008**, *16*, 1777–1785. [CrossRef]
13. Organisation for Economic Co-operation and Development, OECD. *Higher Education for Sustainable Development. Final Report of International Action*; Research Project; OECD: Paris, France, 2007.
14. Hoballah, A.; Clark, G.; Abbas, K. Education: Key to Reaching the Sustainable Development Goals. In *Higher Education in the World. Towards a Socially Responsible University: Balancing the Global with Local*; Guni Series on the Social Commitment of Universities; GUNI Network: Girona, Spain, 2017; pp. 85–99.
15. Commission, E. *Greening the Campus with EMAS. EMAS. Performance–Credibility–Transparency*, 1st ed.; European Commission: Brussels, Belgium, 2013.
16. Alonso-Almeida, M.D.; Marimon, F.; Casani, F.; Rodriguez-Pomeda, J. Diffusion Of sustainability reporting in universities: Current situation and future perspectives. *J. Clean. Prod.* **2015**, *106*, 144–154. [CrossRef]
17. Ceulemans, K.; Molderez, I.; van Liedekerke, L. Sustainability reporting in higher education: A comprehensive review of the recent literature and paths for further research. *J. Clean. Prod.* **2015**, *106*, 127–143. [CrossRef]
18. Ceulemans, G.; Severijns, N. Challenges and benefits of student sustainability research projects in view of education for sustainability. *Int. J. Sustain. High. Educ.* **2019**, *20*, 482–499. [CrossRef]
19. Leal-Filho, W.; Brandli, L.; Kuznetsova, O.; Finisterra do Paço, A.M. (Eds.) *Integrative Approaches to Sustainable Development at University Level. Making the Links*; Springer International Publishing: Cham, Switzerland, 2015.
20. IARU. *IARU Green Guide for Universities*; IARU: Newington, CT, USA, 2014.
21. Soares, N.; Pereira, L.D.; Ferreira, J.; Conceição, P.; Da Silva, P.P. Energy efficiency of higher education buildings: A case study. *Int. J. Sustain. High. Educ.* **2015**, *16*, 669–691. [CrossRef]
22. Barata, E.; Cruz, L.; Ferreira, J.P. Parking at the UC campus: Problems and solutions. *Cities* **2011**, *28*, 406–413. [CrossRef]

23. Nhamo, G.; Ntombela, N. Higher education institutions and carbon management: Cases from the UK and South. Africa. *Probl. Perspect. Manag.* **2017**, *12*, 218–227.

24. Motta-Cabrera, D.F.; Zareipour, H. A Review of Energy Efficiency Initiatives in Post-Secondary Educational Institutes. In Proceedings of the Energy 2011: The First International Conference on Smart Grids, Green Communications and IT Energy-aware Technologies, Venice/Mestre, Italy, 22–27 May 2011.

25. Sukri-Ahmad, A.; Hassan, M.Y.; Abdullah, H.; Rahman, H.A.; Majid, M.S.; Bandi, M. Energy Efficiency Measurements in a Malaysian Public University. In Proceedings of the IEEE International Conference on Power and Energy, Kota Kinabalu Sabah, Malaysia, 2–5 December 2012.

26. CRUE-Sustainability Spanish Universities. CRUE Working Group for Environmental Quality and Sustainable Development 7th Conference of the Permanent Seminar on Environmentalisation "Indicators and Sustainability in Universities". CADEP-CRUE, Santiago de Compostela, Spain, 2007. Available online: http://acreditacion.crue.org/Documentos%20compartidos/Res%C3%BAmenes%20y%20Conclusiones/6.Indicadores_sostenibilidad.pdf (accessed on 29 September 2020).

27. CRUE-Sustainability Spanish Universities. CRUE Working Group for Environmental Quality and Sustainable Development. Renewable Energy, Savings and Energy Efficiency at Spanish Universities. In Encuentro del Grupo de Trabajo de la Crue para la Calidad Ambiental, el Desarrollo Sostenible y la Prevención de Riesgos; 8th Seminar of the CADEP-CRUE. CADEP-CRUE: Salamanca, Spain, 2009. Available online: http://acreditacion.crue.org/Documentos%20compartidos/Res%C3%BAmenes%20y%20Conclusiones/10.ENERGIAS_RENOVABLES_UNIVERSIDADES.pdf (accessed on 29 September 2020).

28. Ull, M.A.; Aznar-Minguet, P. *Red Española Interuniversitaria de Docencia e Investigación en Educación para la Sostenibilidad*; Conference Paper; WEEC: Torino, Italy, 2015.

29. Jorge, M.L.; Madueño, J.H.; Cejas, M.Y.C.; Peña, F.J.A. An approach to the implementation of sustainability practices in Spanish universities. *J. Clean. Prod.* **2015**, *106*, 34–44. [CrossRef]

30. Larrán-Jorge, M.; Andrades, F.J. Implementing Sustainability and Social Responsibility Initiatives in the Higher Education System: Evidence from Spain. In *Integrative Approaches to Sustainable Development at University Level. Making the Links*; Springer International Publishing: Cham, Switzerland, 2015.

31. Aznar-Minguet, P.; Martínez-Agut, M.P.; Palacios, B.; Pinero, A.; Ull, M.A. Introducing sustainability into university curricula: An indicator and baseline survey of the views of university teachers at the University of Valencia. *Environ. Educ. Res.* **2011**, *17*, 145–166. [CrossRef]

32. GreenMetric. *UI GreenMetric World Universities Ranking*. Several years. Available online: http://greenmetric.ui.ac.id/ (accessed on 29 September 2020).

33. Coolmyplanet. *Universities Ranking 2016*. 2016. Available online: http://coolmyplanet.com/universities/ (accessed on 29 September 2020).

34. UAH. *Ecocampus. Office of Participation, Analysis and Environmental Initiatives.* Available online: https://www.uah.es/es/conoce-la-uah/compromiso-social/sostenibilidad-medioambiental/ecocampus/ (accessed on 29 September 2020).

35. Gas Natural-Fenosa-UAH. *University of Alcala's Carbon Footprint-2011 Report*; UAH: Alcalá de Henares, Spain, 2013; Available online: http://www1.uah.es/sustainability/docs/informe_huella_carbono.pdf (accessed on 29 September 2020).

36. Gas Natural-Fenosa-UAH. *University of Alcala's Carbon Footprint–2014 Report*; UAH: Alcalá de Henares, Spain, 2016; Available online: https://www1.uah.es/sostenibilidad/docs/informe_huella_carbono.pdf (accessed on 29 September 2020).

37. UAH. *Integral Action for Enhancing Energy Efficiency in the field of Lighting in Buildings, Facilities and Roads of the UAH*; UAH: Alcalá de Henares, Spain, 2013; Available online: https://www1.uah.es/sostenibilidad/docs/proyecto-iluminacion.pdf (accessed on 29 September 2020).

Publisher's Note: MDPI stays neutral with regard to jurisdictional claims in published maps and institutional affiliations.

MDPI

St. Alban-Anlage 66

4052 Basel

Switzerland

Tel. +41 61 683 77 34

Fax +41 61 302 89 18

www.mdpi.com

Sustainability Editorial Office

E-mail: sustainability@mdpi.com

www.mdpi.com/journal/sustainability